D1516597

HAZARDOUS WASTE MANAGEMENT FOR THE 80's

HAZARDOUS WASTE MANAGEMENT FOR THE 80's

Edited by

Thomas L. Sweeney
Harasiddhiprasad G. Bhatt
Robert M. Sykes
Otis J. Sproul

ANN ARBOR SCIENCE
THE BUTTERWORTH GROUP

mc

Copyright © 1982 by Ann Arbor Science Publishers
230 Collingwood, P.O. Box 1425, Ann Arbor, Michigan 48106

Library of Congress Catalog Card Number 82-71532
ISBN 0-250-40429-X

Butterworths, Ltd., Borough Green, Sevenoaks
Kent TN15 8PH, England

FOREWORD

One of the great environmental challenges of the 1980's will be learning to better manage the generation, transportation and disposal of hazardous waste.

As Administrator of the Environmental Protection Agency, and manager of the two primary federal laws for dealing with hazardous waste, I can assure the engineering community that the Reagan Administration has made the control of hazardous waste its chief environmental concern.

A document of benefit to all those who work in this field is *Hazardous Waste Management for the 80's*. This book makes available the latest thinking of academic and professional experts in the field of hazardous wastes. Readers can zero in on the critical issues in hazardous waste management: groundwater and seepage; treatment, storage and disposal; site cleanup; and safety and legal considerations.

Anne M. Gorsuch

PREFACE

This book is an outgrowth of the Second Ohio Environmental Engineering Conference held in Columbus, Ohio, in March 1982. The conference, an annual event, was sponsored by The Ohio State University Department of Civil Engineering and the Central Ohio Section of the American Society of Civil Engineers.

It was the intent of the 1982 conference to provide practical and useful information on hazardous waste management to engineers, scientists, lawyers, governmental officials and those in responsible positions in waste-generating organizations.

The editors are grateful to those conference speakers who prepared chapters for the book, as well as to those speakers who were unable to do so because of other professional obligations. The editors also acknowledge the invaluable assistance of the sponsoring organizations, the conference organizing committee and, especially, Marcia Gibson of The Ohio State University and James E. Hays of Malcolm Pirnie, Inc.

Thomas L. Sweeney
Harasiddhiprasad G. Bhatt
Robert M. Sykes
Otis J. Sproul

Thomas L. Sweeney received his BS, MS and PhD degrees from Case Western Reserve University. After working for The Standard Oil Company in Research and Development and Commercial Development, he joined the faculty of The Ohio State University, where he is now Professor of Chemical Engineering. In addition to teaching conventional Chemical Engineering courses, Dr. Sweeney has taught courses in environmental regulation, environmental science and technology, and environmental pollution abatement at Ohio State. He has presented or published a number of papers and has been a consultant to more than 50 industrial organizations, governmental agencies and legal firms. He is a registered Professional Engineer in Ohio and is admitted to the practice of law in Ohio and United States courts. Dr. Sweeney has been active in various professional and technical organizations and is a member of the American Institute of Chemical Engineers, the American Chemical Society, the American Society for Engineering Education, and the Columbus, Ohio State and American Bar Associations.

Harasiddhiprasad G. Bhatt is a Senior Project Engineer with Malcolm Pirnie, Inc. He holds BE and ME degrees in Civil Engineering from M.S. University of Baroda in India and an MSCE from Michigan State University. Mr. Bhatt has had more than 25 years of varied experience in undergraduate and graduate teaching, project management, process design for municipal and industrial wastewater treatment, facilities planning and design, and construction administration of water and wastewater treatment facilities. He is a registered Professional Engineer and has authored several technical articles on land application of wastewater and sludge, industrial pretreatments, acid mine drainage abatement, and operation of wastewater management facilities. He is actively associated with several national and international professional societies and is a member of the American Society for Testing and Materials Committee D-34 on Hazardous Waste Disposal.

Robert M. Sykes is Professor of Civil Engineering at The Ohio State University in Columbus, Ohio, where he teaches environmental engineering. Before joining the OSU faculty in 1972, he pursued undergraduate studies at Northeastern University in Boston, Massachusetts, and graduate studies at Purdue University in West Lafayette, Indiana. His consulting and research has focused on the biokinetic aspects of water quality modeling and wastewater treatment. These efforts have led to developments in activated sludge theory, low-temperature anaerobic digestion, landfill gas generation and dynamic water quality modeling. Dr. Sykes is a member of the American Society of Civil Engineers, the Water Pollution Control Federation, the American Water Works Association, the International Association on Water Pollution Research and the American Society for Limnology and Oceanography.

Otis J. Sproul is Professor of Civil Engineering and Dean of Engineering and Physical Sciences at the University of New Hampshire (UNH) in Durham, New Hampshire. Before joining the UNH faculty he taught environmental engineering for 14 years at the University of Maine in Orono, and was Chairman of Civil Engineering for 5 years at The Ohio State University in Columbus. He received his BSCE and MSCE at the University of Maine and his ScD at Washington University in St. Louis. His consulting and research activities include work on detection and elimination of viruses and pathogenic protozoa, on ozone as a disinfectant and on industrial wastewater treatment. He participated in the development of regulations for air resources and solid waste disposal in the state of Maine and was one of the founders of the International Ozone Institute. Dr. Sproul is a registered Professional Engineer in the state of Maine and is a member of the American Society of Civil Engineers, the American Society for Engineering Education, the American Water Works Association, the American Association of University Professors, the International Association on Water Pollution Research, the International Ozone Institute and the Water Pollution Control Federation. He is past Editor of the *Journal of the Environmental Engineering Division, ASCE* and in 1971 he received the ASCE's Rudolph Hering Award for excellence in research.

CONTENTS

xi

CHAPTER 1

A SERVICE INDUSTRY PERSPECTIVE

Richard L. Hanneman
   National Solid Wastes
   Management Association

The question is often put to me as director of government and public affairs for the trade association representing the hazardous waste management industry in North America:
What government policies do we need to manage these hazardous
wastes? The answer, of course, is that we need absolutely no
government policies to manage these wastes!

The real question is: How can we improve the manner in
which we manage these hazardous wastes to provide protection
to public health and the environment? And, what government
policies will most effectively promote improved management of
hazardous wastes?

The waste service industry which I represent offers a
perspective, and a program, for the proper management of
hazardous wastes. To understand the requirements of that program, we must understand the reasons why we have not done
better in the past so that we can map out a workable program
for the present and so avoid, for the future, the deficien-
cies apparent to us today. In truth, the problems of the
past are the cause of our present situation. Among these, I
would single out three for special mention: lack of public
knowledge and understanding; lack of technical knowledge on
proper hazardous waste management; and lack of incentives for
proper hazardous waste management.

By and large, decisions in this country are made
democratically. Political decisions and economic decisions,
both are the result of mass decision-making. And, when the
polity is uninformed and the marketplace is uninformed,
uninformed decisions are the result. A generation ago, the
public had absolutely no conception of hazardous waste, the
magnitude of its production, and its potential for public
health damage and environmental destruction. Things have
changed. Repeated bombardment from the mass media has

raised the level of public consciousness. But, I think a good case could be made that this process of public education has been biased and uneven so that still today the general public has no true understanding of the basic questions involved: the reasons why we have hazardous wastes, the amounts of hazardous wastes, the potential for damage from these wastes, the protections and the limitations of our technology to prevent damage, and the present means by which these wastes are being managed.

A generation ago, many of those who produced hazardous wastes did not, themselves, understand fully the potential which these wastes had for damage and destruction. Some, perhaps most, of the responsible generators employed or contracted for state-of-the-art disposal techniques, many of which have proved fully adequate to the task. Unfortunately, some other generators were less careful or outright irresponsible. Wastes were dumped in the water. They were dumped on the land. They were abandoned in rusting containers in the hope that there would be no problems. Often they were entrusted to unknown and unregulated haulers charged simply to make the wastes "go away." In this past generation, we have learned many lessons on the technical requirements of proper hazardous waste management. No longer can ignorance be offered as an excuse for disposal practices which threaten public health and the environment. Technically, we know how to manage hazardous wastes and we have known for some years.

PROBLEMS OF THE PAST

Which brings us to the last of the problems we inherit from our past; namely, the lack of incentives for proper hazardous waste management. A generation ago, virtually every government adopted that management option which I mentioned at the outset, that is, no government policy on hazardous waste management. We know now what they may have known then, that hazardous wastes were being produced. And we know now, all too well, that in the absence of market incentives for proper disposal and in the absence of government regulation requiring proper disposal, there is no incentive to properly manage hazardous wastes. Clearly, the judicial process of tort recovery is insufficient to discourage improper disposal of wastes. In addition to the incomplete technical understanding of the requirements of proper hazardous waste management, generators who produced hazardous wastes a generation ago found that the more expensive special care required for proper management practices put them at a severe competitive disadvantage in a marketplace where less responsible or even unscrupulous competitors were disposing of hazardous wastes with improper, but bargain basement price, methods.

Fortunately, we are entering a new era. Public concern has created a government mandate to regulate management of hazardous wastes. Regulatory programs are being developed. Enforcement programs are being implemented. We are confident that we are well along on the road to assuring proper management of industrial hazardous wastes.

Far from the caricature of corporate disregard for the environment and opposition to governmental regulation, the waste service industry supports a strong program of government involvement in the regulation of hazardous waste management. We believe that governments should develop specific standards to protect the public and the environment against unreasonable risk. Equally important, we support funding necessary to assure full and strict enforcement of those regulations. Not only can the waste service industry survive in the face of strict government environmental regulation but, without it, we believe the entire program will falter and the public will be forced to turn, instead, to government itself to provide facilities necessary for proper management of hazardous wastes. This is an eventuality which no one should welcome for reasons which I will get into later.

In our opinion, there are five basic requirements for a successful hazardous waste management program. First, recognizing our limited resources, we must prioritize our efforts. Second, we must develop specific and reasonable standards for managing hazardous wastes. Third, we must assure that all those who would manage hazardous wastes are capable of discharging their financial responsibilities for liabilities incurred as managers of these wastes during both transit and facility operation and in perpetuity following facility closure. Fourth, we must insist on full and fair enforcement of facility standards and operators' performance. And fifth, we must devise mechanisms that assure construction of the new facilities which meet these standards where they do not presently exist.

PRIORITIES

First of all, we need to recognize that we have only limited resources to devote to proper management of hazardous wastes. Although it may seem difficult when one examines how much we each pay in taxes, it is nevertheless true that governments have limited resources to manage their programs. Likewise, in the private sector. Whether a corporation lacks an absolute ability to raise capital or faces competitive pressures that will not allow it to recover its pollution abatement costs in the marketplace, there is a limit to how much a private company can spend on proper hazardous waste management.

On the other hand, the job of proper hazardous waste management must be done. Other priorities must give way to the imperative to devote capital and human resources in both the public and private sectors to the job of managing hazardous wastes properly. Indeed, the entire credibility of the government's program to regulate proper hazardous waste management depends on giving waste management that high priority. And the waste service industry supports that priority. Without a credible government program, it is impossible for our industry to get its job done.

It is also possible and, in fact, highly desirable, that priorities be set among the various functions which might be encompassed in a government hazardous waste program. Clearly, some things are more important than other things. The basic reason for the slow start of the federal hazardous waste management program in the States has been the inability of the U.S. Environmental Protection Agency to prioritize. In framing the scope of its program in the broadest possible manner, the EPA found itself without sufficient resources to get that enormous job done. Dedication of the same resources to a more manageable job description would see that program much further down the road today. Specifically, we are encouraged that the new Administration at EPA is committing resources to developing a degree of hazard classification system which segregates wastes into several classes according to their appropriate management technology. This is step one.

STANDARDS

Step two is for the government to establish clear and specific rules which define permissible management practices. We now have final RCRA standards for all portions of the program with the exception of the most difficult and important land disposal facility standards. Development of these standards has been a long and hard-fought exercise. In December, 1978, EPA published a proposed set of design standards that the regulated community found unrealistic. EPA then went back to the drawing boards. As the Carter team left office, they published a completely different approach, one based on very vague standards and "best engineering judgment" which placed virtually all permitting decision-making responsibility in the hands of the permit writer. Industry again objected that each application would be judged on an ad hoc basis against an invisible yardstick. With the enormous risk of investing the huge amounts of capital required to build these facilities, industry requires a greater degree of certainty. The final land disposal standards must set targets for an applicant to shoot at.

Finally, under enormous pressure from the federal courts, EPA has proposed a "basic standard" and two options for those wishing RCRA land disposal facility permits. Public comments having been received, EPA has promised the courts that these standards will be published within the next month or so.

The waste service industry is concerned that in allowing distinctions between new and existing facilities the final standards may tilt so far in accommodating existing disposal practices, even where those practices may represent substantial risk or actual harm to the environment, that the regulations will remove incentives to upgrade disposal practices. RCRA was never intended to create markets for the waste service industry by imposition of unnecessarily stringent disposal standards; however, RCRA certainly anticipated closure of some existing waste management facilities which cannot measure up to tougher environmental standards. We must not back away from that commitment. Development of specific and reasonable facility standards should be done quickly and represents the second step in a responsible hazardous waste management program.

FINANCIAL RESPONSIBILITY

The third step in a successful hazardous waste management program is to assure that all those who incur liabilities in handling or disposing of hazardous wastes are able to meet their financial responsibilities. Clearly, not everyone with a truck should be allowed to haul hazardous wastes. And, not every land owner should be permitted to own and operate a hazardous waste management facility even if it can be designed and operated to meet environmental standards. Good public policy demands not only expert engineering and qualified operators, but a buffer of protection for the public against any accidents which may occur even with the best of engineering design and the most capable management practices.

What the public needs is assurances of the financial capacity of those allowed to manage hazardous wastes. Such assurances need not be limited to insurance provided by a commercial insurance company; it might be self-insurance, a surety bond, a trust fund, a test of net corporate assets, or some other means of determining that the corporation engaged in managing hazardous wastes can fulfill its obligations. In recent years, operating facilities have been able to qualify for increasingly broad insurance coverages and this trend seems likely to continue. There should be a clear policy reserving permission to transport, process or dispose of hazardous wastes only to those financially capable of meeting their obligations.

In this regard, we are pleased that EPA's financial responsibility regulations for required closure and post-closure activities will become effective on April 13. We are very concerned, however, that EPA may follow through on its threat to withdraw federal financial responsibility regulations for operators of RCRA facilities during their operating lifetimes. We see such a federal requirement as absolutely essential to maintaining public confidence that waste management operations will be conducted safely by responsible firms. We are concerned that without a federal requirement the growing interest of insurance companies in offering environmental impairment liability insurance, an interest born of the expectation since 1976 that such would be the federal requirement, will result in flagging insurance industry interest, unavailability of insurance or less competitive policies and premiums. Should this happen, private companies seeking to purchase insurance might find it difficult to do so and states seeking to impose such requirements might find it impossible because of the unavailability of such insurance.

With these provisions in place, only one loophole remains. While an operator can be required to set aside funds for proper closure and routine post-closure care, he cannot be presumed to know whether any unanticipated problem may occur at the facility requiring remedial expenditures for incurring liabilities for personal injury or property damage. It would be infeasible to require an operator to set aside funds sufficient to meet a "worst case" disaster. Typically, such risks are borne by commercial insurance. Unfortunately, no insurance company in the world today offers insurance to cover such an eventuality. The only types of commercial policies available are of the claims-made variety which are cancellable upon notice. Thus, there is no "insurance of insurance availability" during the possibly-perpetual period during which a facility might conceivably cause problems. Recognizing this in the States, the Congress enacted, late in 1980, legislation establishing a federal government-managed post-closure liability fund created by a special tax on all hazardous waste disposal. This fund pays all court judgments or settlements associated with qualifying facilities including cleanup expenditures, property damage compensation and personal injury claims. Facilities must be properly designed, operated and closed before their operators qualify to have their liabilities managed by the fund. Otherwise, the financial responsibility requirements of operating companies are available to satisfy any damages. Thus, operator financial responsibility is assured and represents the third component of a successful hazardous waste management program.

ENFORCEMENT

Unfortunately, identifying the problem and laying out the ground rules for its solution does not get the job done. The fourth part of a successful program is strict enforcement of these rules. There is a tendency in government to gesture dramatically in the direction of looming problems, proclaim loudly the enactment of new legislation or adoption of new regulations or inauguration of new programs to address these problems, and then quietly shortchange the less visible enforcement of those statutes and programs. The job is not done when the ground rules have been set. Full and fair enforcement is imperative. Again, it may sound strange that an industry should so loudly protest a possibility that government inspectors might not come calling at its door. The fact of the matter is, our concern is that the government inspectors will not be calling at the door of all our competitors. The rules should be the same for off-site facilities and on-site facilities, for government facilities and private facilities. It should be of general concern that everyone involved in managing hazardous wastes is required -- absolutely -- to live by the same set of rules. Different rules for different participants bias and distort the marketplace. They require either overly expensive costs on some facilities and those who use those facilities, or they require inadequate environmental protection of competing facilities which may provide customers of those facilities with temporary lower prices but expose those customers to enormous long-term liabilities associated with improper disposal. A budgetary commitment to effective enforcement is wise public policy.

I think it is safe to predict that our luncheon speaker today will document a record of progress in developing new regulations and reiterate her often-voiced warning to the regulated community to expect vigorous enforcement. We have heard that message. We would like to believe it. Unfortunately, we are continually reminded that actions speak louder than words. The Administrator has thrown her full and vocal support to the efforts of the Bush Task Force on Regulatory Reform which has promised a complete review and likely overhaul of the basic hazardous waste management program. The herky-jerky pattern of reorganization after reorganization in EPA's enforcement staff has suggested, hopefully erroneously, that the Administration is more concerned with preventing harassing suits than in prosecuting those who abuse waste management regulations. The 1983 budget recommendations are widely viewed as cutting muscle as well as fat at the EPA. Although there is no doubt that the Administration is firmly committed to a policy of state primacy in the hazardous waste management program, recommended cuts in state program grant assistance suggests to some a federal "copout."

In complete fairness, few if any of the threatened "regulatory relief" measures have been adopted, enforcement activity does seem to be vigorous (at least in terms of the off-site disposal industry), overall cuts in the EPA budget were largely directed at programs other than hazardous waste which retains a high Agency priority, and EPA is taking steps to encourage states to develop their own tax resources to support their hazardous waste management programs.

Unfortunately, there are times when reality is not as important as public perception. Whatever the truth about the commitment of the "twelfth floor" at the EPA to retaining the essence of a strong environmental protection program while trimming the fat, whatever improvements in enforcement may result from focusing scarce resources on perhaps fewer but more winnable actions, whatever the success of "getting more effective regulation from less resources," the stark fact remains: the public believes that the leadership of the Agency is not committed to a strong program of environmental protection. The public believes that the Agency is sympathetic to polluters, not those who might suffer the effects of pollution. Like the proverbial customer, it matters not whether the public in this case is right or wrong, but only what the public believes.

Restoring public confidence and reestablishing the credibility of EPA (and the state environmental agencies as agents of the federal EPA) is the greatest single challenge facing the EPA today. Rather than virtually eliminating the EPA Office of Public Awareness, the efforts of that office should be redirected. For four years, EPA steadily fashioned a climate of public concern, in some cases near-hysteria, in an effort to build a constituency for enactment of Superfund. Now that Superfund is law, it is time to substitute rational environmental education for the counterproductive "horror stories" of the recent past.

But public relations will not be enough. Besides showing that the EPA program is replacing unsound environmental practices with those which protect the public, the EPA must move quickly to enforce against violators against hazardous waste regulations and shorten, not lengthen, the permitting process. It was not too long ago that the EPA told us that it would take fully 3-4 years before all interim status hazardous waste facilities would be reviewed for their Part B RCRA permits. We were upset that the process would take too long. I'm sure that the public was disappointed that some possibly sub-par facilities would remain in operation for as long as four more years. Now, however, some have suggested that it may be as many as 8-10 years before some Part B applications are called. This is unconscionable. Facilities which have not been judged by RCRA standards will be allowed, if this is true, to continue and complete their entire planned operating lives without being subjected to the new standards. This is not what the public wants. EPA should not wait until it is forced to compress this schedule but should act to accelerate it. We recommend that when the final facility standards become effective, EPA immediately call all Part B applications and dedicate sufficient resources to processing them within 18 months. Thus, effective enforcement and public perception of effective enforcement is the fourth requirement of a successful hazardous waste management program.

SITING MECHANISM

The fifth and final imperative recognizes that no matter how successful the allocation of scarce resources, no matter how specific and reasonable the facility standards, no matter how effective the safeguards requiring only responsible operators to handle hazardous wastes, and no matter how effective the enforcement, there must be some facility to take hazardous wastes where they can be managed properly. I suspect that you agree with our companies who feel that they provide an essential service to society and to industry. Like traveling salesmen, they take their product from community to community trying to line up orders. And yet, in state after state, waste service companies have been given only two orders: "Get out and stay out!" It is all well and good to regulate hazardous wastes from cradle to grave, but what if there are no gravesites? The elaborate regulatory program I've described is essential in an effective hazardous waste management program, but it alone is not enough. The "siting problem" for hazardous waste facilities is the most serious single obstacle remaining to be overcome. And yet, its solution depends importantly on the success of the earlier steps. Unless the public is convinced that the government hazardous waste management program is effective in preventing operation of environmentally destructive facilities and fly-by-night operators from handling hazardous wastes, it is small wonder that community after community voices concern, to choose a mild word, over the proposed location of facilities within their borders.

On the other hand, all evidence and experience to date indicates that it may not be possible for a proposer to find a willing host community. What local politician would lay down his career to attract a hazardous waste management facility to his town? Local politics are virtually impossible. And yet, there is no place in the world that is not "local" to someone. And were we able to find such a place, its very remoteness would impose huge costs associated with distance from its customers.

The basic problem is that, because of the huge investment necessary to construct a modern hazardous waste management facility, the site must necessarily serve a regional market. So, on the one hand, the benefits of a proper hazardous waste management facility are regional in nature, while on the other, the perceived costs are local.

What can be done? We do not believe that it is desirable -- or even possible -- to "buy off" local opposition. Rather, we believe that the general public which benefits from proper disposal should do everything in its power to expend the necessary resources to operate an effective hazardous waste management program. If the waste management program itself provides every reasonable expectation that the wastes can be managed safely, and if provision likewise is made that someone outside the community, either a facility operator or the government, will pay for injuries or damages associated with the facility, we believe the community should be required to accommodate a hazardous waste facility if that facility is necessary to meet the state's waste management needs. After all, to return to my original point, the wastes are there and they are being managed one way or another today. Until we get new facilities on-line, the wastes will continue to go to less adequate alternative disposal which may become a problem later.

One mechanism devised to balance the broader public requirements for the construction of these facilities against local concerns about their safety and environmental impact is that of the siting board. Michigan is the model for many states, Ohio among them, in establishing a state siting board with the authority to overrule any local zoning decisions aimed at frustrating establishment of a desirable and technically sound hazardous waste management facility. The Michigan statute has the longest operating history and has already produced one successful siting. The State of Massachusetts has enacted legislation requiring facility developers to compensate host communities, a concept which has, to date, been entirely successful. Wisconsin is on the verge of enacting a new hybrid siting program. In the Wisconsin model, the state establishes a forum for negotiations between facility developers and prospective host communities and a state arbitrator makes a final decision. Another administrative option is that employed by the State of Arizona which has attempted to preclude privately-owned hazardous waste facilities by establishing a state-owned facility (on federal land) and short-circuiting the public hearing process to secure its environmental permit. The planned facility will be privately operated. Finally, the State of Illinois has come up with a unique approach. The State Legislature established specific criteria against which a proposer's application would be judged for acceptability. Rather than establish a state siting mechanism, however, the Illinois plan would have the local governmental unit conduct the hearing to determine if the proposal met the state criteria. This would seemingly fly in the face of the well-documented record of local opposition to hazardous waste facility siting, a fact attested to even by such ardent environmentalists as Environmental Action, Inc. But there is method in the Illinois madness. Having established a definite criteria, the Legislature also provided for expedited judicial review of the local decision by the appeals-level courts. It is too early to know if this will be a successful approach. We are more certain about the likelihood of success of the programs now in place in Alabama and Kentucky. Both provide for a vote of approval by the State Legislature before facility siting can be approved. (Alabama's single commercial hazardous waste facility was established before this law became effective in late 1981.)

Time after time, I'm asked: Why haven't private companies responded to this need? How can we be assured

11

that private companies will build the necessary facilities in time? Shouldn't we consider a government-owned facility? As you know, this debate has raged widely over the past year. We in private sector feel unfairly condemned for failure to provide more of the solution than we already have, given the uneven record of providing the rudiments of an effective program such as I've described. Yet, the repeated frustrations of private companies to respond to this need speaks well for the continued interests of our member companies in constructing those facilities that will be necessary once the groundwork is laid.

The program which I've outlined provides the missing ingredient that explains why private companies have not succeeded in the past: risk management. This program reduces risks. It reduces risks to the public by assuring that improper facilities and irresponsible operators are not allowed to handle hazardous wastes. But it also reduces risks to companies by assuring that if they comply with rigid standards, if they provide access to financial resources, and if the government, in effect, provides the markets by denying improper disposal alternatives, and if the siting procedure offers some prospect of converting good proposals into actual facilities -- if the government provides this program, private industry will provide the sites.

In conclusion, we have had problems in proper management of hazardous wastes in the past because the public has not been properly educated about the real situation in hazardous wastes, because in the past we lacked the technical expertise to manage these wastes properly, and because there existed no incentives for proper management of wastes. We are making progress on all fronts. A proper hazardous waste management program, however, demands that we prioritize our scarce resources, establish specific and reasonable facility standards, require all those authorized to handle hazardous wastes to be financially responsible for their actions, demand strict enforcement of regulatory standards, and establish a siting mechanism to bring into existence the necessary new facilities required to get the job done. The government should exercise its regulatory role vigorously and, if it does, it can expect that the waste service industry will provide the necessary and proper facilities to manage industrial hazardous wastes properly.

CHAPTER 2

OHIO'S HAZARDOUS WASTE PROGRAM

Wayne S. Nichols
    Director, Ohio Environmental Protection Agency

I am pleased to present the State of Ohio's perspective on hazardous waste management. I know that all of you recognize the importance of sound management practices. In looking over the program for this conference, I believe it is a tremendous opportunity for all of us to broaden our knowledge in this area.

In Ohio, we have placed a top priority on hazardous materials management. We believe we are making significant progress toward effective control of hazardous materials, and I'd like to briefly describe our approach.

Ohio's hazardous waste program has three main parts: the cleanup of abandoned hazardous waste sites, a regulatory program to control the disposal of hazardous wastes now and in the future, and a siting program to provide Ohio's industry with additional, environmentally sound disposal facilities it must have in order to comply with those regulations and safely get rid of hazardous wastes.

Our cleanup program has to be one of the best in the nation, especially considering the financial constraints under which we must operate. We have twelve sites, our clean dozen if you will, that have been or are being cleaned up without the use of State tax dollars. We've accomplished this through negotiations with the generators and/or the parties responsible for the sites, and through an aggressive pursuit of Superfund dollars.

The largest cleanup is underway now at the Summit National site in Portage County. A $2.5 million project for total surface cleanup is being paid for by several of the waste generators at that site with Goodyear Tire and Rubber, Goodyear Adhesives, and Morgan Adhesives being the primary contributors.

Since the operation began in December, 1981, over 100 truck loads of waste have been removed for proper disposal. We hope to see the surface cleanup completed during 1982.

Ohio EPA and the Attorney General recently reached agreements with five more generators, totalling $49,000. We are continuing legal action against about thirty other parties. Any money that is gained through that effort will be used to begin a study of buried waste and water and soil contamination.

We have another large project underway at the Chem-Dyne site in Butler County. More than 4,700 drums were removed in 1981, and another 473 so far this year. In all, the generators have removed about two-thirds of the waste from that site. We've recently received an emergency Superfund grant to clean up the loading dock area at Chem-Dyne. This remains our top priority site for remedial Superfund money for long term cleanup.

The Chem-Dyne grant makes the sixth emergency grant Ohio has received from Superfund. The grants total over $700,000, and we're very grateful to U.S. EPA for this help in eliminating our problem sites.

The second part of our program is our regulatory effort. Ohio has hazardous waste regulations parallel to P.CRA. We plan to file amendments to bring us up to date with the federal rules through January 1, 1982, in the very near future. Of course, these amendments will not change our prohibition concerning the landfilling of liquid hazardous wastes, even though the federal ban has been temporarily lifted.

We've applied to U.S. EPA for Interim Authorization for our hazardous waste program. Our staffing and training is nearly complete, and we are moving full speed ahead in this program area.

The third part of our program is our permitting process which, again, we believe to be one of the most progressive in the nation. We have already issued 336 permits to existing TSD facilities, and we have applications from roughly a dozen proposed facilities.

I would like to explain briefly what the permit process involves. Ohio EPA receives the permit application and reviews it thoroughly. The staff inspects the site, and makes a recommendation. Then the application is transferred to the Hazardous Waste Facility Approval Board.

The Board is separate from the Ohio EPA and was created by the General Assembly specifically for the purpose of issuing hazardous waste facilities permits. I am its chairman. The Board's staff will conduct its own technical review, as well as a review of the company's compliance with environmental laws and regulations.

The law requires that the Board hold a public hearing in the county in which a facility wishes to locate. This hearing must be scheduled sixty to ninety days after the Board receives the application. The members of the Board place great importance on this opportunity for the public to express their opinions.

Following the public hearing, the Board will schedule an adjudication hearing in Columbus. This is a formal proceeding at which those who feel they would be adversely affected by a facility may present evidence to a hearing examiner. The law requires that the chief municipal official and the county commissioners participate, to guarantee the public proper representation.

Using all of the information it has collected, the Board will make a decision on the application. Each permit granted will be tailored specifically for the facility which receives it. This means that in addition to meeting all state and federal hazardous waste regulations, each facility will have specific limits on what it can do and what it cannot do. The permit will state what types of waste the facility can accept, how much waste, and what processes the facility may use. In addition, Ohio EPA will conduct regular inspections to make sure that these requirements are being followed.

As I noted earlier, we believe ours is one of the most advanced permit programs in the nation. We place great importance on it for two reasons. One is that no one wants to create new hazardous waste problems to add to the old ones we're trying to solve. We must make sure that the hazardous waste we are generating today is properly disposed of, and regulatory control of all TSD facilities is crucial if we are to succeed in that.

The second reason is that Ohio, like many other states, does not have enough hazardous waste facilities to handle the waste it generates. We estimate that we need one or two additional secure chemical landfills, several major treatment and neutralization facilities, and more on-site facilities to meet Ohio's needs. Without these facilities, Ohio's industry is faced with costly transportation bills to properly dispose of its waste.

Ohio does have the means, through the Ohio Water Development Authority, to construct and operate its own waste facility. However, we believe that the development of such facilities properly belongs in the private sector. We have pledged ourselves to a strong enforcement program, which will protect the people and the environment of Ohio and at the same time provide a demand for approved disposal facilities. We also support the increased availability of low-interest loans to stimulate private investment in hazardous waste facilities.

While the development of additional hazardous waste facilities is important, there are other things we can do to help solve the problem of insufficient treatment and disposal capacity.

One is to encourage waste recovery. Certain material that is waste in one industry can be raw material in another. Many solvents can be recycled and used again. Used automotive oil can be recycled. We must begin to seriously pursue this option.

We must also do more in the area of waste reduction. Industry must receive incentives to reduce the amount of waste it generates. Clearly, the less we generate, the less we must worry about when it comes to disposal.

We are presently working to develop a computerized data bank of information about hazardous materials management in Ohio. When it is completed, this system will help us to advise industries about waste reduction and waste exchange options, to identify specific regional problems and needs, and to better utilize the resources we already have in Ohio.

Conferences such as this one play a valuable role in bringing together those who must deal with hazardous waste management. We have some good technical and non-technical sessions that will help give us an overview of where we are.

We want to work as closely as we can with everyone involved in hazardous waste management, from U.S. EPA to state and local government officials and from industry representatives to concerned citizens. We are all faced with the problem, and we must all participate in the solution.

16

CHAPTER 3

THE HAZARDOUS WASTE MANAGEMENT TRIANGLE

Peggy J. Vince
    Ohio Hazardous Waste
       Facility Approval Board

Over the past twenty years, there has emerged a national trend in participatory democracy concerning governmental regulation. This has generated definitive, vocal interest groups representing the various aspects of any one issue.

In my business, hazardous waste management, there are three distinct participants in every project we handle; industry, government and the citizens. I'd like to call this the hazardous waste management triangle. To those of us involved in the hazardous waste management triangle, it sometimes seems like the one down by Bermuda--once you get in, you never get out.

Miller B. Spangler of the Nuclear Regulatory Commission has put forth a theory about those faced with energy decisions, and his theory can easily be extended to hazardous waste decisions.[1] He identifies "syndromes," which are basically combinations of emotional feelings about good and bad and thought patterns that predispose an individual to make decisions in a certain way. Three of Spangler's syndromes correspond well to the three basic parties involved in the hazardous waste issue. They are the "Yankee Ingenuity Syndrome," the "Trustee Syndrome," and the "People are Wisest Syndrome."

The "Yankee Ingenuity Syndrome" best exemplifies the attitude industry sometimes adopts. According to this thought pattern, private enterprise with limited government intervention is responsible for the high standard of living we enjoy today.

The "Trustee Syndrome," which belongs to government, reasons that complex technical issues cannot be understood by the public, so those with expert knowledge must act as trustees of our environmental resources.

The "People-are-Wisest Syndrome," the province of citizen groups, holds that the people are the best judges of their own interests and that technological decisions should be subjected to democratic rule.[2]

Obviously, not all industries, government officials, and citizen groups will fall into one of these three categories. But in general, I think they help us to see how each side operates out of a mind-set that is at least somewhat at odds with the fundamental belief of the others. And that is where working together to resolve hazardous waste issues becomes a challenge.

It would be helpful if each side recognized the premises under which the others operate, and the weakness in each position, including their own. With that understanding in common, perhaps other common ground could be discovered. But this does not usually happen.

For example, the "Yankee Ingenuity Syndrome" fails to recognize that technology sometimes has unexpected social consequences. And it is those social consequences that citizens are most concerned about. A company that can put aside for a moment its all knowing attitude may be able to negotiate with a citizens group such issues as truck routes and the adequacy of the community's fire-fighting equipment. Unfortunately, some companies have taken the attitude that they don't have to deal with citizens.

For example, one company planning a hazardous waste facility in Ohio attempted to quietly purchase land in a rural county. No announcement was made to the local citizens or to the regulatory authority, Ohio EPA. When the citizens learned of the proposal, they were outraged and organized a citizen group to protest the proposal. While there is no legal requirement, we strongly encourage industry to inform the public as well as the regulatory agencies of any proposal. It is the responsibility of the industry to work cooperatively with the communities. Ignoring the citizens desire for input serves no purpose, and is usually very detrimental.

While this particular proposal would not have been permitted by Ohio EPA for technical reasons, the industry's credibility was damaged.

In the past and sometimes presently, government, operating on the theory that the people are unqualified to make complex technical decisions, has been guilty of failing to take citizens' opinions as seriously as they should. The environmental movement of the 60's and 70's has helped to change that. Public participation is now a recognized right of every citizen, and is included in the rule-making process as standard procedure. Beyond this, the subject matter citizens may address has been broadened. No longer is testimony restricted to technical matters; social and economic impacts are also reviewed. This is not to say that government does not still have its quirks that sometimes make it frustrating for both citizens and industry to deal with us. For example, a newspaper reporter asked a government employee who was an expert about the type of facility proposed for a certain community if radioactive waste could be deposited there. The employee answered "Yes, it could." Of course, the next day's headlines carried the "news". The phone calls poured in. When the government employee was asked why he hadn't pointed out that nuclear waste disposal at such a facility is illegal in Ohio, he answered, "They asked me if it was physically possible, not if it was legal."

In fact, the company had no desire to deal with radioactive waste, and would never be permitted to do so. Company officials were understandably dismayed at the negative publicity this item generated. The members of the community were misinformed and needlessly alarmed, and all because the quintessential government bureaucrat insisted on responding to the reporter's question in its narrowest possible sense. To add insult to injury he vowed never to deal with that reporter again because, "he doesn't check his facts." Other examples can be found where the bureaucratic procedures helped to delay and frustrate those involved.

While we can't eliminate "bureaucrat-itis" across the board, I can assure you that my staff and I try hard to be responsive in a timely manner.

The "People-are-Wisest Syndrome," reasons that the people can decide their future. What it fails to admit, is that lay people may in fact lack the specialized training to decide knowledge-ably about technical issues. Because of the localized nature of the issues as most citizens groups confront it, they fail also to accept the scope of the problem and the need for solutions. "Put it in Texas," is a convenient argument for local use (unless you're in Texas), but it merely passes the buck and denies the fact that those who benefit from technological advancements must also share the burden of responsible management of its by-products.

Some citizen action groups involved in hazardous waste issues exhibit certain general characteristics.

1)   They are formed in reaction to the announcement, usually by the press, of some hazardous waste activity in their community.
2)   They are generally led by one or two charismatic individuals. A good example is Lois Gibbs, whose role at Love Canal was recently portrayed by Marcia Mason in a TV special.
3)   They are not usually members of nation-ally know environmental groups such as the Sierra Club or the Natural Resources Defense Counsel. In fact, they are generally people who have never taken a public stand or been politically active.[3]
4)   Their focus is on the local site rather than on the national scope of the pro-blem, and their approach is often inten-sely emotional.

In addition to sharing these characteristics, some citizen action groups also tend to share a certain set of assumptions.

1)   All hazardous wastes are life-threatening

and cancer-causing, regardless of the
situation.
2) Industry is not to be trusted. A
    National Wildlife Federation handbook
    warns, "One thing citizens are up against
    in the conflict with business is the
    willingness of some businesses to bend
    the truth, or tell half-truths, in order
    to present a convincing case."[4]
3) Government will not be responsive to
    citizens' concerns.

In general, no distinction is made between
illegal dump sites and legitimate TSD facilities.
Rather than seeing properly operated management
facilities as part of the solution, a means of
preventing illegal dump sites, they are perceived
as one and the same thing. As stated, the
solution to the problem is rarely addressed.

Citizens can find plenty of "advice" when it
comes to confronting the hazardous waste issue.
Paperbacks on citizen action are standard fare on
library and dime-store shelves. The National
Wildlife Federation has published, "The Toxic
Substances Dilemma: A Plan for Citizen Action,"[5]
and readers of Jane Fonda's new exercise book will
find a chapter on how to fight hazardous waste
sites inserted between the sit-ups and the
deep-knee bends.

A new field of hazardous waste "specialists"
is also arising and they appear frequently at
public meetings. Popular speakers include key
citizens from other groups, preferably from groups
that have opposed a facility which was never
approved or have lobbied for a cleanup that was
then completed. Former government and industry
officials who consider themselves "whistleblowers"
are also well-recieved.

Actually, a speaker has only to solemnly
announce the fact that hazardous wastes are a problem, and his or her success is virtually
assured. Citizens with no prior experience with
the hazardous waste issue often fail to realize
that this is the one point on which everyone
agrees! Probably nowhere is credibility more
easily achieved than at a public meeting on hazardous wastes.

At one meeting I attended, a woman testified
that toxic emissions from an incinerator had
caused her goat's rectum to turn purple. When I
consulted the attending veterinarian, he stated
that there was no medical evidence to suggest that
the incinerator's emissions were responsible for
this unusual malady, but he found the suggestion
so wildly amusing that he turned purple.

When all theatrics are put aside, however, one
thing that becomes clear from public meetings
about hazardous waste is that not all the parties
are operating under the same set of assumptions or
expectations. As Spangler's syndromes illustrate,
industry may take a "What's good for us is good
for you" attitude, government may patronize concerned citizens, and citizens may feel frustrated
and overwhelmed. Obviously, this makes communication in the citizen, industry and government triangle difficult.

I believe that all parties need to recognize
the valuable role they can play in the hazardous
waste siting issue and devote themselves to making
their most effective contribution.

Industry must, for its part, recognize the
fact that it is part of the community. It may be
a corporate neighbor, but it is a neighbor all the
same, and it shares the responsibility for protecting the community's interests. Providing jobs
and a tax base are good, but they are not enough.
Hazardous waste management facilities must also
respond to questions about the immediate and
long-term welfare of the community. By providing
environmentally sound storage, treatment, and
disposal capacity, these facilities definitely
contribute to the overall well being of Ohioans'
health and the environment. The regulations
require safety plans, closure plans, and financial responsibility to protect the community. But
industries must do more than simply comply with
the letter of the law. Industry has an obligation
to work with the citizens, an obligation that it
cannot afford to shirk. Again, when the parties
to hazardous waste issues work together, everyone
wins in the end.

Government must be sensitive to the fact that
along with its regulatory authority goes a respon-

sibility to provide the people with the facts regarding the issue at hand and to learn from the people any information that can be helpful in the regulatory process. The bureaucratic process must be responsive to the needs of the citizens it serves. It must monitor the "big picture," but it must listen to the valid concerns of knowledgeable people. It must also act in a timely manner.

If industry and government should listen to both sides of the hazardous waste management story, then so too should citizen groups. Unfounded speculation, inaccurate data and thoughts based solely on emotion are not the answer.

Citizens groups can be most helpful by ex- pressing public sentiment, airing public concerns, and addressing those issues that local people are most familiar with. In doing so, the citizens groups can help industry and government representatives to recognize the possible social impacts of a proposal, and discussion can begin on how to mitigate those impacts. The citizens can also make government aware of their chief areas of concern. In fact, they are vital to the regula- tory process. The Ohio Hazardous Waste Facility Approval Board's rules allow for special terms and conditions to be attached to a permit, and this can and has been done to address specific problems the facility posed for the community. In addition, the Board will hold both a public and an adjudication hearing to receive comments before making a permit decision.

When government, industry and citizens work cooperatively in this way, everybody benefits. For example, one group of concerned citizens with whom I am currently working is reviewing a pro- posed hazardous waste facility with an issues oriented approach. Together, we are discussing the site geology, finanical responsibilities of the facility, the suitability of the waste streams, and legal questions about the land owner- ship. This working relationship is beneficial to the citizens in that the Board can provide specific technical data and explain permitting procedures. It is beneficial to the Board in that local concerns can be identified at an early stage and thoroughly addressed. Additionally, the facility representatives have made themselves available by sponsoring frequent question and answer sessions to provide specific information and to listen to citizens' specific concerns.

The responsible attitude of these citizens towards protecting their community is admirable. It takes a great amount of their own personal time and effort to familiarize themselves in the procedural process as well as the specific technical proposal. I feel it is a privilege to be working with these citizens both professionally and personally. And, there has been no mention of purple rectums in any of the meetings.

I have a strong personal belief in, and am regarded by my peers as an ardent supporter of the citizens "right to know" and participate. The government as a regulatory body is inherently responsible to the people--and to protect the health and safety of the public. My particular policy is to attend all public meetings held concerning hazardous waste facility proposals and require the Board staff to be available at all times to the public.

I have also suggested that public participation can be most productive when a responsible, issues-specific approach is taken. Similarly, government and industry also make their most valuable contributions when they take an open, cooperative position. We cannot afford to forget that we're all in this together. After years of little or no regulation in hazardous waste we are faced with the almost overwhelming challenge of catching up and we think that we're doing a pretty good job. Sometimes in the face of that difficult but critical task, it helps if we can laugh a little at our own foibles. But let us not forget that an honest, responsible approach is necessary by all parties towards resolving hazardous waste management problems.

Unlike the Bermuda Triangle, ours does not have to be a threat. If all corners of the hazardous waste triangle do a responsible job the end results will be a better quality of life for all.

24

REFERENCES

1. Spangler, Miller B., "Syndromes of Risk and Environmental Protection: The Conflict of Individual and Societal Values," The Environmental Professional, Vol. 2 (1980).
2. Spangler, P. 289.
3. U.S. Environmental Protection Agency, Community Relations in Superfund - A Handbook (Draft)(Washington, D.C.: U.S. EPA, 1981), Chapter 1, pp. 2-3.
4. National Wildlife Federation, The Toxic Substances Dilemma: A Plan for Citizen Action Washington, D.C.: National Wildlife Federation, undated). Another example is Joseph L. Sax, Defending the Environment: A Handbook for Citizen Action (New York: Vintage Books, 1970).
5. National Wildlife Federation, p. 26.

CHAPTER 4

REUSE AND RECYCLE OF HAZARDOUS
MATERIAL WITHIN THE STEEL INDUSTRY

D. M. Gubanc, P.E.
  Republic Steel Corporation

SUMMARY

The hazardous materials recycle and reuse scenario
within the iron and steel industry is significantly impacted
by the present U.S. EPA and state hazardous waste management
regulations. In general, regulations that control reclamation, reuse or recycle will restrict the application of those
processes.

The following information is presented in this paper:

1. A description of the materials generated by the
    steel industry and presently characterized by
    U.S. EPA as hazardous wastes.

2. The ongoing recycle and reuse programs associated
    with these materials.

3. Recent regulatory action taken by U.S. EPA to
    encourage recycle/reuse. This action has been
    in the form of exemptions, or reduced regulation.

4. The barriers to recycle/reuse that still exist as
    a result of both federal and state hazardous waste
    management regulations.

The following questions will try to be answered:

1. How does recycle and reuse of hazardous materials
    benefit the public and the steel industry?

2. How do the existing rules negatively affect the public and the steel industry?

BACKGROUND

Regulatory Framework

The Resource Conservation and Recovery Act (RCRA) was enacted to ensure that the nation's resources were properly managed. One of the objectives of the act was to promote the recycle, reuse and reclamation of wastes. The encouragement of recycle and reuse is of great public value when the material is hazardous material. Recycle/reuse has the potential to remove the need for disposal or at least reduce the amount to be disposed, and does provide economic resource value. The decision to recycle/reuse versus disposal is basically an economic one. It is important that the cost of recycling and reusing hazardous materials not be allowed to increase due to the implementation of regulatory controls, since this will serve to discourage recycle/reuse.

Steel Industry Material Management

The United States produces approximately 125 million metric tons of raw steel annually. An integrated steel plant, like many other large manufacturing plants, consists of a number of processes that all have incoming "raw" materials, a product, and outgoing by-products. Some of the processes in a steel plant are the blast furnace, basic oxygen furnace, electric arc furnace, continuous caster, rolling mills, strip mills and bar mills.

The production of steel is essentially an extractive process, and therefore each process generates material that is not considered the primary product. For every ton of steel shipped from an integrated plant, one-and-one-half tons of material is produced that is not steel product.

The U.S. EPA conducted a study prior to the effective date of the hazardous waste management regulations to determine the state of the art in resource conservation in the steel industry [1]. The annual quantity of material generated by the steel industry that is not steel product is about 140 million metric tons. About 3.84 million metric tons, or 3%, was estimated to be classified by U.S. EPA as hazardous waste. Figure 1 shows by-product and hazardous waste generation by type. The hatched areas represent that portion of each material that the study indicated was recycled or reused. Recycled slag is used as an aggregate for concrete or road fill. Metallic scrap and some iron

28

oxide wastes are recycled in the steelmaking process for
their iron value. Coke plant by-products are usually sold
to chemical processors, and spent pickle liquor is generally
used as a chemical for municipal water treatment, or recycled
into new acid. Approximately 80% of the steel industry's
by-product materials are recycled or reused.

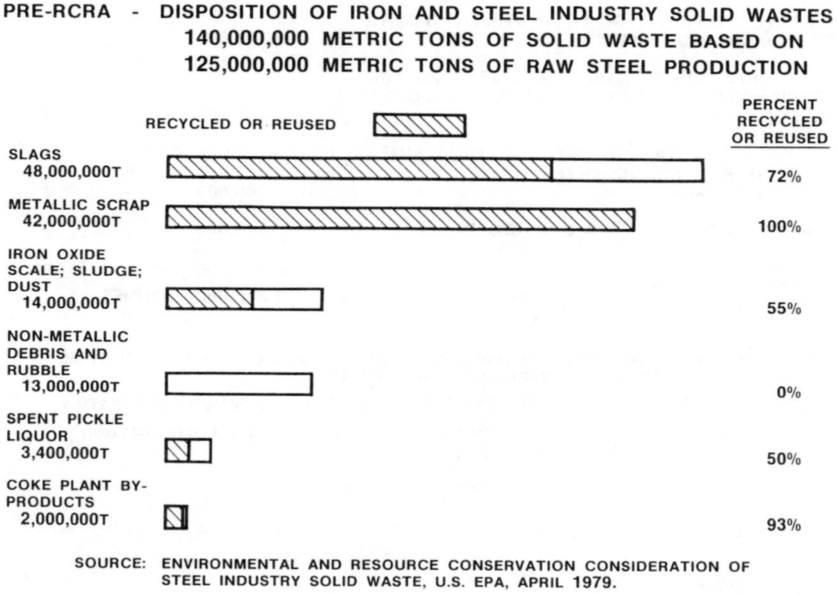

PRE-RCRA - DISPOSITION OF IRON AND STEEL INDUSTRY SOLID WASTES
140,000,000 METRIC TONS OF SOLID WASTE BASED ON
125,000,000 METRIC TONS OF RAW STEEL PRODUCTION

		PERCENT RECYCLED OR REUSED
RECYCLED OR REUSED	⧄	
SLAGS 48,000,000T		72%
METALLIC SCRAP 42,000,000T		100%
IRON OXIDE SCALE; SLUDGE; DUST 14,000,000T		55%
NON-METALLIC DEBRIS AND RUBBLE 13,000,000T		0%
SPENT PICKLE LIQUOR 3,400,000T		50%
COKE PLANT BY-PRODUCTS 2,000,000T		93%

SOURCE: ENVIRONMENTAL AND RESOURCE CONSERVATION CONSIDERATION OF
STEEL INDUSTRY SOLID WASTE, U.S. EPA, APRIL 1979.

Figure 1.   Pre-RCRA disposition of iron and steel industry
solid wastes.

        Based on more recent studies including the industry's
own data, the steel industry generation rate of materials
classified by U.S. EPA as hazardous waste is estimated to
be 3.865 million metric tons [2]. These materials and their
estimated annual generation rates are listed in Table I.
They can be divided into three types: spent acids from
pickling operations (87%); heavy metal contaminated emission
control dusts and sludges (11%); and coke plant wastes (2%).

Steel Industry Hazardous Material Recycle/Reuse Programs

        The recycle and reuse of some hazardous materials have
been performed for several years. Other recycle/reuse programs are rather recent. Some of the ongoing programs are
described below including a description of the material that

Table I. Steel Industry Hazardous Material Production

**U.S. STEEL INDUSTRY HAZARDOUS MATERIAL PRODUCTION
FOR 125,000,000 TONNES (137,800,000 TONS) OF RAW STEEL[1]
BASED ON U.S. EPA HAZARDOUS WASTE REGULATION[2]**

	U.S. ANNUAL PRODUCTION	
	SHORT TONS	PERCENTAGE
SPENT PICKLE LIQUOR FROM STEEL FINISHING OPERATIONS (K062)[3]	3,700,000	87
EMISSION CONTROL DUST/SLUDGE FROM THE PRIMARY PRODUCTION OF STEEL IN ELECTRIC FURNACES (K061)	400,000	9
COKE PLANT WASTES: AMMONIA STILL LIME SLUDGE (K060), DECANTER TANK TAR SLUDGE (K087)	60,000	2
MISCELLANEOUS DUSTS & SLUDGES	100,000	2
TOTAL	4,260,000 TONS	100
OR	3,865,000 TONNES	

(1) ENVIRONMENTAL AND RESOURCE CONSERVATION CONSIDERATIONS OF STEEL INDUSTRY SOLID WASTE, U.S. EPA, APRIL 1979.

(2) 40 CFR 261, 46 FR 4614, JANUARY 16, 1981. FINAL LIST OF HAZARDOUS WASTES.

(3) AISI PUBLIC COMMENTS TO U.S. EPA ON RECYCLE, REUSE, RECLAMATION, AUGUST 18, 1980.

---

is recycled or reused.

Spent Pickle Liquor (SPL)

The largest portion of hazardous material generated by the steel industry is spent pickle liquor (SPL). This material is generated in the pickling process, which is basically "acid cleaning" of steel. SPL is the acid that has been weakened through use to the point of no longer being able to remove surface scale from the steel. The continuous strip pickler shown in Figure 2 has an annual capacity of one million tons. The steel in the form of coils is loaded at one end of the line, unwound and passed through a series of acid tanks. Continuous pickling means that new acid is added and spent acid is removed on a continuous basis. Pickling can also be performed in batch tanks, where the acid is added all at once, used and then dumped into a spent acid tank.

Spent pickle liquor is normally held in tanks such as those shown in Figure 3. These storage areas normally have some containment system external to the tank, and a personnel safety station in the acid transfer area.

Figure 2.   Continuous strip pickling line.

        Pickling acids can be hydrochloric, sulfuric or
nitric-hydrofluoric. SPL contains anywhere from 10% to 30%
iron salts and 1% to 8% acid, depending on the type of acid
and type of steel pickled.

        Because there is iron in the SPL, some of this mate-
rial can be used as a replacement for water treatment
chemicals in water treatment plants. The most common use
of SPL is as a replacement for ferric chloride in municipal
sewage water treatment plants. Figure 4 is a picture of a
flocculation clarifier which is typical of the type of equipment used in sewage treatment plants. A study performed for
the U.S. EPA identified the demand for the chemicals available in SPL for the Great Lakes Region municipal sewage treatment market to be several times the total U.S. steel industry
production [3]. It is expected this type of reuse will
continue as chemical costs rise and experience with spent
pickle liquor as a water treatment chemical continues. The
New York Environmental Facilities Corporation (EFC) recently
announced that it is evaluating the use of SPL at some of its
facilities in hopes of reducing operating costs and encouraging reuse and recycle of hazardous materials [4].

Figure 3. Acid and waste acid storage tanks.

Figure 4.   Flocculation clarifier.

Electric Arc Furnace (EAF) Dust/Sludge

The second largest quantity of hazardous material generated by the steel industry is emission control dust and
sludges from primary steelmaking in electric arc furnaces
(EAF). An EAF melts scrap to make steel, and a typical furnace
is shown in Figure 5.   The dust which is removed in a collection device contains between 40% and 60% iron, 20% to
30% calcium, and smaller percentages of heavy metals such as
chrome, nickel, zinc, lead and cadmium.   The concentration
of these metals are somewhat higher than can be found in
other steelmaking process dusts because the feed material in
an EAF is scrap.

A small portion of the EAF dust can be recovered for
its iron content.   These materials are normally mixed with
other iron bearing materials and fed to a sinter plant.
Figure 6 is a picture of a sinter plant, which is a major
iron unit recovery operation for the steel industry.   The
resultant iron bearing pellets, or sinter, are suitable
for charging to the blast furnace.

As the heavy metal content in the dust increases, iron
recovery becomes more difficult and other metal recovery
(i.e. nickel, chrome and zinc) becomes more feasible.   Some

Figure 5.   Electric arc furnace.

Figure 6.   Sinter plant.

EAF dust is recovered for these metals. Most of this recovery is performed at non-steelmaking facilities. The bulk of EAF dust not recovered or recycled is either not suitable for iron recovery or does not contain enough non-iron constituents to be economically recovered for those metals, and is therefore stored and disposed.

Several joint government/industry research programs are proceeding to perform research and develop technology that will enhance the reuse or recovery of these EAF dusts and sludges that have heavy metal and iron recovery potential.

Steelmaking Dusts and Sludges

Other processes in steelmaking have the potential for generating dusts or sludges that may be characterized by EPA as hazardous. One type of steelmaking dust is produced during the pouring of steel into molds. Molten steel is poured into molds and allowed to solidify into ingots. While the steel is liquid in the mold, chemicals are sometimes added that will give the steel the metallurgical characteristics desired by the customer. This operation is called mold additions. One of the materials added to steel in this manner is lead. Some of this lead oxidizes and is emitted as a dust and drawn into a baghouse as it leaves the edge of the mold. The dust accumulates on filter bags which are mechanically shaken and the dust can drop into drums. This leaded steel dust can contain 50% to 80% lead oxides and is kept covered in drums and stored in a hazardous material storage facility. Figures 7 and 8 are pictures of the teeming aisle where the leaded steel dust is generated, and the baghouse where it is collected and stored.

A usual method of managing this material was to dispose of it in a hazardous waste landfill. In mid-1979, a lead smelter was located that could agglomerate the dust and recast it into lead ingots. Figure 9 is a picture of that facility, which is located in Canada.

Coke Plant Wastes

The last significant category of hazardous material which has been listed as hazardous waste is associated with cokemaking. Raw flushing liquor from the coke ovens which contains the volatiles driven off from the coal is processed in a by-product plant. A tar decanter is one of the units used to remove heavy organic material from the process stream, and the scraping from the bottom is called tar decanter tank sludge (tar sludge).

Figure 7. Teeming aisle.

As seen in Figure 10, tar sludge contains coal fines, coke particulate and tar and has a consistency which makes it difficult to handle. The material has heat value greater than coal and considerable resource value due to its high carbon content. Therefore, a significant effort is being made to recycle this material and avoid hazardous waste disposal costs and associated packaging and transportation problems.

One recycle system that has been successfully used for tar sludge consists of a steam heated box and auger type feed device that places the sludge back onto the coal going to a coke oven. Figure 11 shows the box with associated valving, and Figure 12 shows the chute positioned over the coal conveyor. Tar sludge is dumped in the box and the auger moves the material out of the box onto the coal. The coal is enroute to the hammer mill and then charged to the oven.

CONCLUSIONS

Impact of U.S. Steel Industry Hazardous Material Recycle/ Reuse on Hazardous Waste Landfill Capacity

Increased activity in the recycle and reuse of the steel industry's U.S. EPA characterized hazardous wastes

Figure 8. Leaded steel dust baghouse.

Figure 9. Leaded steel dust agglomeration facility.

Figure 10. Tar decanter tank sludge leaving the tar decanter

Figure 11. Tar decanter tank sludge recycle system–steam
heated box.

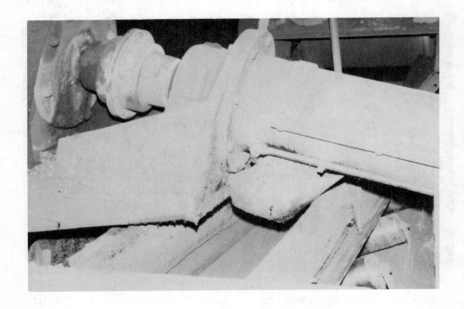

Figure 12. Tar decanter tank sludge recycle system–auger
chute over coal conveyor.

will reduce the hazardous waste landfill capacity presently
needed for these materials. Table II presents the four
significant steel industry generated hazardous wastes
and the estimated quantities of material presently not
recycled. The column entitled, "Without Additional Recycle/
Reuse" assumes that recycling does not expand beyond what
was estimated to exist as of November 1980. The land disposal capacities that are required each year for the entire
U.S. production are calculated for each waste type. The
number under the column entitled, "Area of 25-foot Landfill
Site" is an estimate of the area required for a landfill
with a 25-foot layer of waste. The spent pickle liquor has
been converted to neutralized sludge and is assumed to be
hazardous, which may or may not be the case. Under this
limited recycle/reuse scenario, 75 acres of hazardous waste
landfill capacity would be needed each year to dispose of
the U.S. steel industry's non-recycled solid hazardous waste.

If recycle projects continue forward in the future as
they have in the past, the hazardous waste disposal scenario
will change significantly as shown under the column entitled,
"With Maximum Recycle/Reuse." The average 80% recycle ratio
is assumed to hold for the industry on iron bearing materials
and coal tar sludge. Use of spent acids as a replacement for
municipal wastewater treatment plant chemicals and increased
use of acid regeneration technology is assumed to continue.
The minimum annual quantities of non-recycled wastes are
estimated using the assumptions listed in Table II. Required
annual landfill capacity will be reduced from 75 acres to 22
acres. Most of this capacity will be located in the Midwest,
generally where steel is made.

Impact of Regulations on Recycle/Reuse

It is important to note that the federal hazardous
waste management regulations tend to encourage the recycle
and reuse of some hazardous wastes and discourage the
recycle of other hazardous wastes. The section in the regulations that establishes this dual system is entitled,
"Special Requirements for Hazardous Waste Which is Used,
Reused, Recycled or Reclaimed" (40CFR Part 261.6). Every
hazardous material that is recycled or reused is identified
in Paragraph (a) or Paragraph (b) of Section 261.6.
Paragraph (a) provides that the recycling of the material
and storage, accumulation and transportation prior to recycling will be exempt from all waste management and
permitting regulations. Paragraph (b) places most of the
management and permitting regulations on the storage,
accumulation and transportation of the recycled materials.
With regard to steel industry hazardous materials, spent

41

TABLE II

U. S. STEEL INDUSTRY ANNUAL
HAZARDOUS WASTE DISPOSAL QUANTITIES

GENERATED HAZARDOUS WASTE IDENTIFIED AS NOT RECYCLED	WITHOUT ADDITIONAL RECYCLE/REUSE ANNUAL DISPOSAL REQUIREMENTS			WITH MAXIMUM RECYCLE/REUSE ANNUAL DISPOSAL REQUIREMENTS		
	(TONS)	(ACRE-FT)	AREA OF 25 FT HIGH LANDFILL SITE	(TONS)	(ACRE-FT)	AREA OF 25 FT HIGH LANDFILL SITE
SPENT PICKLE LIQUOR 1,850,000T	2,300,000[1]	1,400	60 ACRES	750,000[2]	500	18 ACRES
ELECTRIC FURNACE DUST 380,000T	380,000	250	10 ACRES	80,000[2]	50	2 ACRES
COKE PLANT WASTES 60,000T	60,000	40	2 ACRES	6,000[2]	5	< 1 ACRE
OTHER DUSTS & SLUDGES 100,000T	100,000	70	3 ACRES	20,000[2]	15	< 1 ACRE
TOTAL	2,840,000	1,760	75 ACRES	856,000	570	22 ACRES

(1) NEUTRALIZED SLUDGE TO WASTE PICKLE LIQUOR RATIO OF 1.25LB./LB. BASED ON ACTUAL OPERATING EXPERIENCE, 1981

(2)   a. 600,000 TONS OF SPENT HF NITRIC DISPOSED, ALL SULFURIC AND HYDROCHLORIC REUSED/RECYCLED.
     b. 80% RECYCLE OF EAF DUST AND OTHER DUSTS & SLUDGES
     c. CONVERSION TO CAUSTIC AMMONIA STILLS
     d. 80% RECYCLE OF TAR SLUDGE

pickle liquor that is reused as a wastewater treatment chemical in NPDES facilities is listed in Paragraph (a), while spent pickle liquor that is regenerated into new acid, coal tar sludge that is recycled in the coke plant, electric furnace dust that is reclaimed off-site and leaded steel dust that is reclaimed for the lead metal are covered in Paragraph (b). Until September 8, 1981, all of the previously discussed steel industry hazardous materials that were recycled were covered in Paragraph (b).

By including the spent pickle liquor exemption in Part 261.6(a), U.S. EPA has recognized that regulation of materials that are recycled/reused will inhibit that activity. Furthermore, U.S. EPA has made a clear statement that beneficial reuse and recycle of hazardous materials is a specific objective of RCRA.

State Versus Federal Handling of Recycle/Reuse

Some states have in place waste management laws and rules which inhibit the timely recycle and reuse of hazardous materials. All of the states that have rules that control resource recovery with greater stringency than the federal regulations claim that these rules protect the public. This perception is not true. Rules that discourage recycle, encourage land disposal. In addition to the possibility of negative impact on public health arising from land disposal (which EPA rules will attempt to prevent), such disposal results in the increased use of land for hazardous waste treatment and disposal facilities. The impact of recycle and reuse of steel industry wastes on land resources is shown in Table II. Regulation of recycle and reuse therefore threatens the misuse of two resources -- land and usable material (or energy).

References

1. Research Triangle Institute, "Environmental and Resource Conservation Considerations of Steel Industry Solid Waste," U.S. Environmental Protection Agency, Washington, D.C., April 1979.
2. American Iron and Steel Institute, "Comments Before U.S. EPA on Recycle and Reuse of Hazardous Waste," Washington, D.C., August 1980.
3. Bhattacharyya, S., "Steel Industry Pickling Waste and Its Impact on Environment," U.S. EPA, IIT Research Institute, Chicago, Ill., July 1979.
4. New York Department of Environmental Conservation, "New York Environmental News," Albany, N.Y., Volume VIII, Issue 13, Pg. 3, October 15, 1981.

CHAPTER 5

CLASSIFICATION AND MANAGEMENT OF
PROCESS WASTES FOR THE TENNESSEE
SYNFUELS ASSOCIATES COAL-TO-
GASOLINE FACILITY

Robert W. Rittmeyer
   Environmental Research &
   Technology, Inc.

John P. Fillo
   Environmental Research &
   Technology, Inc.

Douglas P. Malik
   Koppers Company, Inc.

Arch J. Merritt
   Koppers Company, Inc.

James P. Brennan
   Environmental Research &
   Technology, Inc.

Gerard A. Sgro
   Environmental Research &
   Technology, Inc.

INTRODUCTION

Tennessee Synfuels Associates (TSA), a joint venture partnership of Koppers Synfuels Corporation and Citgo Synfuels, Incorporated, proposes to construct and operate a commercial facility that will produce gasoline from locally available bituminous coal. The facility will be built in 10,000 barrel per day (bpd) modules, with an ultimate design capacity of 50,000 bpd of liquid products. Major byproducts that will be produced by the facility include liquid propanes and butanes, elemental sulfur and anhydrous ammonia. The first module could be in operation before the end of 1986, and the ultimate facility could be in operation by 1995.

The proposed site for the facility is located at the western limits of the City of Oak Ridge, in Roane County, Tennessee (Figure 1). It consists of approximately 1500 acres, and is situated on the south side of the Clinch River (River Mile 9 to 14) across from the Oak Ridge Gaseous Diffusion Plant. The site is presently owned by the U.S. Department of Energy, which proposes to transfer 1217 acres of the site to the City of Oak Ridge for subsequent resale to TSA. Construction of the first module could begin in 1982.

One of the primary media areas which must be addressed for commercial-scale synthetic fuels facilities is the environmental impacts associated with the generation and management of process wastes. Although coal-derived waste is the predominant class of wastes expected from the proposed facility, a wide variety of other wastes/waste classes are also produced. These wastes, non-hazardous or hazardous, must be identified, classified and managed according to federal and state regulations that govern such activities. In support of the preparation of an Environmental Impact Report (EIR), TSA commissioned the development of an inventory, preliminary classifications and management options for process wastes expected from the facility. Permitting activities associated with the management of nonhazardous and hazardous wastes is currently underway in accordance with the Resource Conservation and Recovery Act (RCRA) and Tennessee Department of Public Health (TDPH) regulations and will continue into the detailed design phase.

BACKGROUND

Regulatory Overview

The regulations arising from the Resource Conservation and Recovery Act (RCRA) of 1976 establish criteria which

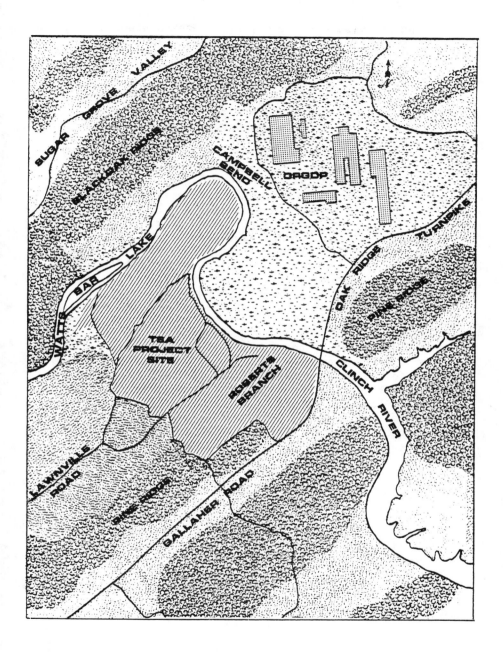

Figure 1. Proposed Oak Ridge plant site

classify certain wastes as hazardous and establish standards governing the generation, transportation, and management (i.e., treatment, storage, and disposal) of hazardous wastes. As part of the EPA Hazardous Waste Regulatory Program, any state may apply for authorization to establish and manage its own hazardous waste program provided that the program is equivalent to, or more stringent than the EPA program. The State of Tennessee developed its own program, and received Phase I authorization from EPA on January 16, 1981 to manage the program with respect to the identification, transportation, and management of hazardous wastes at existing facilities (Tennessee Department of Public Health, Division of Solid Waste Management Rules, Chapter 1200, Hazardous Waste Management; Effective March 2,1981).

In regard to the TSA environmental, health, safety and socioeconomic program, one of the more critical determinations of the RCRA regulations is the criteria pertaining to the classification of wastes (40 CFR Part 261, "Identification and Listing of Hazardous Wastes"). The corresponding Tennessee regulations, as they apply to the TSA project, are essentially equivalent except in one area: RCRA regulations exempt coal combustion wastes (bottom and fly ash) from all requirements, while the Tennessee regulations exempt these wastes from all requirements except notification and hazardous waste determination requirements (TDPH 1200-1-11.02(3)(ii)). A generator of such wastes must notify the Tennessee Department of Public Health that these wastes are produced and must test the wastes to determine if any hazardous characteristics are demonstrated, as defined in the regulations. For convenience, RCRA regulations are discussed and applied in the remaining portion of this discussion.

RCRA Part 261 regulations define four characteristics which trigger a hazardous classification, and identify as hazardous specific wastes known to demonstrate these characteristics. The four hazardous characteristics are ignitability, corrosivity, reactivity, and EP (extraction procedure) toxicity, and are defined as follows:

o   Ignitability (40 CFR 261.21): (1) a liquid, other than an aqueous solution containing less than 24 percent by volume alcohol, with a flash point less than 140°F; (2) a solid which is capable of causing fire at standard temperature and pressure; (3) an ignitable compressed gas per 40 CFR 173.300; or (4) an oxidizer per 49 CFR 173.151.

o   Corrosivity (40 CFR 261.22): (1) an aqueous
solution with a pH less than, or equal to, 2 or
greater than, or equal to, 12.5; or (2) a liquid
which corrodes SAE 1020 steel at a rate greater
than 0.250 inch/year at 130°F.

o   Reactivity (40 CFR 261.23): (1) a waste which
can undergo violent change, with or without
detonation, at standard temperature and pressure,
when exposed to water, or when heated under
confinement; (2) a waste which generates toxic
gases when mixed with water; (3) a cyanide- or
sulfide-bearing waste which, when exposed to pH
conditions between 2 and 12.5, can generate toxic
gases; or (4) a forbidden, Class A, or Class B
explosive per 49 CFR 173.51, 49 CFR 173.53, or
49 CFR 173.88, respectively.

o   EP toxicity (40 CFR 261.24): a waste which, when
subjected to the extraction procedure of 40 CFR
261 Appendix II, produces an extract which
contains any of the contaminants listed in
Table I at a concentration equal to or greater
than those listed. The concentrations listed are
100 times the U.S. Interim Primary Drinking Water
Standards.

After defining hazardous criteria, the RCRA
regulations list wastes known to demonstrate these
characteristics. Wastes are listed generically (i.e.,
regardless of source) and by industry/process source. In
addition, a series of commercial chemical products which,
if disposed of in "pure" form constitute hazardous wastes,
are listed. In all, 13 wastes are listed generically
(e.g., spent 1,1,1-trichloroethane used in degreasing), 75
wastes generated by 12 industries are identified (e.g.,
API separator sludge from the petroleum refining
industry), and 653 commercial chemical products are listed
(e.g., carbonyl chloride).

Without considering the many exceptions and
exclusions, a waste is hazardous if: (1) it is, or
contains, a listed waste, (2) it demonstrates any of the
four hazardous characteristics, or (3) it is otherwise
capable of causing environmental or health damage if
disposed of improperly. The last criterion places the
burden of proof on the generator. Knowledge of waste
characteristics and engineering judgement must be used to
ensure that a waste known to present environmental or
health risks is not managed as a non-hazardous waste

Table I. Maximum Concentration of Contaminants for
Characteristic of EP Toxicity (40 CFR 261.24)

Contaminant	Maximum Concentration, mg/1
Arsenic	5.0
Barium	100.0
Cadmium	1.0
Chromium (VI)	5.0
Lead	5.0
Mercury	0.2
Selenium	1.0
Silver	5.0
Endrin	0.02
Lindane	0.4
Methoxychlor	10.0
Toxaphene	0.5
2,4-D	10.0
2,4,5-TP Silvex	1.0

simply because it does not meet either of the first two criteria. To aid the generator in this judgement, EPA published a list of 374 toxic organic and inorganic compounds in Appendix VIII of 40 CFR Part 261. It should be noted that Appendix VIII of the regulations was provided for informational purposes only, and strictly speaking has no independent regulatory force.

Process Overview

The proposed TSA Coal-to-Gasoline facility is designed to produce unleaded, regular gasoline from coal using proven, commercial technologies (Figure 2). Coal is converted to synthesis gas using the Koppers-Babcock & Wilcox (KBW) gasification process. The synthesis gas is then converted to methanol using the Imperial Chemical Industries (ICI) Methanol Synthesis process, and the methanol is converted to gasoline using the Mobil Methanol-to-Gasoline (MTG) process. The ultimate facility will produce 50,000 bpd of liquid products from approximately 29,000 tpd of bituminous coal. A summary of resource requirements and products for the first module and the ultimate facility is presented in Table II.

Coal Preparation and Gasification

A high-temperature, atmospheric-pressure, entrained-flow KBW process is used to convert bituminous coals to synthesis gas. The plant is designed to use coal from the eastern Tennessee and southeastern Kentucky coal fields with the following average characteristics: a heating value of 11,560 BTU/pound, sulfur content of 1.9 wt percent, and an ash content of 14.3 wt percent. The coal is crushed, dried and pulverized from nominally 2" x 0" (i.e., capable of passing through a screen with 2-inch square openings) to a fineness with approximately 88 percent of the particles less than 0.10 mm in diameter (160 mesh). The KBW process gasifies the coal with steam and oxygen at temperatures in excess of 3000°F, thus decomposing essentially all tars, oils and phenols in the raw synthesis gas. The synthesis gas, products and waste water effluents from the plant are therefore cleaner from an environmental and health standpoint than those typically associated with low-temperature gasification or direct liquefaction processes for coal conversion.

Gas Processing

Raw synthesis gas from the gasifiers is cleaned and the composition is adjusted to enable the subsequent production

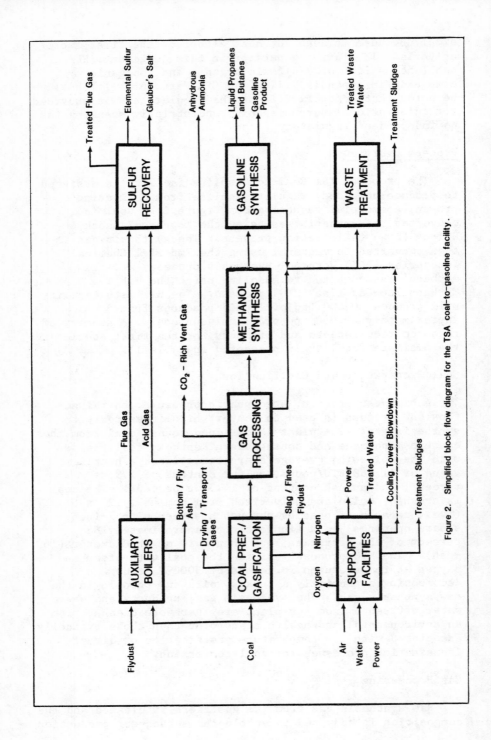

Figure 2. Simplified block flow diagram for the TSA coal-to-gasoline facility.

Table II. Summary of resource requirements and products for the TSA coal-to-gasoline facility

	10,000 barrels/ day module	50,000 barrels/ day facility
Resource		
Coal	5,800 tons/day	29,000 tons/day
Limestone	100 tons/day	500 tons/day
Oxygen	4,200 tons/day	21,000 tons/day
Water	11,360,000 gals/day	56,800,000 gals/day
Electricity	38 MWE	190 MWE
Products		
Gasoline	8,500 barrels/day	42,500 barrels/day
Propane	600 barrels/day	3,000 barrels/day
Mixed Butanes	900 barrels/day	4,500 barrels/day
Sulfur[a]	110 tons/day	550 tons/day
Anhydrous Ammonia	5 tons/day	25 tons/day

Footnote
(a) Based on an average feed coal sulfur content of 1.9 wt percent.

of methanol. Cleaning and quenching to remove entrained particulates and other potential pollutants (e.g., H_2S, NH_3, HCN) is accomplished in the following manner: dry cyclones for removal of gasifier flydust (~50 wt percent carbon), wet quenching for removal of potential pollutants, and wet electrostatic precipitators for removal of fine particulates. The synthesis gas is compressed and processed through (1) CO shift reactors to raise the H_2 to CO ratio for subsequent methanol production, and (2) COS hydrolysis reactors to convert COS to H_2S. The gas is processed through an ammonia scrubber and the selective, Selexol acid gas removal process to remove NH_3, and H_2S and CO_2, respectively, from the process gas. A guard bed of zinc oxide is used as the final gas treatment step to remove trace quantities of sulfur compounds from the gas and thus prevent fouling of the methanol synthesis catalyst.

Methanol Synthesis

Carbon monoxide and hydrogen are converted to methanol using the low-pressure, ICI Methanol Synthesis process. The primary $CO-H_2$ reaction is supplemented by reaction of CO_2 and H_2 to produce CO and H_2O (i.e., reverse water-gas shift reaction). Secondary reactions result in the formation of byproducts including dimethyl ether, higher alcohols, aldehydes, butanes, wax, carbonyl compounds and light-ends products. Effluent from the methanol converters is separated, gas is vented to the plant's fuel gas system, and the crude methanol sent to gasoline synthesis.

Gasoline Synthesis

Gasoline is produced from methanol using the fixed-bed Mobil MTG process. The methanol is partially dehydrated to form a mixture of methanol, dimethyl ether and water. This mixture is subsequently converted to hydrocarbons and water through olefin intermediates. Gasoline components each contain less than 10 carbon atoms due to the unique zeolite catalyst of the process. Depending upon seasonal and market variations, the heavier gasoline may require hydrotreating to reduce levels of durene. Gasoline product is of higher quality with an 88 average octane rating, and negligible amounts of sulfur, nitrogen or oxygenates that are typically found in petroleum-derived gasoline. An HF alkylation process is being considered for the production of additional gasoline product from propylenes and butanes for the ultimate 50,000 bpd capacity.

Major Auxiliary Facilities

A number of auxiliary facilities are required to support operations for the facility. An air separation plant provides 21,000 tpd of 99.5 percent purity oxygen for the gasifiers, and nitrogen for pneumatic coal transport and purging operations. An auxiliary boiler system, which uses coal and gasifier flydust as fuel and is rated at 5,400 MM BTU/hr fired duty, produces process steam for the facility. Raw river water is treated prior to use in the process and process waste water and facility runoff is treated on site prior to discharge to the Clinch River.

Provisions are made for the recovery of elemental sulfur and anhydrous ammonia as byproducts. An integrated system is used to recover sulfur, and consists of a modified Claus plant (fed with the H_2S-rich acid gas from the Selexol process), and a Citrate flue gas desulfurization process. During the gasification of coal, both hydrogen cyanide and ammonia are produced from the coal-bound nitrogen. The hydrogen cyanide is removed from the process gas and hydrolyzed to ammonia, and all the ammonia is recovered as commercial-grade, anhydrous ammonia using the U.S. Steel PHOSAM process.

Finally, an on-site, non-hazardous waste landfill is maintained for disposal of the majority of process wastes that are produced from the facility. Conceptual plans for disposal indicate that 12.2 million cubic yards of waste can be disposed at the Oak Ridge site: 3.6 million cubic yards in a 46 acre area at Roberts Branch and 8.6 million cubic yards in a 111 acre area at Campbell Bend. The combined capacity of these embankments provides sufficient disposal capacity for a single module with a 30-year design life, after which off-site disposal would required. All hazardous wastes are recycled or disposed off-site at an approved hazardous waste facility.

METHODOLOGY OF WASTE CLASSIFICATION, SEGREGATION AND MANAGEMENT

Three distinct activities comprised the waste investigations performed on the TSA project: (1) development of a waste inventory, (2) characterization/classification of wastes, and (3) development of waste management scenarios.

Development of the Waste Inventory

In order to classify wastes and develop management scenarios for them, the following information is required:

o   process source (including treatment systems preceding ultimate disposal),

o   generation rate (quantity/time), and

o   generation schedule (continuous or intermittent).

The process source is an important input to waste characterization/classification because it forms the basis for engineering judgement regarding potentially hazardous waste properties (especially when waste samples are not available) and because it allows a comparison against industry/process source hazardous-waste listings in the RCRA regulations. Waste characterization/classification, generation rate and schedule are all necessary to develop a feasible and acceptable management system.

Detailed process flow diagrams for the TSA project were first reviewed to develop a listing of potential sources of process waste. Wastes generated by both process units and by in-plant treatment units (i.e., raw and treated wastes) were considered. Process designers were then consulted to refine this listing and to develop generation rates and schedules for the final set of process wastes. In many cases it was not clear whether a process source would produce a waste, or what its generation rate and schedule might be because the facility has only selected commercial counterparts which are currently operating. In these cases, engineering judgement and experience with similar processing units in analogous industries were utilized to develop waste inventory data.

Waste Characterization/Classification

As previously discussed, a waste is hazardous if:

1)   it is, or contains, a listed waste;
2)   it demonstrates any of the four hazardous characteristics (ignitability, corrosivity, reactivity, or EP toxicity); or
3)   it is otherwise capable of causing environmental or health damage if disposed of improperly.

These criteria are useful for an existing facility whose wastes are specifically listed (criterion 1), where wastes can be sampled and tested against the hazardous criteria (criterion 2), or where waste characteristics are known (criterion 3). Unfortunately, these descriptors do not apply directly to the TSA facility. No wastes are presently listed in RCRA regulations for synthetic fuels processes, wastes from the facility cannot yet be sampled and tested, and limited representative characterization data are available because the facility has only selected operating counterparts.

Given these limitations, process wastes generated by the proposed facility were characterized/classified based on available design information, literature data, and engineering judgement. Hazardous classification criteria were revised into the following criteria:

1) Is the waste, or its source, similar to a waste listed for an analogous industry (e.g., petroleum refining); or

2) Will, or could, the waste exhibit the characteristics of ignitability (flash point below 140°F), corrosivity (pH less than 2 or greater than 12.5), reactivity (appreciable cyanide or sulfide content), and/or toxicity (appreciable levels of heavy metals included in EP toxicity criteria or metals or organics included in Appendix VIII).

Wastes were classified by mapping these criteria against waste characteristics, either documented or suspected on the basis of engineering judgement. There were three outcomes from this exercise: (1) the waste is clearly or probably hazardous, (2) the waste is probably non-hazardous, but additional data are required to make a final judgement, or (3) the waste is clearly non-hazardous. Therefore generated wastes were classified as potential hazardous, probable non-hazardous, and non-hazardous. It should be noted that these classifications are preliminary and will be confirmed by analyses to be performed on all wastes that are generated by the facility. Indeed, the exercise outlined herein could be viewed as the precursor of a waste analysis program.

Development of Management Scenarios

For each process waste that was identified in the waste inventory, probable and alternate management scenarios were developed based on engineering design, anticipated waste characteristics, and knowledge of the

57

capabilities of standard management techniques. Management techniques that were considered included landfill disposal, landfarming, incineration, physical/chemical treatment, resource recovery, and sales.

Landfill disposal was selected for wastes anticipated to have little recovery value, high ash content, low BTU content, or wastes not amenable to physical/chemical treatment. This option was further divided into disposal in the on-site, sanitary (non-hazardous waste) landfill or an off-site, secure (hazardous waste) landfill. Landfarming was specified principally for organic sludges similar to those presently disposed of in this manner, but was not identified as a probable management technique for any waste because present facility design does not include an on-site landfarm. Incineration was considered for organic wastes with low ash content and high BTU content. Incineration could be accomplished by directly feeding a waste to the facility's auxiliary boiler or by blending a waste into the plant fuel oil pool (for in-plant use). Treatment was evaluated for wastes amenable to elementary physical/chemical methods of reducing hazardous properties (e.g., neutralization) or reducing volume (e.g., dewatering). Resource recovery was specified for wastes whose reclamation would be attractive because they contain commercially valuable materials or are of a proprietary nature. Sales was selected for materials which would be marketable byproducts.

Specified management scenarios are preliminary and are based on the best information available at this stage of engineering design and engineering judgement. As additional waste characterization data becomes available, the set of management techniques will be refined as appropriate. For example, data may show that a waste presently classified as potentially hazardous is actually non-hazardous. In this case this waste could be disposed of in the on-site sanitary landfill instead of an off-site secure landfill.

RESULTS OF ANALYSIS

Tables III, IV, and V present by plant operating area generated wastes, generation rate and schedule, potential hazard classification (or, for non-hazardous wastes, the basis for this classification), and proposed management techniques for potential hazardous, probable nonhazardous, and non-hazardous wastes, respectively. Figures 3 and 4 are schematics of the waste management techniques proposed for the facility. The information that is presented represents best estimates based on current

58

Table III. Summary of Potential Hazardous Wastes and Management Techniques for the TSA Facility

Process Area	Source	Waste Description	Rate	Potential Hazard Classification	Proposed Management
Secondary Gas Compression and Methanol Synthesis	Wax Strainers	Wax	25 ft^3/yr (intermittent)	Toxic	Plant Fuel
HF Alkylation	Acid Rerun Column	Tar	1,500 ft^3/yr (continuous)	Toxic, Corrosive	Landfill [a]
	Propane Defluorinator	Spent Catalyst	6,500 lb (every 6 months)	Toxic	Landfill
	Spent Caustic Neutralizer & Precipitation	Sludge	3 tpy [b] (intermittent)	Toxic	Landfill
Product Blending and Storage	Storage Tanks	Bottom Sludges	23,400 ft^3/yr [c] (intermittent)	Ignitable, Corrosive, Toxic	Landfill
Wastewater Treatment	HCN Hydrolysis Reactor	Spent Catalyst	25 tons (every 6 months)	Toxic, Reactive	Landfill

Table III (Continued)

Process Area	Source	Waste Description	Rate	Potential Hazard Classification	Proposed Management
Wastewater Treatment	Process Area Runoff Collection Basin/API Separator	Skimmed Oil	ND[d]	Toxic	Plant Fuel
Auxiliary Service Systems	Seal Oil System	Waste Oils	65,000 gallons (once a year)	Toxic	Plant Fuel

Footnotes

(a) All materials that are determined to be hazardous will be disposed of off-site in an approved hazardous waste landfill.

(b) Rate reported on a dry basis.

(c) Basis: 1 ft. of sludge accumulates in each tank every five years.

(d) ND = no data currently available. However, skimmed oil from the process area runoff basin will be minor due to good housekeeping of the process area.

Table IV. Summary of Probable Non-Hazardous Wastes and Management Techniques for the TSA Facility

Process Area	Source	Waste Description	Rate	Potential Hazard Classification	Proposed Management
Coal Feeding & Gasification	Slag Quench Tank	Gasifier Slag	493,000 tpy[b,h] (continuous)	Toxic, Reactive	Landfill
Gas Cooling and Cleaning	Cyclones	Gasifier Flydust	1,454,000 tpy[h] (continuous)	Toxic, Reactive	Auxiliary Boiler
	Clarifier	Filter-Cake Solids	269,000 tpy[c,h] (continuous)	Toxic, Reactive	Landfill
Primary Gas Compression and CO Shift	CO Shift Converters	Spent Catalyst	490 tons (every 3 to 4 years)	Toxic	Return to Vendor
	COS Hydrolysis Reactors	Spent Catalyst	640 tons (every 3 years)	Toxic	Return to Vendor
Secondary Gas Compression & Methanol Synthesis	ZnO Guard	Spent Catalyst	920 tons (every 2 years)	Toxic, Reactive	Landfill

Table IV (Continued)

Process Area	Source	Waste Description	Rate	Potential Hazard Classification[a]	Proposed Management
Secondary Gas Compression & Methanol Synthesis	Methanol Reactor	Spent Catalyst	1,630 tons (every 3 to 5 years)	Toxic	Return to Vendor
Gasoline Synthesis	DME Reactors	Spent Catalyst	8,450 ft^3 (once every 2 years)	Toxic	Return to Vendor
	MTG Reactors	Spent Catalyst	32,400 ft^3 (once every year)	Toxic	Return to Vendor
HF Alkylation	Feed Driers	Spent Desiccants	2550 ft^3 (every 2 to 3 years)	Toxic	Landfill
	Acid Oil Washer	Spent Caustic	8,250 gal/yr (every 20 days)	Toxic[d]	Neutralizer and Precipation; Wastewater Treatment

Table IV (Continued)

Process Area	Source	Waste Description	Rate	Potential Hazard Classification	Proposed Management
Sulfur Recovery	Claus Plant	Spent Catalyst	375 tons (every 3 years)	Toxic	Landfill
	Citrate Unit	Glauber's Salt	11,300 tpy (intermittent)	Toxic	Sales
	Claus Plant	Sulfur		Toxic	Sales
	Citrate Unit	Sulfur	182,000 tpy (continuous)	Toxic	Sales
Wastewater Treatment	Coal and Slag Pile Runoff Collection Basin	Solids 850 tpy[e] (intermittent)		Toxic	Return to Coal Pile
	Biological Treatment	Sludge Wastage	190,000 tpy[f] (continuous)	Toxic	Dewatering
	Dewatering	Dewatered Sludge	120,000 tpy[g] (continuous)	Toxic	Landfill

Table IV (Continued)

Footnotes

(a) Characteristics that could result in hazardous classification for each respective waste

(b) Rate reported on a wet basis @ 15% water

(c) Rate reported on a wet basis @ 35% water

(d) Spent caustic is neutralized prior to disposal, hence it will not be corrosive

(e) Rate reported on a dry basis

(f) Rate reported on a wet basis @ 99% water

(g) Rate reported on a wet basis @ 87% water

(h) Rate based on plant feed coal containing 20 wt percent ash.

64

Table V. Summary of Non-Hazardous Wastes and Management Techniques for the TSA Facility

Process Area	Source	Waste Description	Rate	Potential Hazard Classification	Proposed Management
HF Alkylation	Butane KOH Treater	Spent Caustic	240 gal/yr (intermittent)	Neutralized prior to discharge. Contains only KF and KOH.	Neutralizer; Wastewater Treatment
	Propane KOH Heater	Spent Caustic		Neutralized prior to discharge. Contains only KF and KOH.	Neutralizer; Wastewater Treatment
	Acid Relief Neutralizer	Spent Caustic	7,860 gal/yr (every 3 weeks)	Neutralized and precipitated prior to discharge. Contains only inorganic salts (NaCl, CaCl2) and water.	Neutralizer & precipitation; wastewater treatment.

65

Table V (Continued)

Process Area	Source	Waste Description	Rate	Potential Hazard Classification	Proposed Management
Auxiliary Steam Generation	Auxiliary Boiler	Bottom Ash	1,321,000 tpy[a] (continuous)	Exempt per 40 CFR 261.4	Landfill
	Auxiliary Boiler	Fly Ash		Exempt per 40 CFR 261.4	Landfill
Makeup Water Treatment	Raw Water Clarifier	Solids	495,000 tpy[b] (continuous)	Suspended inert solids from raw river water	Dewatering[c]
Wastewater Treatment	Utility Wastewater Sedimentation	Solids	212,000 tpy[b] (continuous)	Inert solids/ salts	Dewatering[c]
Auxiliary Service Systems	Instrument Air Driers	Spent Desiccants	150 ft^3 (every 2 to 3 years)	Saturated with water only	Landfill

Footnotes

(a) Rated reported on a dry basis, and based upon plant feed coal containing 20 wt percent ash

(b) Rated reported on a wet basis @ 98% water

(c) Solids are dewatered in Area 16 and ultimately landfilled (see Table IV)

66

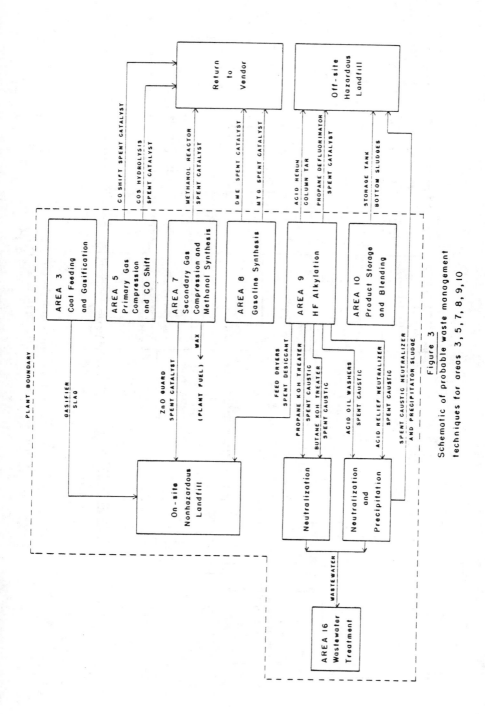

Figure 3

Schematic of probable waste management
techniques for areas 3, 5, 7, 8, 9, 10

67

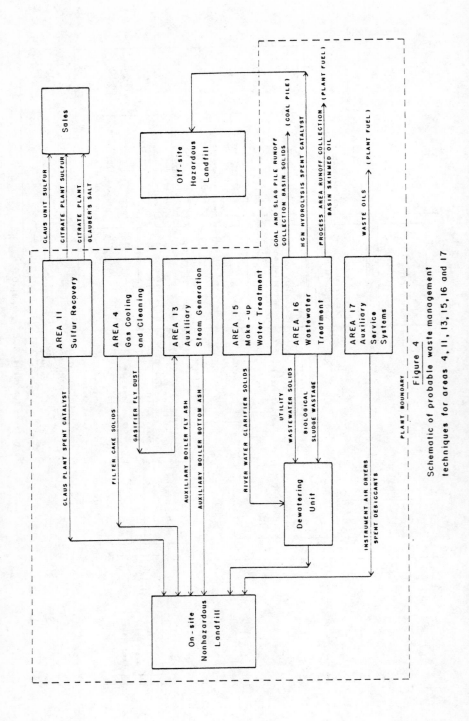

Figure 4

Schematic of probable waste management
techniques for areas 4, 11, 13, 15, 16 and 17

68

process design, available waste characterization data, and engineering judgement.[1]   This information will be updated as the process design develops further, and will be verified once plant operation is initiated.

The facility (i.e., 50,000 bpd capacity) will generate approximately 4.6 million tons/year of raw wastes (i.e., before in-plant management). These wastes can be divided into five broad categories:

o   coal derived wastes (e.g., gasifier slag) which account for about 77 percent of total raw wastes,

o   organic wastes (e.g., methanol synthesis wax) which account for approximately 4 percent of total raw wastes,

o   byproducts (e.g., Glauber's salt) which account for roughly 4 percent of total raw wastes,

o   spent catalysts (e.g., shift catalyst) which account for about 0.06 percent of total raw wastes, and

o   other wastes (e.g., spent desicants) which account for approximately 15 percent of total raw wastes.

In-process management scenarios shown on Figure 3 (e.g., firing gasifier flydust in the auxiliary boilers), reduce the amount of wastes actually leaving process units from 4.6 to 2.4 million tons/year, a reduction of 48 percent. After in-process management, generated wastes are comprised of 87 percent coal-derived wastes, 0.6 percent organic wastes, 8 percent byproducts, 0.1 percent spent catalysts, and 5 percent other wastes. Of the wastes ultimately leaving process units, 0.03 percent are classified as potential hazardous, 45 percent as probable non-hazardous and 55 percent as non-hazardous.

DISCUSSION

The following discussion details the process of waste characterization/classification and management scenario development for three categories of wastes generated by the facility: coal-derived wastes, organic wastes, and spent catalysts. Development of the waste inventory is not addressed because in all cases this information was developed following the generic methodology previously outlined and was accomplished in conjunction with process designers.

Coal-Derived Wastes

Coal-derived wastes generated by the facility include gasifier slag, gasifier flydust, clarifier filter-cake solids, and boiler bottom and fly ash. Classification and management scenario development for these wastes follows.

Classification

As indicated in Table IV, gasifier wastes (slag, flydust, and clarifier filter-cake solids) were classified as probable non-hazardous, but could possess characteristics ultimately resulting in a toxic and/or reactive wastes classification. Gasifier slag could be considered toxic due to metals present in feed coal not volatilized in the gasifier, and/or due to trace organics deposited by raw gas contact cooling/cleaning water used for slag quenching. Gasifier slag could also be considered reactive if it contains appreciable levels of calcium sulfide (CaS) formed from the gasifier limestone flux. Gasifier flydust and clarifier filter cake solids (essentially equivalent wastes except for particle size) could be considered toxic due to metals present in feed coal, or reactive due to calcium sulfide.

Although these wastes could demonstrate the hazardous characteristics outlined above, review of data on leaching behavior of coal-derived wastes (i.e., char, ash, slag) from other coal conversion facilities and conventional coal combustion facilities suggests that these wastes will not be classified as toxic due to metals content. Summaries of leaching data versus EP toxicity criteria (40 CFR 261.24) are provided in Table VI for coal conversion facilities and Table VII for coal combustion facilities. These data indicate that in all cases the concentration of toxic metals in waste leachates are well below existing EP toxicity criteria. Although these data do not strictly apply to the TSA facility due to process differences (which could affect waste composition) and because they do not in all cases represent rigorous EP toxicity analyses as prescribed by RCRA (which could affect leachate composition), they do indicate that gasifier wastes will probably not exhibit the characteristic of EP toxicity. Analyses of these wastes, when available, will provide a more substantive data base to establish their final classification, and will also address the potential for toxic organic deposition on gasifier slag from quench water and the potential reactivity of all three wastes.

Table VI.   Composition of Coal Conversion Char/Ash/Slag
Leachates Versus EP Toxicity Criteria[a]

Constituent	Range of Leachate Conc. mg/l (b)	Maximum Conc. for EP Toxicity, mg/l
Arsenic	< D[c]	5.0
Barium	< D – 40.5	100.0
Cadmium	< D – 0.27	1.0
Chromium	< D – 0.18	5.0 (Chromium VI)
Lead	< D – 1.8	5.0
Mercury	< D	0.2
Selenium	NR[d]	1.0
Silver	NR	5.0

Footnotes

(a) 40 CFR 261.24; TDPH Rule 1200-1-11-.02-12.
(b) Abstracted from references 2,3 and 4. Analyses run on
    char/ash/slag from H-Coal, Lurgi (Westfield), Hygas,
    and Synthane gasifiers firing Illinois No. 6 coal.
(c) Below detection limits.
(d) Not reported.

Table VII

Composition of Conventional Coal Combustion Wastes
Versus EP Toxicity Criteria(a)

Constituent	Bottom Ash Leachate(b)	Concentration mg/l Fly Ash Pond Overflow(c)	Sluice Pond Overflow(d)	Maximum Concentration for EP Toxicity
Arsenic	0.065	0.01-2.5	0.002-0.084	5.0
Barium	<1.0	0.3-1.2	0.3-40.0	100.0
Cadmium	<0.01	0.06	0.001-0.01	1.0
Chromium	<0.02	0.15	0.001-1.0	5.0 (Chromium VI)
Lead	<0.04	NR(e)	0.0027-0.024	5.0
Mercury	0.0006	NR(e)	0.003-0.015	0.2
Selenium	0.02	0.1(e)	0.0005-0.47	1.0
Silver	NR(e)	NR	NR(e)	5.0

Footnotes:

(a) 40 CFR 261.24: TDPH Rule 120-1-11-.02-12
(b) Multiple power plants were investigated, see Reference No. 5.
(c) Two power plants were investigated, see Reference No. 6.
(d) Multiple power plants were investigated, see Reference No. 7.
(e) Not reported.

Boiler wastes (bottom and fly ash) are exempt from RCRA requirements per 40 CFR 261.4.* Aside from the regulatory exemption, the data presented in Table VI also suggest that these wastes would not be toxic due to residual metals content. Therefore they were classified as non-hazardous.

Management

Coal-derived wastes are generated in large quantities: they account for approximately 2 million tons/year of the 2.4 million tons/year of wastes ultimately leaving the process plant. Gasifier slag, boiler bottom ash, and boiler fly ash share similar characteristics with regard to management scenario development. These wastes are largely inert solids (ash) and are not amenable to landfarming, incineration, or treatment. The wastes also contain only minute amounts of commercially valuable materials, and therefore are not candidates for resource recovery or sales. Landfill disposal was therefore specified as the management technique for these wastes. Due to the very limited options to dispose of these wastes, alternate management techniques were not developed.

Gasifier flydust and clarifier filter-cake solids demonstrate most of the characteristics previously discussed for gasifier slag and boiler ashes with the exception that these two wastes do contain appreciable levels of residual carbon (~50 percent) and therefore possess reclaimable heating value. The gasifier flydust is collected in a dry state (via cyclones) and can be conveyed to the auxiliary boiler for destruction. The flydust will supply approximately 14 percent of the auxiliary boiler's fuel. Clarifier solids, however, contain approximately 35 percent water which heavily penalizes the recoverable heat value and handling and firing logistics. Incineration was therefore not selected for clarifier solids, and this waste will be landfilled. Landfill disposal was also selected as the alternate management technique for gasifier flydust because it is the only viable option other than incineration. Due to the limited options for disposal of clarifier solids, no alternate management technique was developed for this waste.

---

*As discussed earlier, Tennessee regulations exempt these wastes from all except Notification and hazardous waste determination requirements.

Organic Wastes

Organic wastes generated by the facility include methanol synthesis wax, HF alkylation tar, storage tank bottom sludges, biological treatment sludge wastage, API separator oil skimmings, and waste seal oil system oils.

Classification

Methanol synthesis wax is generated by polymerization reactions occurring in the methanol synthesis reactors, and is removed from reactor effluent by in-line strainers. This waste could contain complex, toxic organics formed by polymerization reactions. For this reason methanol synthesis wax was classified as a potential hazardous waste due to toxicity.

HF alkylation tar is essentially still bottoms from the acid rerun column, and could contain complex, toxic organics formed by polymerization reactions during alkylation. This waste could also contain residual hydrofluoric acid (HF), which is a listed hazardous waste (due to toxicity and corrosivity (40 CFR 261.33(f)) and is also listed as a toxic constituent in 40 CFR 261, Appendix VIII. Hence HF alkylation tar was classified as a potential hazardous waste.

Storage tank bottom sludges are generated from the settling of contaminants from intermediates, products, and other plant stock stored in bulk form (e.g., gasoline additives). These wastes could be ignitable, corrosive, or toxic depending upon characteristics of the material from which they settle. All storage tank bottoms were classified as potential hazardous wastes to be conservative because the exact characteristics of materials are not available at this time.

Biological treatment sludge wastage is comprised of excess microorganisms used to treat plant wastewaters. A portion of plant wastewater entering biological treatment is used to cool and clean raw product gas and could contain appreciable levels of toxic metals and organics. These toxic materials could subsequently concentrate in the biological sludge. Although this is possible, biological sludge samples taken from four operating refineries and subjected to EP toxicity analyses yielded extracts with metals concentrations far below the EP toxicity criteria presented in Table I.[8] These data do not necessarily address the potential toxicity of TSA biological sludge due to process differences (e.g., feedstock) and because organics content of the refinery sludges was not investigated, but do suggest that this waste could ultimately be classified as non-hazardous. Biological treatment sludge wastage was classified as probable non-hazardous[*] because there is reason to believe this waste will be non-hazardous, but additional data are required to allow a final judgement.

API separator oil skimmings consist of organics and other materials contaminating process area runoff. This waste could contain toxic organics, and is generated by the same process unit that generates a listed hazardous waste for petroleum refining, API separator sludge. For these reasons oil skimmings from the API separator were classified as a potential hazardous waste.

Waste seal oils, lubricating/seal oils which are periodically replaced with fresh oils, will contain organics and metals (additives) themselves and may also absorb metals and organics during use. Because these organics and metals may be toxic, and also because EPA intends to repropose the listing of waste oils as a hazardous waste[9], this waste was classified as a potential hazardous waste.

Management

Methanol synthesis wax, API separator oil skimmings, and waste seal oils all possess reclaimable heating value and could easily be blended into the plant fuel oil pool for in-plant use. Therefore, the primary management technique selected for these three wastes was used as in-plant fuel. The alternate management technique chosen for these wastes was landfill disposal, but methanol synthesis wax and API oil skimmings could be managed at an off-site waste incinerator, if available, and waste seal oils may be reclaimable.

HF alkylation tar and storage tank bottom sludges may also contain reclaimable heating value, but will probably be contaminated by non-combustible or inert materials (e.g., storage tank sludges typically contain high levels of tank metal corrosion products) to a degree which will make them unsuitable for use as in-plant fuel. These wastes are also not amenable to physical/chemical treatment, and contain little if any valuable materials

---

*Dewatered sludge, which is comprised of three raw sludges including biological treatment sludge wastage, was similarly classified as probable non-hazardous.

which would qualify them for resource recovery or sales.
Hence these wastes will probably be landfilled.
Alternatively these wastes could be incinerated in an
off-site waste incinerator, if available, because these
facilities are typically equipped to handle wastes with
appreciable levels of non-combustible materials.

Biological treatment sludge wastage contains approximately 99 percent water as generated, making it a viable
candidate for dewatering (physical treatment). Dewatering
of biological sludge to roughly 87 percent water reduces
the volume for ultimate disposal by 93 percent. Ultimately
dewatered sludge, a combination of dewatered biological
sludge and two other dewatered sludges (raw water clarifier
and utility wastewater sedimentation sludges) which consist
primarily of inert materials, will be disposed of in a
landfill. Should biological sludge not be dewatered, it
could alternatively be landfarmed as are many wastewater
treatment sludges that are generated by the petroleum
refining industry and municipal treatment works.

Spent Catalysts

The TSA facility will generate spent catalysts from the
following processes: CO shift, COS hydrolysis, zinc-oxide
(ZnO) guard, methanol synthesis, gasoline synthesis (DME
and MTG catalysts), propane defluorinator, Claus plant, and
wastewater HCN hydrolysis.

Classification

Carbon monoxide shift, COS hydrolysis, ZnO guard, and
methanol synthesis catalysts all process gasifier product
gas. Product gas could contain toxic metals and organics
which could subsequently plate out on these catalysts
causing them to be hazardous wastes due to toxicity.*
However, levels of metals and organics in product gas
feeding these units are expected to be minimal because
these units are downstream of gas cooling and cleaning,
where the bulk of metals and organics are expected to be
removed from product gas. These materials were classified
as probable non-hazardous wastes because there is reason to
believe that these catalysts will be non-hazardous, but
additional data are required to allow a final judgement.

---

*ZnO guard catalyst could also be reactive due to sulfides
content. It should be noted that no catalyst to be used
at the TSA facility is toxic in and of itself.

Propane defluorinator catalyst is used to remove residual fluorine from product propane leaving HF alkylation. Since fluorine is a listed hazardous waste due to toxicity (40 CFR 261.33(e)) and also is listed a toxic constituent in 40 CFR 261 Appendix VIII, this waste was classified as a potential hazardous waste due to toxicity.

Claus catalyst processes H_2S-rich acid gas removed from gasifier product gas by the Selexol unit. Gasifier product gas could contain toxic metals and organics which could adsorb onto Claus catalyst and cause it to be hazardous due to toxicity. However, levels of metals and organics in the gas streams feeding the Claus unit are minimal because these materials are expected to be removed from gasifier product gas during gas cooling and cleaning and, if necessary, by contact with Selexol solution. Spent Claus catalyst was classified as probable non-hazardous because there is reason to believe that it will be non-hazardous, but additional data are required to allow a final judgement.

Hydrogen cyanide hydrolysis catalyst is used to convert HCN removed from plant wastewater to ammonia for subsequent recovery via the PHOSAM process. Plant wastewater is used to cool and clean raw product gas, and may contain appreciable levels of toxic metals and organics. These materials could subsequently adsorb onto the HCN hydrolysis catalyst, causing it to be hazardous due to toxicity. In addition, residual cyanide compounds if present on the spent catalyst would cause it to be reactive. For these reasons, spent HCN hydrolysis catalyst was classified as a potential hazardous waste.

Management

Of the set of six management options considered, essentially only two apply to spent catalysts: resource recovery and landfill disposal. Resource recovery (noted in Tables III, IV, and V as return to vendor) was considered for catalysts containing valuable metals (e.g., cobalt, platinum, chromium) or those of a proprietary nature. Catalysts for which resource recovery was selected as the proposed management technique include CO shift catalyst (cobalt), COS hydrolysis catalyst (platinum), methanol synthesis catalyst (proprietary), and gasoline synthesis catalysts (DME and MTG: proprietary). In all cases where resource recovery was selected as the probable management technique, landfill disposal was chosen as its alternate.

Landfill disposal was chosen as the primary management technique for propane defluorinator, Claus, and HCN hydrolysis catalysts because the catalysts are all alumina-based. Landfill disposal was also selected for ZnO guard catalyst because, although zinc recovery itself may be economically viable, the cost of equipment to control large sulfur dioxide emissions associated with reclaiming spent guard catalyst probably would make its reclamation uneconomical. For all of these catalysts, resource recovery was specified as the alternate management technique.

CONCLUSIONS

The TSA facility will generate approximately 4.6 million tons/year of raw wastes (i.e., before in-plant management) consisting of coal-derived wastes (77 percent), organic wastes (4 percent), byproducts (4 percent), spent catalysts (0.06 percent), and other wastes (15 percent). In-process management of selected raw wastes reduces the amount of wastes actually leaving process units from 4.6 to 2.4 million tons/year. Of the 2.4 million tons/year of disposed wastes, 87 percent are coal-derived, 0.6 percent are organic, 8 percent are byproducts, 0.1 percent are spent catalysts, and 5 percent are other wastes. Disposed wastes were classified as potential hazardous (0.03 percent), probable non-hazardous (45 percent) and non-hazardous (55 percent).

Information that is presented herein represents best estimates based on current process design, available waste characterization data, and engineering judgement. This information will be updated as the process design develops and will be verified once plant operation begins. Initially all wastes generated by the TSA facility will be sampled and analyzed to determine their appropriate hazard classification (as required by 40 CFR 262.11). Once all wastes have been classified, their physical and chemical characteristics will be monitored as described below.

Wastes classified as hazardous will be subjected to routine analyses according to a general waste analysis plan to be developed for the facility (as required by 40 CFR 264.13). The plan will specify (1) parameters for analysis, (2) test methods, (3) sampling methods, and (4) the frequency with which analyses are to be repeated. Items (1) through (3) are highly site and waste specific and cannot be presented at this time. The frequency of analyses will parallel waste generation frequency. Wastes which are continuously generated will be analyzed at least semianually, or more often if any process, operational or other change occurs which could affect waste characteristics. Wastes which are generated infrequently will be analyzed when they are generated.

Non-hazardous wastes will be analyzed at least annually, or more often if any process, operational, or other change occurs which could significantly affect their characteristics. This action will satisfy the hazardous waste determination requirements of 40 CFR 261.11. In addition, leachate collected from the on-site, sanitary (non-hazardous) landfill will be analyzed to ensure that the landfill is not receiving and subsequently discharging hazardous constituents.

ACKNOWLEDGEMENTS

The work presented herein reflects substantial cooperative efforts between Tennessee Synfuels Associates and their subcontractors Environmental Research & Technology, Inc. and M.W. Kellogg. The authors wish to thank Koppers' Engineering and Construction Group and M.W. Kellogg's Process Design and Systems Engineering Department for their continued cooperation and support throughout the course of this work. The authors also wish to acknowledge the contributions of staff from ERT's Pittsburgh Engineering Division.

REFERENCES

1. Tennessee Synfuels Associates, "Environmental Impact Report: Coal-to-Gasoline Facility, Oak Ridge, Tennessee," Prepared by Environmental Research & Technology, Inc., (1981).

2. Griffin, R.A., R.M. Schuller, J.J. Suloway, S.A. Russell, W.F. Childers and N.F. Shimp, "Solubility and Toxicity of Potential Pollutants in Solid Coal Wastes", Proceedings from the Symposium on the Environmental Aspects of Fuel Conversion Technology, III, EPA-600/7-78-063, (1978).

3. Luthy, R.G. and M.J. Carter, "Experimental Analysis of the Leaching Characteristics of Residual Hygas Coal Gasification Solids," Carnegie-Mellon University, (1978).

4. Neufeld, R.D., J. Bern and M.A. Shapiro, "Environmental Significance of Coal Conversion Residuals", University of Pittsburgh, (1979).

5. Keairns, D.L., C.C. Sun, C.H. Peterson and R.A. Newky, "Fluid-Bed Combustion and Gasification Solids Disposal", Westinghouse Corporation, (undated).

6. Chu, T.-Y.J. and R.J. Ruane, "Characterization and Re-use of Ash Pond Effluents in Coal-Fired Power Plants," Tennessee Valley Authority, (1978).

7. Argonne National Laboratory, "Environmental Control Impacts of Generating Electric Power from Coal", (1976).

8. Environmental Research & Technology, Inc., Client-confidential data, (1980).

9. Federal Register, Volume 7, No. 70, (1982).

CHAPTER 6

DETERMINATION OF DISPOSAL SITES
FOR PCB CONTAMINATED MATERIAL DREDGED
FROM THE HUDSON RIVER BED

J.S. Reed
J.C. Henningson
        Malcolm Pirnie, Inc.

INTRODUCTION

Contaminated sediments may be removed from a river bottom in several situations. Historically, maintenance dredging in navigation channels has been required wherever shoaling occurs, including in contaminated reaches of rivers. More recently, remedial dredging has been proposed in several river systems as a means of removing toxic materials which have accumulated in the river bottom and pose a threat to the aquatic environment.

Both types of dredging serve a useful purpose in our society, but result in the same problem: finding disposal sites for the unwanted, contaminated dredge spoil. Ocean dumping of the spoil is often precluded by high transport costs from inland river dredge sites, as well as the potential harm to the ocean environment. In-river disposal of contaminated materials may reduce transport costs, but incurs similar environmental concerns, especially long-term site stability and the practicality of maintenance and monitoring. State-of-the-art methods appropriate for more concentrated forms of hazardous waste, including microbial decomposition and incineration, have not been demonstrated as practical for dredge spoil.

Upland containment is frequently the preferred alternative, because the material is removed from the aquatic system and placed in a secure containment site. Long-term monitoring and maintenance are facilitated, as is the collection and treatment of leachate. However, different issues arise. The shoreline areas of contaminated river reaches are likely to be heavily populated, making isolated disposal sites difficult to find. Human health issues arise, including protection of surface and well drinking water supplies, air quality, and local agriculture. Land use problems may be created, including conflicts with existing or intended uses, and possible negative impacts on the local tax base or property values.

SITE SCREENING

Given the myriad of environmental, socioeconomic and political issues associated with upland disposal, it is imperative that any site selection process be scientifically valid, objective and thorough. This paper describes a particular site screening procedure which was successfully applied to two dredging projects in the Hudson River, one a remedial dredging project in the Upper Hudson, the other a maintenance dredging project in the Lower Hudson.

The Hudson River received an estimated 500,000 lbs of polychlorinated biphenyls (PCB) over a 25 year period, from two General Electric capacitor plants located approximately 192 river miles north of New York City.[1] In 1974, the New York State Department of Environmental Conservation began investigating the optimal strategy for removing the highly contaminated sediments within a 40-mile section of the upper river, north of Albany (see Figure 1). A series of alternatives was examined. It was determined that dredging of the most contaminated sediment "hot spots" in combination with secure upland containment, offered the most feasible opportunity for long-term maintenance and monitoring of the PCB-contaminated spoil. As a second phase of the project, Malcolm Pirnie was to locate a site or sites suitable for such containment.

More recently, the New York District of the Army Corps of Engineers initiated preparation of a 10-year management plan and environmental impact statement for maintenance dredging of the Hudson River federal channel between New York City and Albany. This lower portion of the river has markedly reduced PCB levels in comparison with the upper Hudson. However, as a contingency against encountering sediments contaminated with PCB or other materials such as heavy metals, the Corps requested Pirnie to identify all disposal sites along actively dredged reaches potentially suitable for contaminated or hazardous dredge spoil.

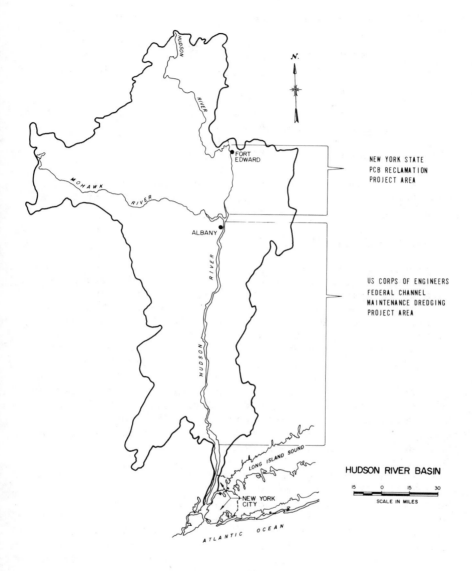

NEW YORK STATE
PCB RECLAMATION
PROJECT AREA

US CORPS OF ENGINEERS
FEDERAL CHANNEL
MAINTENANCE DREDGING
PROJECT AREA

HUDSON RIVER BASIN

15   0   15   30
SCALE IN MILES

Figure I: Location of Upper and Lower
Hudson River Dredging Projects.

PHASED APPROACH TO UPLAND SITE IDENTIFICATION

For both studies, a phased approach to site identification was employed. Generalized screening procedures were followed by progressively more rigorous field investigation as potential sites were identified and, in the case of the PCB study, finalized. The approach comprised five steps:

o   Develop screening criteria.
o   Prepare overlays based on the criteria.
o   Delineate specific sites.
o   Conduct preliminary field visits.
o   Conduct detailed field visits.

In addition, public participation played a key role throughout the site identification process. Each of these phases is described in detail below.

Development of Siting Criteria

The critical first step in the screening process is to develop siting criteria which satisfy regulatory requirements as well as good engineering and scientific judgment. The criteria must then be applied to the entire study area without preconception as to the best site locations. Such objectivity is imperative in anticipation of public opposition to any hazardous waste disposal site, and the inevitable questions "Why here?", "Why not over there?".

Federal and New York State laws pertaining to hazardous waste disposal were researched to determine characteristics which would exclude an area from further consideration. In several cases, as deemed necessary, criteria more stringent than the regulations were developed.

For the remedial dredging project in the Upper Hudson, sediment PCB levels commonly exceed 50 ppm PCB and are therefore classified as a toxic material under 40 CFR Part 761.2 In the lower river, the majority of the channel sediments are sandy and "clean", with the potential for reuse as fill, road sand, etc. The hazardous sites were delineated as a contingency; sites for clean material and for slightly contaminated material were found as well, but will not be discussed in this paper.

As shown in Table 1, the criteria for maximum hydraulic conductivity is 0.10 μm/sec. Section 360.8 of New York State's "Solid Waste Management Facilites" requires that the soil underlying a hazardous waste site must meet or exceed this standard, and that an impermeable barrier (either

Table 1. Preliminary Site Screening Criteria-
Hazardous Dredged Material

Parameter	Unacceptable	Reference
Soil	Permeability greater than 1×10^{-5} cm/sec	Part 360 NYCRR[3]
	Less than 3 ft thick in situ.	Malcolm Pirnie
	Class I or II agricultural soils.	Part 360 NYCRR
Slope	Deep gullies, slope over 15%.	Malcolm Pirnie
Surface Water	Closer than 300 ft to any pond or lake used for recreational or livestock purposes, or any surface water body officially classified under state law.	Malcolm Pirnie
	In 100-year floodplains.	40 CFR Parts 257[4], 761[2]
	In official state-designated wetlands	Malcolm Pirnie
Bedrock	Closer than 30 ft to highly fractured rock or carbonates. Closer than 10 ft to all other rock.	Malcolm Pirnie
Ground Water	Closer than 10 ft to ground-water table.	Part 360 NYCRR[3]
	Closer than 500 ft to any water supply well or recharge area.	40 CFR Part 250[5]
Committed Land	Located in designated agricultural districts or closer than 1000 ft to parks, residential areas, historic sites, reservoirs.	Malcolm Pirnie
Biologically Sensitive Areas	Endangered plant or animal habitats, unique or regionally significant habitats.	Malcolm Pirnie

natural or synthetic), having a hydraulic conductivity of
1.0 ηm/sec or less must exist between the waste and the
underlying soil. Because the barrier requirement can be
"engineered away" by constructing a liner, 0.10 μm/sec for
the underlying soils was adopted as the criteria.

The minimum depth to ground water is 10 ft, based upon
Section 360.8. In addition, the site must be at least 500
ft from a water supply well or recharge area, as specified
in 40 CFR Part 250, "State Hazardous Waste Programs,
Proposed Guidelines."

The minimum depth to bedrock is 10 ft, as also mandated
by Part 360.8. An additional precaution was added: depth to
bedrock must be at least 30 ft in highly fractured rock or
in carbonate rock, where movement of leachate would be
facilitated.

Preparation of Overlays

A series of overlays was prepared on transparent sheets
using USGS topographic maps (at a scale of 1:24,000) as a
base. This widely used overlay technique was first developed by Ian McHarg at the University of Pennsylvania.
Computer data entry and graphics may be substituted for hand
drafting; however, unless the data base is to be continually
updated and re-used, hand drafting of the overlays may be
considerably less costly.

For both studies, the overlays covered a land area 2
miles on either side of the Hudson River. This distance was
determined by hydraulic dredge pumping capabilities. Primary references were employed to determine areas of unacceptable conditions on each overlay. These references included:

o Soils surveys, general soils maps, and preliminary
   field maps.
o County ground water bulletins.
o New York State significant habitat reports.
o New York State freshwater wetlands maps.
o Floodplain maps.
o New York State Land Use and Natural Resources
   Inventory (LUNR).
o Aerial photographs.

Because data gaps were encountered with respect to
several key criteria, final site determination was deferred
until after the detailed field investigation. For example,
detailed soils maps were unavailable for portions of the
study areas. Judgments based upon general maps were re-

86

quired, to be confirmed later by borings. Minimum depth to bedrock could be mapped at 5 ft, although the criteria specified 10 ft. Depth to ground water was considered too variable to be mapped at all, based upon scattered well information. In addition, inconsistencies among the various maps required constant checking. A small shift in map scale or alignment could radically alter the acceptability of an area.

Figure 2 illustrates how the individual overlays were utilized. Areas which did not satisfy the criteria were shaded for each overlay. When the overlays were placed together, a composite was formed in which the clear areas or "windows" represented broad areas of environmental suitability. The result of the overlay procedure was a substantially reduced land area in which to concentrate the subsequent site screening procedures, thereby saving labor and expenses. For example, in the Upper Hudson PCB-related study, approximately 3,200 ac consisting of 40 parcels were found to be acceptable, out of approximately 100,000 ac in the original study area.

Delineation of Specific Sites

Specific sites within the environmentally acceptable "windows" were identified by using secondary criteria. The secondary criteria consisted of engineering and economic factors, such as:

o Adequate size for the projected disposal needs (200 ac for the Upper Hudson PCB study; 20 ac for the Lower Hudson).
o Proximity to the Hudson River and to the anticipated dredge area.
o Low elevation and associated transport costs.
o Site accessibility (presence of nearby roads).
o Absence of obstacles (utilities, pipelines, and other structures).
o Sufficient screening from residential areas and other sensitive land uses.
o Site ownership (one or two owners as opposed to multiple owners).

These criteria were applied to the windows in an iterative manner, with the aid of recent aerial photographs and USGS topographic maps. In the Upper Hudson, where the ultimate objective was to locate a single suitable site of 100 ac or more, the secondary screening process reduced the number of candidate sites to four. In the Lower Hudson, where the objective was to identify all potential sites of 20 ac or more, 133 candidate sites were delineated.

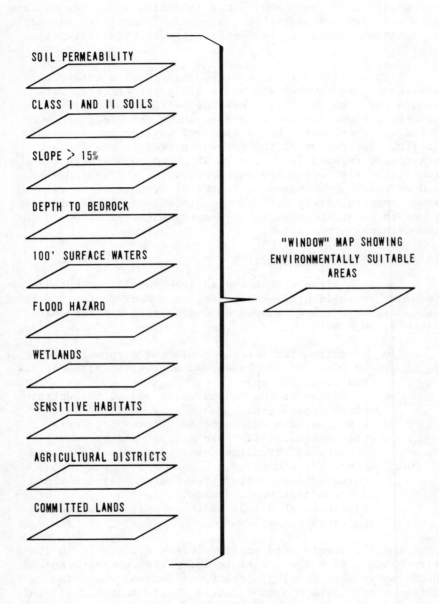

SOIL PERMEABILITY

CLASS I AND II SOILS

SLOPE > 15%

DEPTH TO BEDROCK

100' SURFACE WATERS

FLOOD HAZARD

WETLANDS

SENSITIVE HABITATS

AGRICULTURAL DISTRICTS

COMMITTED LANDS

"WINDOW" MAP SHOWING
ENVIRONMENTALLY SUITABLE
AREAS

Figure 2: Representation of Overlay Technique

Preliminary Field Investigation

The objective of the preliminary field investigation was to confirm and refine the in-house screening results and to eliminate obviously unacceptable sites. Two tasks were performed at each site: a preliminary soils investigation, and a site walkover. For the Upper Hudson PCB study, backhoe pits were excavated to a depth of 10 ft and laboratory analyses performed on the soils. For the Lower Hudson maintenance dredging study, hand auger samples to a depth of 4 to 5 ft were collected at a number of locations within each site. The soils were visually examined for clay content. Portions of the sites which did not have a clayey subsoil, or where bedrock was encountered at shallow depth, were eliminated from further consideration.

Based on visual observation, areas exhibiting rock outcrop were eliminated. Aquatic and terrestrial habitats were subjectively evaluated in order to rank otherwise similar sites. Land use was checked, particularly recent changes which did not appear on the maps. Distances to surface waters or nearby wells were determined, and wetland vegetation recorded.

A conflict arose among three criteria: soil permeability, depth to groundwater, and wetland vegetation. In the Northeast, clayey soils with the required permeability (0.10 μm/sec or less) are generally saturated at depths very close to the surface. Depth to ground water is, therefore, considerably less than 10 ft as specified by New York State law. Furthermore, these saturated soils may support wetland vegetation, a second violation of the siting criteria.

In practice, however, water movement in these tight soils is extremely slow to nil. The low yield is not characteristic of a aquifer. Wetland vegetation, if present, would likely be typical of a wet meadow. A priority was therefore assigned to soil permeability.

As a result of the preliminary field visits, three of the four Upper Hudson sites were eliminated from further consideration. The fourth underwent detailed field investigation as described below. In the Lower Hudson study, 62 of the 133 initial sites underwent preliminary field investigation. Thirty one were eliminated, leaving a total of 102 sites. For these sites, a detailed field investigation may be performed in the future when and if: (1) maintenance dredging is required in the vicinity of a given site; and (2) sediment testing indicates that the material to be dredged would be considered hazardous by EPA.

Detailed Field Investigation

The preferred site in the Upper Hudson study area com-
prises 250 ac of active and abandoned farmland in the
Village of Fort Edward, New York. In the winter and spring
of 1980, soil borings, a resistivity survey, and test pit
investigations were conducted on the site. Twenty borings
were initially conducted. Five encountered bedrock at less
than 30 ft; the remainder were driven to 30 ft. Six borings
were added during the course of the investigation to better
verify rock levels. In addition, 10 undisturbed samples
were taken and 5 temporary piezometers installed. The
borings indicated varved Lake Albany clays at all depths.
The varving is in a horizontal plane, and does not affect
the vertical permeability of the clays in-situ.

After several days, ground-water levels as measured in
the piezometers, were 3 to 4 ft below the surface. An
artesian condition was encountered in one of the borings
which extended to bedrock. Water was observed flowing
slowly out of an auger which had remained in the hole over-
night. This artesian condition indicates that groundwater
in the bedrock aquifers is under pressure, which should act
to limit any movement of leachate from the overlying clays
into the bedrock aquifers. If the site leachate collection
system should unexpectedly fail, water could accumulate
within the site and negate this pressure differential.

Surface resistivity readings were made at a total of 19
stations to provide additional depth to bedrock information.
The results indicated that depth to bedrock across the site
ranged from near the surface, to 75 ft. The site alignment
was shifted slightly to assure that minimum depth to bedrock
was 10 ft as required by State regulations.

Eight new test pits were excavated, in addition to
those undertaken previously as part of the preliminary field
work. The pits were excavated to a depth of 9 to 13 ft,
soil morphological descriptions were provided for all but
two of the pits. The predominant soils types were Kingsbury
silty clay, Covington silty clay loam, and Vergennes silty
clay loam, all poorly drained with seasonally high water
tables. The estimated permeability of this association was
$2.6 \times \eta m//sec.$

The field investigations showed that with the exception
of depth to groundwater as strictly interpreted, the site
met or surpassed regulations and/or guidelines established
by State and Federal agencies (see Table 2). It was concluded that, with proper design, construction, and maintenance procedures, the site would provide a satisfactory

Table 2. Comparison of Regulatory Requirements

Factor	40 CFR 761 [2]	40 CFR 250 [5]	6 NYCRR 360 [3]	PCB Feasibility Study-"Ideal" [6]	Actual Site Characteristics
Permeability	$\pm 1 \times 10^{-7}$ cm/sec	$\pm 1 \times 10^{-7}$ cm/sec	$\pm 1 \times 10^{-7}$ cm/sec	$\pm 1 \times 10^{-7}$ cm/sec	2.6×10^{-7} cm/sec
Thickness	In-situ 4 ft	5 ft	case by case	in-situ > 10 ft	> 30 ft
Passing #200 sieve	> 30%	> 30%	-	> 30%	> 90%
Plasticity Index	> 15	> 15	-	> 15	15-37
Liquid Limit	> 30	> 30	-	> 30	41-85
Depth to Bedrock	-	-	> 10 ft	> 50 ft	> 30 ft
Classification	-	CL, CH, SC, OH	-	-	CL, CH, ML
pH	-	> 7.0	-	-	6.9-8.3

91

containment area for the proposed PCB hot spot dredging
program. Hearings were recently completed by the New York
State Facilities Sitings Board, to determine whether the
site could be approved. If the permit is issued, site
preparation is anticipated in 1983 as the first step in the
Hudson River PCB reclamation project.

PUBLIC PARTICIPATION AND RESPONSE

In the maintenance dredging study the public was
actively involved from beginning to end. Five public meetings were held, to review the screening criteria, evaluate
the "window" maps and the potential site maps, and finally
to comment on the draft EIS. In addition, a Coordinating
Committee was formed comprised of representatives from
environmental and business groups and government agencies.

The public participation program improved communications and relations in general between the Corps and the
residents along the river. However, fear persisted regarding PCB's and the possible impacts of upland disposal.
The principal objections were not technical in nature. They
tended to involve subjective judgments related to effects on
property values, and on the perception of the community as a
desirable location to live or work. Requests were received
to delete certain sites. Each of these sites was re-
examined with respect to the original criteria. If all
criteria were satisfied, the site remained on the maps.
Some sites were deleted as a result of this process.

In the Upper Hudson PCB study the public was not involved from the onset. However, a technical advisory committee was set up as a result of the settlement between the
New York State Department of Environmental Conservation and
General Electric. Public participation was initiated after
the site screening had been completed but prior to field
investigations. During the meetings and hearings there was
considerably more emphasis on technical issues, particularly
ground-water and air quality impacts, than observed during
the meetings on the Corps of Engineers project. It is
difficult to determine whether the emphasis on specific
technical issues was a result of the focus on a few specific
sites, or a lower degree of confidence in the technical data
because the public was not involved in the early stages of
the project.

These experiences indicate that the early exposure of
the public to the siting process prior to the identification
of specific potential disposal sites may be desirable. Such
a program may increase the public's understanding of the
technical issues. This in turn may simplify the site

evaluation process by focusing on the socioeconomic factors which appear to be the most significant public concern.

CONCLUSION

The need for upland disposal of contaminated dredge spoil will likely continue as a result of both river reclamation efforts and, more commonly, channel maintenance projects. Disposal site location must be determined through a procedure which is both legally and scientifically valid, to assure maximum protection to the environment and to gain public trust. The site screening approach described in this paper has two principal advantages: 1) the overlay technique is objective and applies the criteria equally to all lands; and 2) the phased approach is economical and permits the determination of the level of effort as the study objectives evolve. In the Upper Hudson PCB study, carrying this approach through the detailed field investigation stage resulted in a site which meets or exceeds both State and Federal standards.

REFERENCES

1.   Tofflemire, T.J., L.J. Hetling, and S.O. Quinn. "PCB in the Upper Hudson River: Sediment Distributions, Water Interactions and Dredging." New York State Department of Environmental Conservation Technical Paper No. 55. January 1979, p.1.

2.   40 CFR Part 761, "Polychlorinated Biphenyls (PCBs) Manufacturing, Processing, Distribution in Commerce, and Use Prohibitions," Subpart E, Annex II, Section 761.41.

3.   New York Compilation of Rules and Regulations, Title 6, Chaper 360, "Solid Waste Management Facilites".

4.   40 CFR Part 257, "Criteria for Classifcation of Solid Waste Disposal Facilities and Practices", Section 257.3.

5.   40 CFR Part 750, "State Hazardous Waste Programs, Proposed Guidelines".

6.   Malcolm Pirnie, Inc. "PCB Hot Spot Dredging Program Upper Hudson River, Containment Site Investigations Program Report No. 1", for New York State Department of Environmental Conservation, May 1980.

CHAPTER 7

HAZARDOUS WASTE MANAGEMENT AT
PPG'S DEEP LIMESTONE MINE, NORTON, OHIO

R. J. Samelson,
    Manager, Environmental Programs, PPG Industries, Inc.

T. A. Zordan,
    Project Manager, D'Appolonia Consulting Engineers, Inc.

I. INTRODUCTION

For nearly 40 years, PPG extracted limestone from a
deep formation at Norton, Ohio to provide feedstock to soda
ash operations at the nearby Barberton plant. Since the
mid-1970's, when mining operations stopped, PPG has sought
beneficial uses for this mine. With the advent of RCRA
regulations, the potential for use of the mine for hazardous
waste management was evaluated. Recently, PPG has publicly
announced its intent to proceed with the permitting, design,
and operation of this limestone mine as a hazardous waste
management facility.

This facility possesses several unique features which
allow novel and innovative approaches to hazardous waste
management with unparalleled safety. The geological, hydrological, and structural aspects of the mine are discussed
with respect to environmental impacts from waste management
operations. The operational plans incorporate several
innovative measures which provide for unprecedented measures
of safety and allow for future recovery should the need
arise.

II. SITE CHARACTERISTICS

The location of any hazardous waste management facility
depends upon the complex interaction of many environmentally
related areas. Factors such as physiography, demography,
transportation, climatology, and the nature of the waste
material itself all play important roles in helping assess
whether a given site is suitable for its proposed use. In
addition, when deep underground facilities are considered

for hazardous waste management activities, the geology and
hydrogeology of the site play an even more important role
than they do in near-surface facilities.

The PPG limestone mine (Figure 1) is located near the
western boundary of the glaciated Allegheny Plateau Subpro-
vince of the Appalachian Plateau Province.

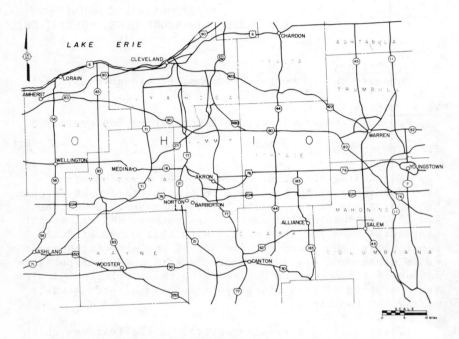

Figure 1.   Location of PPG's Deep Limestone Mine.

It is in an area characterized by gently rolling hills
and intervening valleys, representing a glacially modified
stream-dissected plateau. The preglacial land surface in
Summit County had a relief of several hundred feet which was
subsequently reduced as a result of erosion and deposition
associated with the Pleistocene glaciations. The glaciations also had a profound effect on the drainage system
in the site area as demonstrated by the relatively poor
drainage conditions that are characteristic of glaciated
regions.

In terms of geological factors, the PPG limestone mine
is located in the Central Stable region of the United States
which is characterized by broad low structural arches and
intervening basins with the stratigraphy throughout this
region being consistent. The depth of the Precambrian Basin

in Ohio varies from 2,800 feet to approximately 13,000 feet. The overlying Paleozoic sedimentary rocks which consist of limestone, dolostone, sandstone, and shale form a wedge-shaped deposit that thickens to the east and southeast.

The information on the stratigraphy at the PPG limestone mine has been collected from both drilling data and from data collected during sinking of the shafts. The mine itself is approximately 2,200 feet below the surface and is located in a thick limestone layer known regionally as the Onondaga. The mined-out volume exceeds 300 million cubic feet and, at a total extraction of 53 percent, the mined-out area underlies approximately 450 acres. The mine horizon is penetrated only by the two shafts; one for personnel and equipment, and the other for product transfer. During the early stages of production, the final entry and room configuration was established on 75-foot centers with both room and entry width at 32 feet (Figure 2). Initial room heights were established at 17 feet but, in significant portions of the mine, mining height was increased to either 28 or 45 feet by stoping.

Figure 2. Typical Haulways and Room Entries.

The mine horizon is in the upper 46-foot interval of the Columbus limestone which dips at a low and fairly uniform rate of 30 feet per mile to the southeast. The lithology of the Columbus limestone is a bluish-gray, partly crystalline limestone, bedded in layers, separated by several shale partings, varying in thickness from trace to over one inch (Figure 3). In several horizontal beds, chert nodules are concentrated in the upper layers of the limestone. Shale partings and cherty horizons are consistent throughout the mine.

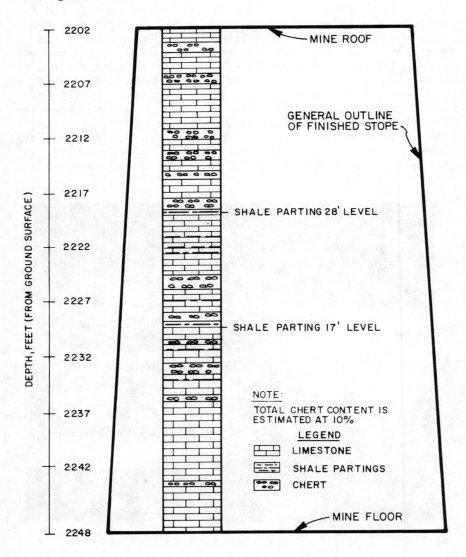

Figure 3.   Typical Geologic Strata in Mine Horizon.

Structurally, the mine is in a strong, consistent, unfractured strata. The generalized physical and mechanical properties of the Columbus limestone are given in Table I. With the exception of a single vertical joint observed in the westernmost portion of the mine, there are no vertical joints or cleats anywhere in evidence in the parts of mine inspected. The absence of fractures is a characteristic that makes this mine unique; it is evidence that the formation has been stable through geologic time. Modern records show the mine to be located in a region of very low seismic activity and risk and may be further insulated from seismic events by the relatively thick shale layers that both underlie and overlie the Onondaga.

Table I: Physical and Mechanical Properties
Columbus Limestone

Specific Gravity	2.69
Apparent Porosity (%)	0.72
Compressive Strength	
Perpendicular To Bedding	18,000 psi
Parallel To Bedding	12,000 psi
Modulus of Rupture (psi)	2,300 psi
Young's Modulus (psi)	$7.5 \times 10^6 - 8.0 \times 10^6$
Modulus of Rigidity	3.33×10^6
Poisson's Ratio	0.20
Velocity of Sound	15,400 ft/sec
Hardness	
Scleroscope, scleroscope units	58
Dorry Abrasive, lb/sq in./rev	0.10×10^4
Page Impact Toughness, cm	8.6

Based on reports of seismic activity from 1776 to 1977, no damaging earthquakes have been reported in the Barberton area near the site. Minor earthquake activity has been reported from the Cleveland area further to the north with earthquakes ranging from Modified Mercalli Intensity IV to Modified Mercalli Intensity VI or VII. Earthquake activity in this area is generally of low intensity and diffuse throughout the region and has not been associated with faulting (Figure 4). In a probablistic sense, the site is located in a Zone 1 seismic area which indicates that the seismic risk is minor.

Overlying the Onondaga limestone formation is approximately 1,500 feet of shale. Known regionally as the Ohio shale strata, there are five lithographically recognizable units. This great thickness of shale is an important feature as it isolates the mine horizon. It precludes the

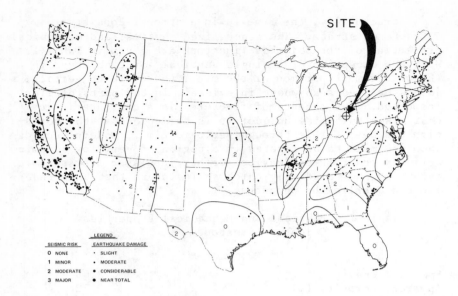

Figure 4. Seismic Risk Map.

penetration of water into the mine from above and is of utmost importance as a geologic barrier to prevent the upward migration of any waste materials placed in the mine (Figure 5). This 1,500-foot shale layer is overlain by an interbedded sequence of sandstone and shale of Mississippian and Pennsylvanian Ages. Finally, this strata is mantled by glacially derived sediments and alluvial deposits. It is in these geological strata overlying the shale that the characteristics of surface water and ground water are exhibited.

The land surface above the mine lies within the Hudson Run watershed. Surface runoff from this watershed drains through Hudson Run, through a short run of Wolf Creek, through the Tuscarawas River, which is a tributary of the Muskingum River, and from the Muskingum River finally into the Ohio River. Near the surface facilities for the mine, and connected to Hudson Run, are two lakes: Lake Dorothy and Columbia Lake. The average annual flow rate of Hudson Run is estimated to be 11 cubic feet per second. Preliminary analyses indicate the the probable maximum flood will have a flow rate of about 56,000 cubic feet per second at the mouth of Hudson Run where it joins the Tuscarawas River. This flood could raise the Lake Dorothy pool level up to 12.5 feet (Elevation 1025.5 above sea level). Since the Shaft No. 1 elevation is 1040 feet, no direct inflow into the mine through the shaft is possible during probable maximum flood.

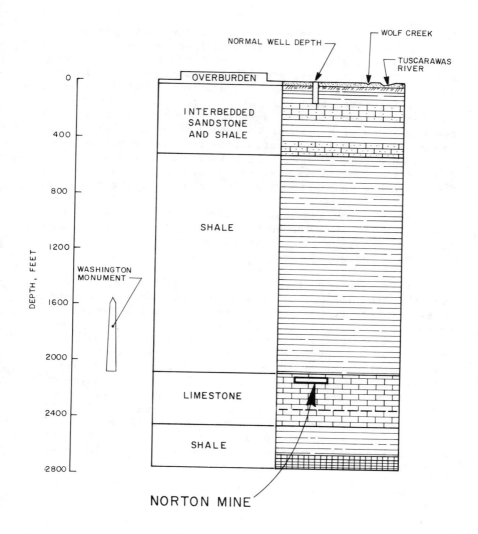

Figure 5.   Typical Geologic Strata.

The municipal water supply for the city of Barberton is from a surface reservoir on Wolf Creek, in southern Copely Township, outside of the Hudson Run Watershed.

Industries in Barberton pump about 2 billion gallons of ground water each year from a major aquifer that underlies Wolf Creek and a large part of the city. Recharge to the aquifers at Barberton is principally from local precipitation. Because of the fine-grained material associated with the creek bed, recharge by infiltration from Wolf Creek is probably small. The typical well depth in the area is 150 feet. In terms of the ground water, the stratigraphic unit present in the area can be grouped into subdivisions: (1) consolidated sedimentary layers of Devonian, Mississippian, and Pennsylvanian Ages; and (2) unconsolidated surficial deposits of Pleistocene Age. The unconsolidated Pleistocene glacial deposits cover most of the area under consideration for regional ground water and are the more important of the two divisions with respect to ground water availability. The mine horizon is separated from the overlying aquifers by approximately 1,700 feet, the majority of which is 1,500 feet of the impervious Ohio shale, which isolates the mine from penetration of the water from the overlying aquifers.

The Oriskany sandstone located approximately 200 to 300 feet below the mine has been reported to yield saline water. It is not anticipated that these formations would be utilized as a source of potable water in the future.

It is seen from the geological and hydrological data that the mine is located in a competent, consistent layer of limestone that is unfractured and has been unaffected by seismic activity over geologic time. Further, the presence of approximately 1,500 feet of impermeable shale overlying the mine horizon provides an extremely effective barrier in isolating the mine from those ground water aquifers important to surface activities. The saline aquifers which exist below the mine are not suitable for use as a water supply and have been isolated from the mine horizon by 250 to 300 feet of limestone and shale. The mine itself is both structurally stable and dry. These characteristics combine to allow consideration of the use of the mine for innovative operations in hazardous waste management.

III. OPERATIONAL CHARACTERISTICS

The unique features of the mine itself make it imperative that this facility be operated in a unique manner. The typical materials handling techniques employed in surface storage facilities or landfills are not applicable 2,200 feet underground. This hazardous waste management facility will be operated more like a modern warehouse than any typical surface disposal facility.

Long before any wastes get placed in the mine, several important steps will have occurred between the time that the waste is generated and is actually delivered to the mine site. The operation of this mine will be under permit conditions from the Ohio EPA and the U.S. EPA. The type of wastes which PPG will accept will be reviewed and approved by those agencies prior to acceptance of the material. PPG will develop a list of acceptable wastes during this permit application and review period. This list will be based upon several important physical and chemical characteristics including low vapor pressure, compatibility with container materials, compatibility with other wastes, and compatibility with the host rock. No radioactive, explosive, or flammable wastes will be accepted at the facility.

All wastes must be containerized at the point of shipping. Shipment of wastes to the facility site will not be mixed with other commercial cargoes. The containers will be designed and chosen to help assure long-term stability in contact with the waste material and in the mine conditions. Both the regulatory agencies and PPG will require that the containers be accurately labeled and that the required manifests accompany the shipment.

When a shipment of waste containers is received at the facility, they will be off-loaded into a protected staging area for temporary storage. During this period, both the manifests and the labels will be checked by sampling and analyzing the contents of the waste containers. Once the contents have been confirmed, the waste containers will then be cross-matched and sorted for underground storage with containers of similar compatible wastes. The actual cataloging and scheduling activities will be handled by an automated system similar to modern airline scheduling technologies. Records will be maintained of who generated the waste and shipped it to the facility, what the contents of the waste container are, and the storage location in the mine so that the potential for retrieval or recovery of the material well into the future is assured.

Once a waste container is accepted for underground storage, it will be transported to the mine horizon by the use of a dedicated elevator. Once in the mine itself, the waste containers will be transported to the their ultimate storage locations using modern materials handling technologies. Both the underground transport and storage operations will be automated to the maximum extent practical. It is an objective to minimize the actual hands-on operation requiring human contact with the waste containers.

The waste containers will be taken to an appropriate storage room in the underground space. Once there, they will be stacked in a structurally stable configuration with other containers of compatable waste. Once a storage room has been filled, that storage room will be sealed in order to isolate it from continuing operation in the balance of the mine. A manway will be provided in the sealing wall to

allow controlled access and inspection in the future.
Similarly, provisions will be made through the sealing wall
to allow sampling of the atmosphere in the room without the
requirement for human entry. The operations in the mine
will proceed in a sequence of steps like these. There will
be a relatively small number of active rooms at any one
time. All previous activities will have been completed and
isolated from those on-going activities by erecting isolation walls.

IV. CONCLUSIONS

The PPG limestone mine is a unique facility. It is
stable, strong, and it is isolated from interaction with the
environment immediately accessible to man. The mine itself
and the modern materials handling techniques which will be
used in the operations provide the potential for the use of
this facility for the next 50 years in a manner which gives
the highest assurances possible that unacceptable environmental events will not occur. This facility will go through
the entire permit process required of hazardous waste management facilities. It is being designed and will be operated much in the same manner as a modern warehouse operation
rather than a conventional surface landfill. These features
combine to lead to the conclusion that this facility and its
proposed mode of operation will provide an unprecedented
degree of protection to the environment for this type of
hazardous waste management activity.

CHAPTER 8

HAZARDOUS WASTE MANAGEMENT
IN THE R&D ENVIRONMENT

Susan Schmidt
    United States Air Force

David R. Lawrence
    United States Air Force

C. David Douthat, P.E.
    United States Air Force

The enactment of the Resource Conservation Recovery Act
(RCRA) has set forth the guidelines by which hazardous waste
materials are managed from "Cradle to Grave". The intent of
this regulatory program is to insure environmentally sound
storage, treatment, and disposal of waste materials. Additional benefits of the Act are to encourage the substitution
of less hazardous materials and processes, and discourage the
development of environmentally unsafe materials. This mandate from Congress has created mixed emotions from private
industry and environmental gorups alike but the goal of all
is the safeguarding of human health and the long term protection of the environment.

The problems normally associated with instituting and
managing new high impact programs such as this have been
greatly reduced by public sensitivity to recent environmental problems. This public awareness has affected both
employee and management concern for environmental protection - no one wants or can afford another Love Canal.

Research and development institutions find themselves in
a difficult position, attempting to comply with regulations
that were written mainly for the large volume/small variety
industrial generator. The vast variety of small quantity
waste generated by an R&D activity tends to complicate the
already difficult task of compliance with RCRA. Only within
the last six months has the EPA looked toward hazardous waste
regulatory reform with the laboratory waste generator in
mind. Relaxing of the ban for landfilling liquid and liquid
ignitable waste lab packs and the shift toward the "degree of

hazard" approach to regulating treatment, storage, and disposal (TSD) facilities are just two examples.

On November 17, 1981, two days before the official ban was
to go into effect, the EPA reconsidered its ban on landfilling lab pack or overpacked drums of liquid and liquid ignitable wastes in secure landfills. This ban would have virtually eliminated hazardous waste disposal options for laboratory waste generators. The degree of hazard approach to
regulating TSD facilities has not yet been adopted by the
EPA. This approach would require facilities which have
greater potential for environmental impact to comply with
more comprehensive regulations. These facilities would be
secure landfills and hazardous waste treatment sites. Storage areas would be subject to substantially less regulatory
requirements. These changes by the EPA will do much to
reduce the burden of compliance with RCRA regulations facing
R&D organizations.

WASTE MANAGEMENT

Effective management is critical to a successful program,
especially when serious environmental damage and personal
liabilities are possible under the law. One way to accomplish this is to have a single responsible person or office
with authority to develop and set up a program, and make
decisions regarding hazardous wastes. Visible support must
come from top level management, including adequate funding
for the program. Also, for a program to be successful in an
R&D organization, the researchers must find it simple and
relatively easy to use, and responsive to their needs (i.e.,
with a feedback system). Written procedures should be
available to the researcher, particularly in large organizations.

One of the major long-term goals of management should be
the reduction of hazardous wastes generated by the organization. The program structure might include a committee that
meets periodically to review research projects, with discussions centering on changes to research procedures in order
to reduce wastes. Also, employee problems with the program
could be aired, so that the committee could serve as an
important feedback loop, and other items such as recycling or
exchanges of wastes could be discussed. Reduction of wastes
could also be enhanced through early identification of future
wastes. For example, in the planning stages of a project,
estimates could be made, if possible, of types and amounts of
chemicals to be used and wastes to be generated. Thought
could then be given to potential alternate lab procedures
which might result in less hazardous wastes or smaller
amounts. Perhaps a form could be used for these estimates,

with review by the responsible hazardous waste office and/or the above mentioned committee, and with periodic revisions as the project takes shape. Early identification can also aid in budgeting for future disposal costs.

Employee participation is critical to the success of a hazardous waste management program. Participation starts with education. The RCRA regulations, both Federal and State, are quite complex, very long, and are changing almost continuously. Researchers cannot be expected to memorize all the regulations, so the hazardous waste managers must be responsible for being familiar with regulations and their particular impact on the R&D organization. The essential requirements of the regulations should be passed on to the researchers. Seminars can be very useful for this purpose, and for encouraging researchers to think more in terms of the wastes generated as integral parts of their projects. Participation also requires the encouragement of employees. Researchers should be urged to find ways (cost effective, hopefully) to reduce hazardous waste generation, such as alternative experiment designs, use of less hazardous materials for a project, or perhaps neutralization or some other in-house treatment to lessen the hazards or amounts. Money not spent for disposal of wastes is money available for further research. Encouragement to reduce wastes, however, should not mean discouragement to use the waste disposal system that has been set up in the organization. The procedures for disposal must be easy to use by the researchers.

A successful hazardous waste management program must include strong top level support, a visible waste management structure, and employee participation, with a view towards not only disposal of wastes, but also the reduction of wastes.

THE SMALL QUANTITY GENERATOR

In reviewing disposal requirements, some consideration should be given to the classification of "Small Quantity Generator". The regulatory exemption for this classification can be beneficial, assuming the organization can consistently maintain the small generator status. This exemption attempts to relieve the small generator of many of the RCRA requirements while still assuring safe disposal of the wastes. With this exemption a generator is not required to register with the EPA. Therefore, the "paper trail" normally associated with assuring waste disposal through approved landfills is not required. The major drawback of the small quantity generator status is the problem of locating a transporter, disposal contractor, or approved landfill that will accept

wastes from a generator not registered. Also, possible
future increases in the quantity of waste generated should be
taken into account. The RCRA guidelines specify maximum
quantity generation permitted to maintain the small generator
status. Changes in activity levels may remove you from this
status. Regardless of the quantity, the burden of assuring
proper disposal is the responsibility of the generator.

INFORMATION SOURCES

Several sources of information are available to the disposal manager for the handling of chemical wastes. The
researcher generating the waste typically has the greatest
knowledge about the waste's properties and should be the
primary source of information. Other researchers within the
same group may have had different experiences with the same
chemical, therefore, it is very important for the hazardous
waste disposal manager to establish a means of communication,
such as the committee noted above, with the research
personnel to discuss disposal techniques. Although research
personnel normally do not know what information is required
for transportation they can provide more than adequate information for proper storage and handling. Unfortunately, many
times the original generator may terminate employment, leaving little or no information behind on the waste chemicals
remaining in the laboratory from the past projects. This is
one reason why good written records can be extremely useful.
Additionally, a check-out procedure for personnel may prove
helpful in minimizing this occurrence.

Chemical manufacturers, or local distributors can be
another source of information. Very often they will be able
to forward Occupational Safety and Health Administration
Material Safety Data Sheets (MSDS), which contain all the
necessary information for storage, transportation, and disposal. Because of the tremendous number of chemical compounds in existence and the new compounds manufactured
yearly, the MSDS is not always available. Most chemical
manufacturers are willing to disclose storage and disposal
methods as long as the discussion does not involve proprie-
tory information. Some manufacturers are willing to take the
waste material back for their own reuse or disposal.

In addition to the above sources, other technical
references often provide the chemical composition, health
related hazards, and usually storage related hazards for a
large number of chemical compounds. Trade chemicals are
usually not included in the reference and with the expanding
number of new compounds manufactured, these references will
not be comprehensive. Regardless which of the above sources
are used, it is very important for the disposal manager to

keep detailed disposal records for future reference. The same problems with gathering information for storage and disposal can be expected to occur routinely and the time and effort is wasted retracing the same steps.

ANALYSIS AND CATEGORIZATION

A problem that frequently seems to be encountered in the R&D environment is the unknown waste. These mysterious chemicals tend to appear in storage cabinets, fume hoods, and benchtops, with no hint as to the contents, or resemblence to any compound known to science. Unfortunately, this problem will continue to exist, and the best hope of minimizing these occurrences lies with the education and cooperation of employees. The waste program managers should keep in mind the analytical costs for identification of unknown chemicals. Most R&D activities are fortunate to already have the in-house expertise to accomplish analysis although additional equipment will probably be necessary. Regardless of whether the analysis is done in-house or by contract, analytical service, the expense of identification of unknowns can quickly exceed the original cost of the chemical(s). Once a waste is analyzed or already known, it must be categorized. Unfortunately, it is common for R&D wastes to not fall into either the listed wastes or the four EPA categories, yet still be hazardous. Most of these seem to be wastes that are toxic, but not EP Toxic. Classification of these types of hazardous wastes should be by Department of Transportation Hazardous Materials Tables (for example, the ORM-E category, Hazardous Waste, liquid or solid, N.O.S.). Care and professional judgement should be used in categorization of those wastes that fall through the "cracks" of the RCRA regulations.

LABORATORY DRAIN DISPOSAL

Drain disposal can often be utilized, with careful consideration to the compounds being disposed and the type of sewer system. Although there are no systematic means for determining chemical suitability for drain disposal, there are some general rules regarding drain disposal which should be followed:

1. "Only water-soluble substances should be disposed of in laboratory sink. Solutions of flammable solvents must be sufficiently diluted that they do not pose a fire hazard.

2. Strong acids and bases should be diluted to the pH 3-11 range before they are poured in the sewer system. Acids and alkalis

should not be poured into the sewer drain at a rate exceeding the equivalent of 50 ml of concentrated substance per min.

3. Highly toxic, malodorous, or lachrymatory chemicals should not be disposed of down the drain. Laboratory drains are generally interconnected; a substance that goes down one sink may well come up as a vapor in another. Sinks are usually communal property, and there is a very real hazard of chemicals from two sources contacting one another; the sulfide poured into one drain may contact the acid poured into another, with unpleasant consequences for all in the building. Some simple reactions can even cause explosions (e.g., ammonia plus iodine, silver nitrate plus ethanol, or picric acid plus lead salts).

4. Small amounts of some heavy-metal compounds may be disposed of in the sink, but larger amounts may pose a hazard for the sewer system or water supply."[1]

Responsibility lies with the hazardous waste manager or other disposal focal point to coordinate any drain disposal with local, State, and Federal authorities to insure compliance with applicable regulations.

WASTE NEUTRALIZATION

Waste neutralization can be an effective means of disposal for small quantity generators with proper planning and consideration for individual wastes. The most obvious application is for acids. Many research facilities are equipped with acid neutralization pits. The use of these pits for neutralization will require compliance with performance criteria established by the EPA. This may require a self monitoring program for pollutants introduced, pH monitoring, sampling and analysis, and periodic pit cleaning. The use of acid pits can be costly and time consuming and the installation of such pits are probably cost effective only when bulk quantities of waste acids are generated. Regardless, small quantities of acids can be neutralized in the laboratory. Whatever neutralizing method is used, consideration should be given to material contaminants such as heavy metals which are not readily neutralized.

SECURE LANDFILLING

Because of the relatively few EPA approved secure land-fills, available space for disposal of hazardous waste is considered a natural resource to be protected and used wisely. This means that only a waste meeting all the EPA criteria of a hazardous waste and for which no other disposal option exists should be placed in one of these areas. These landfills are extensively designed and strictly managed so as not to pollute the surrounding area or groundwater. The generator never loses liability even when pains are taken to insure proper design, and management considerations are met, but the extent of generator liability is reduced because the EPA requires owners of secure landfills to carry liability insurance. Currently, full scale chemical analysis is required prior to acceptance into one of these areas, which makes secure landfilling one of the more expensive disposal options.

INCINERATION

Incineration should be considered if sufficient volume of organic wastes are available. This is one of the more attractive forms of waste removal because the waste is essentially destroyed during the process. Although there are by-products of incineration, such as ash, that require proper disposal, the volumes are significantly reduced and much easier to manage. Incineration is typically less expensive than landfilling and once destruction is complete, little or no liability remains.

SANITARY LANDFILL

Although many chemicals are suitable for disposal in licensed sanitary landfills, it is not the most prudent dis-posal means. The time and money saved by this disposal technique would not be worth the potential adverse publicity for an institution found to be "dumping hazardous waste chemicals" in the community landfill. Landfilling chemical waste on one's own premises should be discouraged. Regula-tory requirements and manpower associated with establishing such a facility for most smaller institutions are cost pro-hibitive. Also, the long term liability and clean-up of a poorly designed facility make this option undesirable.

EXCHANGE/RETURNS

Normally exchange of waste chemicals is not feasible in the R&D community; however, its potential should not be ruled out. All waste chemicals need not be contaminated chemicals and may be useful to other researchers either in-house or at another activity. A good communication system within an organization and between the R&D community including local universities aid the exchange of valuable information and certain chemicals. Also, the State or Federal EPA may maintain a listing of chemical reprocessors and exchange groups.

Manufacturers or distributors of some hazardous materials will accept items for return and this policy may be established prior to purchase. This policy may be particularly useful when gas cylinder disposal is involved. Disposal of full or partially full cylinders containing a hazardous material is relatively complicated and expensive. Small gas cylinders or lecture bottles are very difficult to dispose of because they are often not returnable and disposal guidance has not been established by the EPA. The purchase of returnable gas cylinders should be encouraged even if only a small volume of gas is required.

CONTRACT DISPOSAL

Contract disposal should be approached carefully and thoroughly, since the generator's liability does not end when the wastes leave the generator premises. Federal and State EPAs should be contacted to find out if there are any environmental or legal problems with the potential contractor. The contractor should provide references to the generator and these should be contacted for an indication of satisfaction with the service.

A decision must also be made on what level of service is required. Some disposal contractors provide a full range of services, such as analysis and categorization of wastes, on-site packing, and manifesting. This convenience and saving of labor comes at a cost, of course, and it must be determined if the benefit exceeds the extra cost. It is important to continually monitor your contractor to insure proper disposal of wastes. Site visits would be appropriate before and during contract administration. Continued contact with EPA representatives to keep abreast of contractors compliance during the term of the contract is also advisable.

CONCLUSION

The need for regulation of waste chemicals and their safe disposal is real. With all its preparation and review, the RCRA regulations have many unclear and undefined areas which are particularly troublesome for the R&D community. Effective management of the waste disposal program will require a concerted effort for all. For the R&D community primary emphasis must be placed on the minimization of waste generation through education and support of management and employee alike. This will help reduce the disposal problem to manage-able proportions and allow R&D organizations to continue to function as they should, with minimal adverse impact from RCRA.

Book References

1. National Academy of Sciences, <u>Prudent Practices for Handling Chemicals in Laboratories</u>, (National Academy Press, 1981), pp 231-232.

CHAPTER 9

ROLE OF THE ANALYTICAL LABORATORY
IN HAZARDOUS WASTE MANAGEMENT

V. K. Varma
    Harmon Engineering &
      Testing

B. B. Ferguson
    Harmon Engineering &
      Testing

M. S. Shearon
    Harmon Engineering &
      Testing

The passage and implementation of the Resource Conservation & Recovery Act (RCRA) and the Comprehensive Environmental Response, Compensation & Liability Act (CERCLA or
"Superfund") has converted the field of hazardous waste
management to a high stakes ballgame.

As the regulators and the regulated struggle for the
ballpark turf, the analytical laboratory engaged in
environmental tests takes on an increasingly significant
role. This is true whether the laboratory is part of the
regulated industry, or part of the regulatory apparatus, or
- as is increasingly the case - a commercial professional-
services laboratory.

This paper describes the functions of the analytical
laboratory in hazardous waste management, examines selected
analytical chemistry considerations, presents selected case
studies, and suggests guidelines for obtaining laboratory
services.

THE ANALYTICAL LABORATORY FUNCTION IN HWM

The Hazardous Waste Management (HWM) system of the
Resource Recovery and Conservation Act (RCRA) is based on the
separation of hazardous and non-hazardous waste with a
subsequent interest in the nature, disposal method, environmental fate and health effect potential of the hazardous

portion. Within each step of this "cradle to grave" HWM
system the analytical laboratory has been given a critical
role to play for most industrial wastes. Chemical and
physical analyses establish whether the waste is hazardous;
whether waste constituents have migrated to the environment;
whether these constituents present a health effect hazard.
Conversely, laboratory testing may determine whether a waste
can be removed from the EPA listing of hazardous waste;
whether no potential exists to contaminate the environment
via waste constituent migration; or, perhaps ultimately,
whether wastes have been thoroughly destroyed, immobilized or
removed from a previously contaminated area.

In many aspects of the HWM system, the analytical
laboratory rests at the center of the decision-making
process. It is this vital role that has produced a keen
awareness among the regulated industrial community,
regulatory personnel, hazardous waste engineers, chemists and
analysts that this critical link in environmental protection
can not be overrated.

Within the HWM field the analytical laboratory functions
can be grouped under three categories:

1. Waste identification
2. Waste and waste constituent monitoring
3. Treatment, clean-up and other certifications

Special and specific areas of analytical need are shown
in Table I.

Table I. Analytical Laboratory Functions In Hazardous Waste
Management

Waste Identification
- Determination of hazardous/non-hazardous nature of non-listed wastes
- Location and isolation of waste generation sources
- Delisting of wastes which do not meet the EPA listing assumptions

On-Going Monitoring
- Annual waste characterization
- Groundwater monitoring

Certifications
- Hazardous waste treatment success
- Closure completion
- Spill clean-up completion
- Waste characterization for disposal

Thus, the main function of the analytical laboratory can be said to be an aid in decision-making:

- Do I have a hazardous waste? What is the source of the hazardous component?
- Are waste constituents entering the groundwater or other environmental vectors?
- Can my hazardous waste be treated?
- If so, how?

Figure 1 illustrates the key points in the HWM system in which the analytical laboratory performs a critical part in determining the course of HWM practices for individual industries, situations or waste streams.

In addition to the HWM System regulations of RCRA, Superfund regulations will rely heavily on the expertise of the analytical laboratory. Under this system, waste will need to be located, delineated, characterized, isolated and removed. This presents a greater analytical challenge than under most RCRA situations, since little information will be available regarding types or quantities of waste materials present in a disposal area. Also in accomplishing removal of waste and cleanup of the site, the analytical laboratory will undoubtedly play a major role in answering the "how clean is clean?" question. In all aspects, Superfund analytical programs will be amplified versions of the RCRA program.

LABORATORY PROGRAMS FOR HWM SYSTEM FUNCTIONS

As discussed above, the HWM system has several specifically defined functions for the analytical laboratory. The programs designed to meet the compliance objectives for these areas are discussed below.

Waste Identification

The waste identification task is the most basic aspect of the HWM system: a waste is either hazardous and therefore regulated, or it is non-hazardous. The EPA has "listed" many sources of wastes that are to be considered hazardous unless proven otherwise. Non-listed wastes must be tested for a set of four characteristics so that they can be classified as either hazardous or non-hazardous.

The tests are specified by EPA [1,2] to be the following:

1. Ignitability - A waste which has a flash point below 140°F in a closed cup tester [3] is defined to be

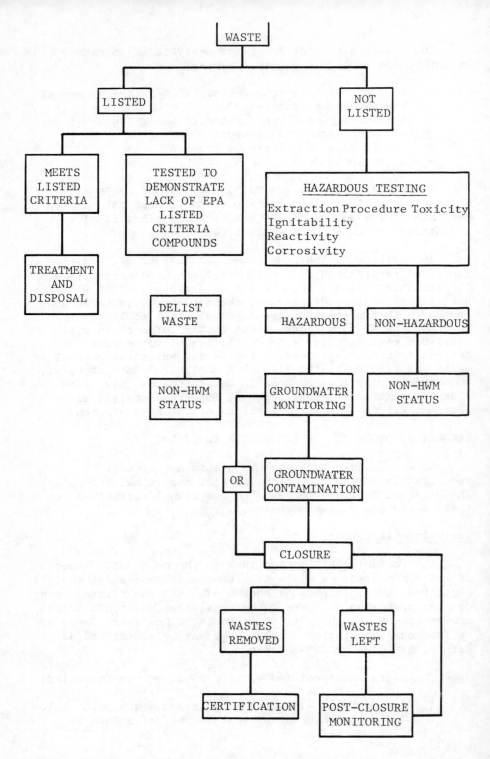

Figure 1. The analytical laboratory as a HWM
decision making tool.

hazardous due to the characteristic of ignitability. Non-liquid wastes are tested for ignitability by slurrying with water and testing by this procedure.

2. Corrosivity – A waste is corrosive if it has a pH of less than 2.5 or greater than 12. Wastes for which pH determinations are irrelevant (e.g. methyl butyl ketone sludges) can be tested for corrosivity by a method given[4] by the National Council of Corrosion Engineers. A waste with the listed pH conditions or capable of eroding a predetermined amount of metal (amount determined as part of method) is listed as hazardous due to the characteristic of corrosivity.

3. Reactivity – The reactivity of a waste is defined as the property of bearing cyanide or sulfide in concentrations "capable of producing effects on human health or the environment". This prose characterization is vague and encompasses many wastes which are innocuous. Many cyanide and sulfide wastes are fully stabilized but remain hazardous because of their cyanide-bearing or sulfide-bearing property. The property is measured[2] by exposing the waste to strong acid conditions and observing or analyzing for the release of sulfide or cyanide. If the concentration of either compound reaches more than a few hundred parts per million the waste may be considered hazardous due to the characteristic of reactivity.

4. Extraction Procedure (EP) Toxicity – The EP test was devised to simulate conditions within landfills by exposing candidate wastes to an acetic acid solution at pH 5.0 $\pm$0.2 to determine "toxic" leachate potential. The filtered leachate is analyzed for selected heavy metals and pesticides[4] identified in the Safe Drinking Water Act (SDWA). If these constituents are found to exceed 100 times the level set by the SDWA, the waste is characterized as toxic due to the presence of the constituent(s) detected.

Various studies are underway within the American Society for Testing and Materials, the Department of Energy, and EPA to devise an improved standard method for producing representative leachate extracts for organic compounds. The development of such a standard protocol will be a tremendous asset to the definition of hazardous wastes.

The identification of hazardous wastes has several applications beyond simple characterization of existing wastes:

1. Development of Treatment Schemes - Hazardous waste
    treatment schemes can be rapidly evaluated as to
    success by the application of the definitive tests.

2. Isolation of Waste Sources - When hazardous wastes
    are identified, it is possible to "backtrack"
    through the waste treatment and production processes
    to determine the origin of the waste's constituents.
    Upon source isolation, new feedstocks, treatment or
    process technologies may be investigated to reduce
    or eliminate hazardous waste generation.

3. Evaluation of New Processes - As with source
    isolation, when new processes, feedstocks or waste
    treatment schemes are in the feasibility stage,
    bench-scale batches of the probable wastes can be
    generated and evaluated to determine the hazardous
    or non-hazardous nature of the proposed waste.

Delisting of Waste Sources and Specific Wastes

The HWM system has provisions for demonstrating that a
listed waste is not of the characteristic for which it was
listed and, therefore, should be "delisted" or removed from
EPA's list of hazardous wastes. The vehicle for delisting a
waste is known as a "Delisting Petition". Among the items
which should be included in a viable petition are the
following:

- description of manufacturing process
- description of waste generation process
- demonstration that a representative sample of the
  waste does not contain the constituents for which
  it was listed
- generator's certification of accuracy
- presentation of analytical and quality control data

ON-GOING MONITORING

The owner-operator of every HWM unit (impoundments,
landfills, etc.) must develop plans to routinely perform
certain analyses. Among these are Waste Analysis Plans and
Groundwater Monitoring Plans. Other plans which may be
required during later phases of the HWM regulations are
analytical plans to demonstrate adequacy of hazardous waste
incineration (incineration efficiency) and soil monitoring
plans at "waste farming" operations. The purpose of these
plans is to set forth the methods and rationale for sample
collection, analysis (including quality control) and
reporting.

Waste Analysis Plan

The waste analysis plan is required to be followed on an annual basis or other schedule necessary to determine the characteristics of the waste being disposed. The principal elements of this plan should include:

1. Sample Collection Method - The EPA has specified several acceptable methods for sampling hazardous waste streams[5]. A specific method should be thoroughly described in a step-by-step fashion.

2. Analytical Method(s) - Describe methods to be used from available standard literature[2,6,7] or describe method specifically (preferable).

3. Parameter Selection Rationale - A description of the waste generation and treatment including feedstocks, intermediate reactions, etc. will show that the selected parameters represent the waste being disposed.

4. Quality Control - The minimum acceptable quality control practices for the laboratory performing analysis must be described. The criteria for accepting or rejecting an analysis based on QC data must be stated.

5. Chain-of-Custody - A specialized form to link a sample, via the signature of all parties involved, from collection to analysis. Figure 2 is a sample.

The waste analysis practice serves a double purpose if a hazardous waste is disposed of via a contract disposer. The information collected during this program is typically required by the contractor for disposal records. The periodic analyses demonstrate the consistency of waste characteristics.

Groundwater Monitoring Plans

The groundwater monitoring regulations of the HWM system specify that those facilities operating a land treatment unit, e.g. landfill, land farm, impoundment or waste pile, must have a groundwater (GW) monitoring system. This GW monitoring system is required to have at least one well capable of demonstrating background (upgradient) and at least three wells downgradient of each HWM unit or area. The GW monitoring system is then to be monitored according to the Sampling and Analysis Plan required for compliance. This plan must contain:

1. Sampling Protocols - These protocols should be simple,

GROUNDWATER MONITORING
CHAIN-OF-CUSTODY DOCUMENT

FIELD DATA

Facility: _____
Location: _____
_____
_____

Contact Person: _____
Phone: _____

LAB DATA

Project No. _____
Date Received _____
Condition Upon Receipt_____
_____
_____

SAMPLE COLLECTION INFORMATION

Well No.	Bail or Pump	*Water Level Pre-Purge	Water Level Pre-Sample	Casing Volumes Removed	Elapsed Time to Sampling	Aliquots Taken	Initials	Date	Time

ANALYSES REQUESTED

☐ pH
☐ Conductance
☐ Total Organic Carbon
☐ Total Organic Halogens
Upgradient Well(s) _____

☐ Chloride
☐ Iron
☐ Manganese
☐ Sodium

☐ Sulfate
☐ Phenols

OTHER

☐ Drinking Water Parameters
☐ _____
☐ _____

SAMPLE TRANSFER INFORMATION

	Relinquished By	Relinquished To	Date	Time	Reason
1.					
2.					
3.					
4.					

Note: First Signature is Sampler or Field Team Supervisor   Final Signature is Authorized Laboratory Representative. Chain-of-Custody complete upon Lab Receipt. A copy will be returned to client.

• Referenced at top of well casing

Figure 2. Sample chain-of-custody form.

explicit and complete "cookbook" type statements.

2. Preservation Techniques – Each parameter to be analyzed should be referenced to the particular containerization and preservation step necessary and this should be referenced to standard accepted techniques $^{5-10}$. Instructions should be included (Table II).

3. Analytical Techniques – The methods to be used (and/or required of a contracting laboratory) must be stated (Table III). Statements to the effect that "Standard Methods will be followed" are not acceptable; many methods are available for each parameter. The specific method must be listed. This documentation allows the data to be of a known quality as regards detection limits, sensitivity, interferences, etc. If any listed method needs to be modified to obtain acceptable results, the modifications should be incorporated into the GW monitoring plan.

4. Quality Control – The description of QC procedures is to be included. In the case of an in-house laboratory, the QC plan should be attached. If a contracting laboratory is utilized, this section should specify the minimum acceptable criteria for the selected laboratory to follow.

    In order to ensure the integrity of all samples collected, the following points should be observed as a minimum during collection and handling samples in the field.

    ●   Utilization of trained field personnel for collection of samples.

    ●   Use of proper collection and containment equipment which has been properly prepared as, for example, in Table II.

    ●   Strict adherence to sampling protocols as pertaining to each specific endeavor.

    ●   Strict adherence to preservation and shipping protocol as specified for specific test parameters.

    ●   Proper completion of chain-of-custody forms including notation of field observations of relevance.

    The validity of all reported data must be documented with the incorporation of a sound analytical QA/QC program. A minimum effort to ensure the integrity of

Table II. Groundwater Preservation and Containerization Techniques

PARAMETERS	CONTAINER	PRESERVATIVE
Groundwater Quality Parameters		
Chloride	plastic	none
Iron	plastic	filter on site, HNO_3, pH <2
Manganese	plastic	filter on site, HNO_3, pH <2
Phenol	glass	H_3PO_4, pH <4, 1 gm $CuSO_4$/liter
Sodium	plastic	filter on site, HNO_3, pH <2
Sulfate	plastic	none
Indicator Parameters		
pH	plastic	none
Specific Conductance	plastic	none
Total Organic Carbon	glass/teflon lined lid	H_2SO_4, pH <2
Total Organic Halogen	glass	none
SDWA Parameters		
Arsenic	plastic	filter on site, HNO_3, pH <2
Barium	plastic	filter on site, HNO_3, pH <2
Cadmium	plastic	filter on site, HNO_3, pH <2
Chromium	plastic	filter on site, HNO_3, pH <2
Lead	plastic	filter on site, HNO_3, pH <2
Mercury	plastic	filter on site, HNO_3, pH <2
Selenium	plastic	filter on site, HNO_3, pH <2
Silver	plastic	filter on site, HNO_3, pH <2
Pesticide	glass/teflon liner	none
Herbicides	glass/teflon liner	none
Radium	plastic	HNO_3, pH <2
Gross Alpha	plastic	HNO_3, pH <2
Gross Beta	plastic	HNO_3, pH <2
Fecal Coliform	sterile glass	none
Nitrate	plastic	none
Fluoride	plastic	none

Table III. Analytical Methods for Groundwater Analysis

PARAMETER	METHOD	REFERENCE
Groundwater Quality Parameters		
Chloride	Mercuric Nitrate, Titrimetric	7,325.3
Iron	AA-Flame	7,236.1
Manganese	AA-Flame	7,243.1
Phenol	4-AAP, Spectrophotometric	7,420.1
Sodium	AA-Flame	7,273.1
Sulfate	Turbidimetric	7,375.4
Indicator Parameters		
pH	Electrometric	7,150.1
Specific Conductance	Wheatstone Bridge	7,120.1
Total Organic Carbon	IR Combustion	7,415.1
Total Organic Halogen	Pyrolosis/Microcolumeter	7,450.1
SDWA Parameters		
Arsenic	AA-Hydride	7,206.5
Barium	AA-Furnace	7,208.2
Cadmium	AA-Flame	7,213.1
Chromium	AA-Furnace	7,218.2
Lead	AA-Furnace	7,239.2
Mercury	AA-Cold Vapor	7,245.1
Selenium	AA-Hydride	7,270.3
Silver	AA-Flame	7,272.1
Pesticides	GC	7,608
Herbicides	Derivatization/GC	11
Radium	Proportional Counter	6,705
Gross Alpha	Proportional Counter	6,703
Gross Beta	Proportional Counter	6,703
Fecal Coliform	Membrane Filtration	6,909
Nitrate	Automated Cadmium Reduction	7,353.2
Fluoride	Selective Ion Electrode	7,340.2
Cyanide	Titrimetric after Distillation	6,413

analytical results obtained in the laboratory will
consist of the following:

- Use of analytical blanks and standards.

- Use of replicate sample analyses.

- Use of spiked samples.

- Use of standard approved methodologies.

- Observation of proper sample holding times.

- Performance of critical analyses, i.e., gas chromatography, by experienced analytical personnel.

- Maintenance and documentation of instrumentation
  service contracts.

- Routine calibration and certification of analytical
  instrumentation.

- Documentation of all analytical procedures and QA/QC
  data.

- Check and acceptance of all data by designated QC
  officer.

5. Chain-of-Custody — As with waste analysis, groundwater
   samples must be transported under chain-of-custody
   procedures. The plan must either specify a chain-of-
   custody form (Figure 2) or require that one be used by
   the sampling personnel.

Groundwater monitoring represents the most elaborate and
potentially the most costly of the analytical programs
required by RCRA. Because of the importance and cost-fac-
tors, the selection of analytical laboratories should be a
cautious undertaking. The results of the analyses on
groundwater samples are the key to demonstrating that land
treatment is not affecting the groundwater environment or,
conversely, what constituents have escaped to the
groundwater. In the event of groundwater contamination,
analyses are again used to determine the concentration,
migration limits and migration rates of the constituents.

CERTIFICATIONS

The HWM system often requires that certain analytical
data be utilized to certify something e.g., a detoxified
waste, a process, or even analytical results themselves. The

certification process typically involves the laboratory and the owner-operator. The laboratory must certify (and demonstrate the rationale of certifying, via QC, qualifications, etc) that analyses are valid. The owner-operator must typically certify that the laboratory has, to the best of the owner-operator's knowledge, gathered data representative of the system being analyzed. Some specific certifications are:

1. Hazardous Waste Treatment Success - Various schemes have been devised to "detoxify" certain sludges by combining the hazardous waste with portland cement or other adhesive solidifiers. Preliminary testing of EP Toxicity values of the sludge are compared to post-treatment EP Toxicity values to demonstrate that the treatment process prevents the leaching of levels of the EP toxic heavy metals in concentrations greater than 100 times the SDWA limits.

   The laboratory must provide certification on the methods, instruments, analysts, QC and related information.

   The owner-operator, through the vehicle of the delisting petition, must certify that the samples analyzed are representative of the hazardous waste treatment process.

2. Site Closure Certification - Under the site closure regulations of the HWM system, the owner-operator must demonstrate that all hazardous waste and hazardous waste constituents have been removed from the site. In the case of a waste pile being "closed" by removal to a secure landfill, the certification scheme would involve taking samples of the soil beneath the old pile site, extracting and analyzing for the hazardous waste constituents of the waste material.

   The laboratory must certify that representative samples of the closed HWM area were sampled and analyzed by techniques to provide representative data on the chemical constituents monitored. Methods, QC, analyst and instrumentation information is required.

   The owner-operator must certify that the closure is complete to the best of his knowledge, based on the laboratory analysis of representative samples.

3. Incinerator Certification - Current proposed incinerator regulations require that 99.99% of the

hazardous waste constituents entering the incinerator be destroyed. For many wastes to be incinerated, current stack gas sampling techniques and analytical techniques of ash residues for trace organic components are not sufficient to meet the certification requirement. In this instance, the regulatory requirement may have exceeded the practical ability of the laboratory to perform the required function.

These areas of the HWM system identify the critical role the analytical laboratory plays in the regulatory process - especially from the viewpoint of the regulated community. The laboratory has many other functions that are peripheral to the HWM system: waste treatability, workplace monitoring for toxic chemicals, discharge monitoring, air pollution monitoring, process evaluation, etc. However, the chief function of the laboratory is to provide decision-making facts to the decision-making process on the nature, quantity and hazard potential of waste materials and environmental media exposed to them.

ANALYTICAL CHEMISTRY ISSUES AND CONSIDERATIONS

While the analytical laboratory plays a vital role in all aspects of RCRA, it presents a tremendous challenge to the analytical chemist. The variety of analyses that must be performed to provide the information about the waste is extensive. The media in which the analyses must also be performed vary from relatively clean groundwater to every type of waste and material imaginable. It is often necessary to perform analyses for trace compounds in the presence of major interferences. In general, interferences are the major problem that must be overcome in analyses related to RCRA.

A laboratory performing RCRA related analyses must provide a variety of analyses. Groundwater samples must be analyzed for several parameters which have been defined to indicate contamination or potential contamination from hazardous waste management facilities. Waste material must be evaluated for the four characteristics of ignitability, corrosivity, toxicity, and reactivity. In addition, often it is also necessary to analyze a waste material for one of the many compounds that have been specifically identified that can cause a waste to be classified as hazardous. The leachate from a variety of wastes must also be analyzed. This leachate may contain one of the parameters which are used to classify the material or it may be necessary to analyze this leachate for some of the priority pollutants. Many times, it is also necessary to analyze an unidentified

waste to determine what its chemical characteristics are
prior to classification. At other times, it will even be
necessary to synthesize and/or generate the wastes prior to
chemical characterization. In cases like this, a new process
may be expected to generate a waste but until the waste is
actually generated on a laboratory or a pilot plant scale,
the characteristics of that waste can not be evaluated. All
of this serves to indicate the diversity, complexity and
critical importance of the analyses that are performed by an
analytical laboratory. The following paragraphs indicate
some of the potential problem areas and solutions that must
be considered during analyses.

Sample Collection And Preservation

Sample collection is the first major step in obtaining
valid analytical results. The sample collected for analysis
must be representative of the entire lot, otherwise the
result may be meaningless. Wastes are many times multiphased and each phase contains different components and
compounds. The ability to obtain the right proportion of
each phase to separate the phases of analyses is a necessity
for valid results. Generally, standard sample collection
procedures have been specified that will work for many
materials. Those procedures should be adhered to when
possible, but many wastes do not lend themselves to standard
collection procedures.

The utensils used for sample collection must be inert so
as not to contribute to any potential contamination. The
collection utensils and sample containers must be contamination free for representative results. This contamination
free sampling is often difficult when the same sampling
equipment is utilized to collect several samples. The sample
composition may vary from one sample to the next and it is
imperative that the sample equipment be thoroughly washed
between the collection of each sample. In the field it is
difficult to determine when the equipment is actually
contamination free and ready for the next sample collection.

At the time of collection, the sample must be properly
preserved to stop any reaction that may be taking place or to
prevent any sample degradation. Sample preservation normally
consists of storage of the material in the dark and cooling
to 4°C. At other times, it may even be necessary to store
the material under an inert atmosphere or to add some
chemical preservative to stop a reaction from occuring.
Procedures must be developed for each type of waste to
prevent degradation problems for that particular waste.
While standard procedures are normally available for sample
preservation, it should be noted that they are not applicable

in all cases.

Analytical Interferences

The major problem encountered with sample analysis is the interferences that are present in the sample. Waste materials are seldom single component mixtures but many times will contain hundreds of compounds, each of which interferes with the analysis of the others. The fact that many waste samples contain multiple phases, organic materials, inorganic materials, inert and reactive compounds, adds to the problem experienced by the analytical chemist. For this reason, normal, routine analytical procedures often are not applicable for use with waste materials.

Many standard analytical procedures utilize colorametric reactions to determine concentrations. Many waste materials yield highly colored solutions that do not lend themselves to colorametric analyses. The color in this solution can, at times, be removed but more often than not, an alternate analytical technique must be chosen. As a general rule, sample preparation of some type is required prior to analysis. Many wastes are semi-solids that must be dissolved, filtered, or extracted prior to analysis. Each of these extra steps in the analytical procedures has the potential for introducing error and producing results that are not accurate.

Separation of organic materials is one of the major efforts that requires the use of analytical separation schemes followed by compound resolution. Analytical schemes are developed based on the chemist's knowledge of the reaction of the compound or compounds of interest in relation to other compounds. Often, an extraction technique is developed based on the pH of the solution or the addition of specific chelating agents for a particular compound. The separation and extraction techniques must be verified utilizing simulated solutions and then by actually spiking samples with the compound of interest to verify the recovery of that compound. The spiking and method of standard additions as a means of overcoming interferences will be discussed later in this paper.

An example of the type of interferences that are experienced is the analysis of cyanide in the presence of sulfide. Normal procedures call for the acidification of the sample and then stripping the cyanide from the sample as HCN. This same procedure strips any sulfide present as H_2S. The stripped cyanide is then trapped in a sodium hydroxide solution for analysis. Analytical techniques that are normally used for cyanide cannot be used in the presence of sulfide. It is therefore necessary to develop additional separation techniques to separate cyanide from sulfide quantitatively at the trace concentrations present. Most standard methods simply do not address this fact. Methods will identify the interferences but solutions must be developed for a particular sample. This type of problem must be routinely solved to obtain valid results.

Instrumentation

Due to the complexity of the waste material and the trace concentrations of contaminants that must be analyzed, sophisticated analytical instrumentation has to be used. Experience has shown that older techniques that were satisfactory to determine concentrations in the parts-per-thousand or the parts-per-million range are simply not applicable with many waste materials. One such analytical tool that has filled a major need is the gas chromatograph-mass spectrometer (GC-MS). The gas chromatograph portion of the instrument is utilized to separate a variety of organic compounds by the order of elution. The mass spectrometer is used to identify those compounds as they elute from the gas chromatograph. Extensive libraries of mass spectral data are available on a modern GC-MS system to enable the chemist to rapidly pick the best fit mass spectral data for a particular group of compounds. Identification of many organic compounds and a complex matrix is simply impossible without the use of this instrument.

High resolution of compounds utilizing gas chromatography has always been the goal of most modern-day analytical chemists. The advent of capillary column technology (wall coated open tubular columns, support coated open tubular column, and fused silica columns) has increased the ability to resolve very similar compounds. The use of this high resolution chromatography with the mass spectral system has aided the chemist to obtain more accurate data. At present, fused silica capillary columns represent the state of the art for resolution.

New detectors for gas chromatography and other instrumentation have been developed specifically to provide results that are compound-specific. These detectors are necessary in order to quantitatively determine concentrations after the components are resolved. Many other analytical instruments are utilized in eliminating interferences. A complete list of modern-day analytical instrumentation would be a long one and will require describing atomic absorption spectophotometry, high performance liquid chromatography, thin layer chromatography and many other techniques. Each technique is used for a specific analysis or group of analyses to provide one bit of information regarding the waste materials. A group of instrumentation and analytical techniques are required to complete the picture regarding the composition of waste and its potential to be hazardous.

Quality Control

The necessity for quality control (QC) should be evident in the analysis of compounds within the analytical laboratory. It is especially critical in the analysis of waste material and groundwater as it relates to RCRA. The interferences that were discussed above and the potential problems make it imperative that each and every result be verified. Normally, the verification of the results is performed by the analysis of blanks, replicate standards and spiked samples. This group of extra analyses provide the analytical chemist with confidence in the results obtained. The documentation of the QC efforts for a particular sample or a lot of samples is actually a part of the result itself. The recovery of spiked samples for the precision of the results should be considered by the user as he interprets the data. Results without the associated QC data are often meaningless.

One of the most versatile tools in verifying results is the utilization of the method of standard additions for each sample. In utilizing this technique, a sample is analyzed and also spiked with two to five different concentrations of the analyte. The sample is then analyzed to verify the recovery of that particular analyte in the sample matrix. This technique is generally very time consuming and expensive since a single sample must be analyzed three to five times. It is also, many times, impractical to do this because every analyte within the sample that is being analyzed cannot be identified prior to analysis. In such cases, a sample is only spiked with a representative analyte to determine the recovery on a single compound. That result may or may not be applicable to all analytes of concern. Analytical judgement must be used every time an analysis is carried out.

Quality control is of such importance that it must be given management backing. Each laboratory performing analyses of solid wastes and RCRA-related substances should have a written quality control plan which has been approved by management for use on all routine samples. This written plan should document the quality control techniques to be used and should also define the acceptance limits for the data.

CASE STUDIES OF THE ANALYTICAL LABORATORY ROLE IN HAZARDOUS WASTE DECISION-MAKING

Three case studies involving the analytical laboratory in the HWM System decision-making process are discussed. The first involves an instance in poor planning of a delisting petition such that expensive, unnecessary laboratory costs were incurred. The second discussion involves the determination of a new recycle technology for waste material which would have been treated as a hazardous waste had no laboratory studies been available to show the utility of recycling. The final case study describes an integrated program of waste analysis, waste disposal site investigations and groundwater monitoring to determine compliance needs.

Case 1: Imcomplete Analytical Program Design

Situation

A pharmaceutical manufacturing facility treats highly organic-loaded wastewaters with biological treatment ending in an activated sludge process. The dewatered sludge from this process was characterized as hazardous since it was a sludge from the treatment of hazardous constituents.

After November 19, 1980, the industry instituted a program to characterize the waste (determine its hazardous or non-hazardous status) in preparation for submitting a delisting petition.

Analytical Program

A series of analyses were planned for characterization of the waste. The analyses were in two groups.

Influent Constituents	HWM Characteristics
Methylene chloride	EP toxicity
Methanol	Ignitability
Toluene	Reactivity
Ethyl acetate	Corrosivity

The sources of each of the constituents were sampled as was the final dewatered sludge.

In designing the analytical program, the majority of the analytical effort was focused on the analysis of the influent constituents. Prior to any actual sampling and analysis, a methods development study was undertaken to devise headspace analysis and solvent extraction gas chromatograph techniques capable of resolving the solvents present in the sludge. Samples were then collected over a six-week period, and analyzed immediately for the influent constituents.

Initial results were promising. Certain solvents appeared to pass through the activated sludge treatment without total decomposition; however, the residual amount appeared to be defensible as not significant enough to cause effects on human health or the environment. Since the major hazardous components did not seem likely to impede the delisting process, the waste characterization was begun.

The sludge showed no EP toxicity values in excess of the limit, either for heavy metals or organics. Corrosivity was performed by the NACE coupon test method and the sludge did not demonstrate corrosive capacity. Solvent levels were well below flash point thresholds by the closed cup testing method. However, when reactivity was tested an unexpected result was obtained.

All samples of the sludge proved to be reactive by the acid exposure method; the sludge had become anaerobic at some point and contained a significant amount of sulfide. This sulfide was liberated at a rate approaching one percent of the wet weight of the sludge.

Thus, the waste treatment process had removed the constituents of concern but had caused another HWM problem.

Comments

The analytical program for this HWM decision-making process was flawed due to the emphasis placed on the analysis of one particular group of waste constituents without the consideration of the total waste classification. A more appropriate scheme would have been:

1. Devise sampling plan to obtain representative waste samples.

2. Scan all necessary waste characterization parameters on a composite of the representative samples.

3. If the scan indicates the potential for successful delisting, proceed with all required analyses.

4. If problems are indicated, rethink analytical program and/or probable success of entire program.

The case study presented here ended by costing the company approximately $50,000 for a data set not useable for the intended goal. The majority of the effort was expended in devising the analytical methodology and QC system for the trace organic analyses. The implementation of the composite scanning approach might have saved well over 80% of this cost had it been implemented at project initiation.

Case 2: Evolution and Development of Materials Recovery Technology

Situation

A very large brass foundry operating at the same site for over fifty years had accumulated a pile of foundry sand (containing organic binders and trace metals) in excess of 100,000 tons. At the inception of RCRA, the company, having determined that the waste would be classified as hazardous due to the EP Toxicity test, initially evaluated two options. One option consisted of constructing an on-site landfill to meet RCRA regulations and the other involved transport and disposal of the waste at an approved off-site facility. After the latter option was chosen, bids were received in the range of $10,000,000.

At this juncture in cooperation with its corporate research and development group and an outside consulting laboratory, the company initiated bench and pilot scale tests to evaluate materials recovery. Ultimately, a system was engineered to recover copper and reuseable foundry sand from the waste. The capital cost of the recovery plant was approximately equal to the value of recovered materials. The disposal cost of $10,000,000 was avoided.

Analytical Program

The analytical program was aimed at analyzing for total lead and leacheable lead in the waste and its intermediates during the pyrolysis-based materials recovery process. The laboratory scheme examined:

- levels of the toxic metal
- behavior of the metal at different treatment levels, e.g. formation of intermediate compounds
- identification of critical treatment parameters
- optimization of operating conditions and development of basis of design
- determination of EP Toxicity at intermediate points in the pyrolytic process
- determination of EP Toxicity of the final, recovered reuseable materials

Comments

This project benefitted from the team work engaged in by

the (i) plant operations personnel, (ii) the subject
company's central R&D group and (iii) the outside
environmental laboratory familiar with environmental
regulations as well as environmental chemistry. Rapid
turnaround of the analyses provided immediate feedback to the
engineers for improved efficiency in decision making.

Case 3: Integrated HWM Survey for Successful Planning

Situation

A wood products company produces creosote and/or pentachlorophenol treated wood products at several plants in the
southeast. Prior to November 19, 1981, a study was
undertaken to identify the following items:

1.   Analysis of existing waste

2.   Prior waste disposal areas

3.   Presence of waste constituents in soil

4.   Presence of waste constituents in groundwater

The overall objective was to establish the status of the
HWM areas and to determine what further actions would be
necessary under the groundwater monitoring program.

Analytical Program

A series of samplings were undertaken to define the
following:

1.   Concentration of major waste constituents in sludges
     and pond waters.

2.   Vertical profile constituent concentrations in soils
     around the plant site.

3.   Upgradient and downgradient determination of waste
     constituents' presence or absence in the groundwater.

The analysis of the waste sources (impoundment sludges
and waters) revealed that significantly high concentrations
of those constituents for which wood preserving wastes are
listed occurred in the pond. Soil borings identified layers
of old sludge material which had been buried outside of the
current HWM areas. Groundwater analyses demonstrated whether
or not any of the waste constituents were leaving the HWM
units and entering local groundwater regimes.

Comments

The integrated nature of this program points out the critical path such investigations should follow systematically. The questions which should be answered are:

1. Are hazardous waste constituents present?

2. Have any hazardous waste constituents entered the soil?

3. Are any non-RCRA regulated sources of similar waste constituents extant such that data regarding the regulated HWM facility is threatened?

4. Do groundwater contamination problems exist and, if so, to what extent?

By utilizing this step-by-step approach several potential problems were avoided. At two sites, locations of previous disposal areas for wood preservation wastes were found. The groundwater monitoring plans for RCRA compliance were designed around these areas so that no "interference" would occur during later groundwater monitoring. At one site, it was determined that a waiver was possible.

The integrated approach defined by these example projects is certainly cost-effective in utilizing the analytical tool as a decision-making device under the HWM regulations.

CONSIDERATIONS IN OBTAINING LABORATORY SERVICES

An industry will generally have three choices in obtaining laboratory services in support of its hazardous waste management program:

1. In-House Laboratory
2. University-Based Laboratory
3. Commercial Environmental Laboratory

In-House Laboratories

An industry's in-house laboratory may be at the plant where the HWM plans are to be implemented, or it may be at a regional or divisional level. If several plants with similar wastes have substantially similar HWM activities, the use of in-house laboratories, if available, will be a cost-effective approach. Many in-house laboratories, however, are used in connection with process control and have neither the flexibility nor the instrumentation to perform the kinds of analyses required by HWM activities.

University-Based Laboratories

In recent years, several universities have started competing with commercial laboratories for routine and specialized analytical services needed for industrial HWM activities. Universities offer highly-skilled professionals and a wide array of excellent instrumentation. They are a good source for solving special methods development problems. The response, however, is often too slow to meet the requirements for industry. Also inter-departmental insularity often prevents a multi-disciplinary focus. The institution's administrative structure may require time-consuming and expensive contracting steps which reduce the flexibility of the user industry.

Commercial Environmental Laboratory

In the past decade the number of commercial laboratories offering a wide variety of environmental testing services has steadily risen. Various state and federal agencies have developed certification programs which require regular inter-laboratory testing. These commercial laboratories offer specialized expertise in solving environmental compliance problems. The "third party" nature of their involvement in HWM compliance activities is valuable in dealing with regulatory agencies. As an outside agent, their involvement is sometimes viewed with concern by user industries in regards to confidentiality of business and process information. Competent commercial laboratories have written protocols for ensuring confidentiality of data and routinely provide such assurances to their clients in written form.

CHECKLIST FOR SELECTING AN ANALYTICAL LABORATORY FOR RCRA RELATED ANALYSES

Personnel

1. Do the key laboratory personnel have sufficient formal education in chemistry or related subjects?

2. Are the analysts experienced in performing analyses of parameters in the media of concern?

3. Does the laboratory maintain an ongoing training program for analysts?

4. What professional development steps are taken to increase the knowledge of analysts and laboratory management personnel?

5. Are the analytical personnel experienced in the use of instrumentation required for the analysis?

Facilities and Equipment

1. Is sufficient laboratory space available to perform all necessary analyses?

2. Is all necessary instrumentation (GC, AAS, Spectrophotometers, TOX, etc.) available to perform routine analyses of RCRA related substances?

3. Does the laboratory meet the criteria established under good laboratory practices for analytical laboratories?

4. Is sufficient logistics (distilled and de-ionized water, air, vacuum, steam, etc.) available for use in the laboratory?

5. Are routine preventive maintenance schedules followed on all instrumentation and are they properly documented?

6. Is the instrumentation to be utilized for the analyses state of the art or is it obsolete?

7. Are the laboratory facilities conducive to performing trace analyses in the proper environment?

Procedures

1. Are accepted procedures (EPA, ASTM, APHA, etc.) used for analysis where applicable?

2. Are nonstandard procedures developed and validated prior to use?

3. How are nonstandard procedures validated?

4. What steps are taken to prevent contamination of samples and sample containers within the laboratory during analysis?

5. What procedures are available for validating standards for the different parameters (are reference standards purchased or are they made within laboratory)?

Quality Control

1. Does the laboratory have a written quality control plan for laboratory operation which has been apprroved by management?

2. What is the frequency of normal sample replication and spiked sample analysis?

3. What is the frequency of the analysis of blank samples with normal samples?

4. How are quality control acceptance limits established for routine samples?

5. What actions are taken if a sample lot is found to be out of control?

6. What action is taken if recovery of an analyte is not demonstrated by the method of standard addition or a spiked duplicate analysis?

7. Who is responsible for the maintenance of quality control within the laboratory?

Cost & Performance Aspects

1. Does in-house performance of analytical work by the industry involve capital expenditures? Additional staff?

2. Can several plants and/or facets of environmental testing be combined to get favorable terms from a contracted laboratory?

3. Does the contracted laboratory have a reasonable turn-around time? Do references confirm adherence to standard turn-around time?

4. Does the contracted laboratory pass on savings derived from large volume of analytical work from same client to that client?

5. Does the contracted laboratory have satisfactory procedures for ensuring confidentiality of information?

REFERENCES

[1] Federal Register. May 19, 1980 Vol. 45-No. 98 40 CFR 260 (Washington, D.C.: U.S. Government Printing Office 1980).

2
U.S. Environmental Protection Agency. Test Methods for Evaluating Solid Waste Physical/Chemical Methods SW-846 (Washington, D.C.: U.S. Government Printing Office, 1980).

3
American Society for Testing & Materials. "Standard Test Methods for Flash Point by Pensky-Martens Closed Cup Tester" ANSI/ASTM D-93-79, (Philadelphia, PA; ASTM, 1979).

4
"Public Health Service Act" as ammended by the Safe Drinking Water Act. Pub. L. 93-523.

5
DeVira, E. R., B. P. Simmons, and Storm, D. L. "Samplers and Sampling Procedures for Hazardous Waste Streams" U.S. Environmental Protection Agency. EPA-600/2-80018 (Washington, D.C.: U.S. Government Printing Office, 1980).

6
American Public Health Association. Standard Methods for the Examination of Water and Wastewater, 15th ed. 1980 (Washington, D.C. APHA, 1980).

7
U.S. Environmental Protection Agency. Methods for Chemical Analysis of Water and Wastes, EPA-600/4-79-020 (Washington, D.C.: U.S. Government Printing Office, 1979).

8
U.S. Environmental Protection Agency. Procedures Manual for Groundwater Monitoring at Solid Waste Disposal Facilities SW-611 (Washington, D.C.: U.S. Government Printing Office, 1980).

9
Scalf, M., McNabb, J., Dunlap, W., and Cosby, R. Manual of Groundwater Sampling Procedures National Water Well Association (Worthington, Ohio: 1981).

10
Annual Book of ASTM Standards. Part 31 - Water (Philadelphia, PA American Society of Testing and Materials, 1979).

11
Environmental Protection Agency. Methods For Benzidine, Chlorinated Organic Compounds, Pentachlorophenol, And Pesticides In Water And Wastewater, 1978.

141

ENVIRONMENTAL "TERMITE" INSPECTIONS

W. A. Duvel, Jr., P.E.
   Environmental Research & Technology, Inc.

WHAT IS AN ENVIRONMENTAL "TERMITE" INSPECTION?

A new concept is emerging in relation to commercial and industrial real estate transactions--the Environmental Termite Inspection. Suppose Party A wants to sell a parcel of industrial property to Party B. Party A is concerned about long term liability for past practices at the site--usually a concern over potentially contaminated soil or ground water as the result of past spills or disposal practices. Party B is concerned about buying any hidden liabilities or problems which might result in regulatory action against him, might present a hazard to future workers, might somehow constrain or limit the use of the property, or might increase the cost of development of the property. Both parties are therefore usually interested in knowing what environmental/health problems go with the property. The Environmental Termite Inspection is simply an evaluation of a piece of property to determine the presence or absence of potential human health and environmental problems resulting from past practices on the property. The Environmental Termite Inspection is to industrial and commercial property what a regular termite inspection is to private homes--an investigation into hidden conditions which might cause problems later.

WHAT GOES INTO THE INSPECTION?

The inspection is a detective story in which you try to answer these questions:

o   What potentially harmful material is present on the property?

o   What migration pathways are available for exposure to humans and the environment? Is the material moving off the property?

o   What population is presently or potentially
     exposed to the material?
o   What are the present or potential future adverse
     health/environmental impacts resulting from
     exposure?

WHAT DO YOU GET OUT OF AN INSPECTION

The direct output from the Termite Inspection is a
health/environmental risk assessment of future use of the
site. However, the Termite Inspection is done in the
context of a real estate transaction and it is used
primarily to evaluate the financial risk or exposure
related to the transaction. In conjunction with other
information, the Termite Inspection is used by both the
buyer and seller to determine whether the property should
be bought (or sold), at what price, and with what
restrictions or exclusions. Because the Inspection can
influence the terms and conditions of the sale, it is
important that the Inspection be conducted properly and
that the buyer/seller paying for the Inspection understand
the significance, implications and limitations of the
Inspection. The Inspection provides the buyer/seller with
information about the site. Armed with that information,
the buyer/seller can make an informed judgement about the
terms of the sale.

HOW DO YOU DO AN INSPECTION?

There is a great deal of flexibility in what can be
done to get information about a site. The important aspect
is to get the right information and the right amount of
information. But what is the right amount of information?
How big should the investigation be? Presumably the more
information a buyer/seller has, the better will be his
ability to make the right decision. Thus, the
buyer/seller's confidence about his position vis-a-vis the
transaction will increase as more information is
developed. However, getting information costs money and
the buyer/seller does not want to spend any more money than
necessary to evaluate the site.

The conceptual relationship between input information
and level of confidence is shown in Figure 1. The slope of
the curve varies with different situations, but the
relationship is the same. You learn the most about a site
with a relatively modest investigation. More information
(i.e., more investigation and more money spent) increases
your confidence about the situation, but you eventually

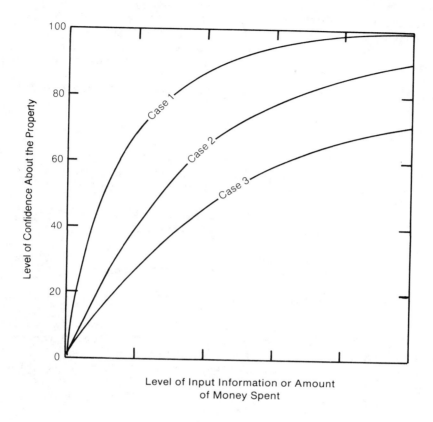

Figure 1. Relationship between input information and level of confidence achieved during an Environmental Termite Inspection.

reach a point of diminishing return. At this point on the curve you are getting more data but it's not telling you anything new. All this is a fairly round about way of making the point that in doing Termite Inspections it is prudent to proceed with the work in small increments, developing information in a stepwise fashion. In this manner you can stop the work when the buyer/seller believes he has enough information to make a decision. The following steps describe the activities to be conducted in the inspection.

SOURCE EVALUATION

The purpose of this activity is to determine what potentially hazardous materials are present on the property. This means you are seeking to find out:

o   What material is present on the property?
o   Where is that material located?
o   How much is there?
o   How did it get there?
o   Is it moving? Where? At what rate?
o   In what form is it?
o   How harmful are these materials?

Get answers to the first of these questions by finding and interviewing the Plant Supervisor, Technical Director, Site Manager, Plant Engineer, or other people who are directly familiar with what went on at the property. Find out from them:

o   How has the site been used?
o   Who had ownership? When?
o   What did the owners make/do/process/manufacture on the site?
o   What processes were involved?
o   Did these processes involve solvents or liquid chemicals? What chemicals?
o   Were there spills? How often? What material? Where? How were they cleaned up?
o   Where were chemicals loaded and unloaded (a good place to look for spills)
o   What happened to the wastewater? Were there lagoons, pits, trenches, or other open conduits? Where? How big? How deep?
o   What happened to the solid waste? Where is the dump? What things went there? How old is it? What are the boundaries? How deep is it?
o   Are there underground storage tanks? What's in them? How old are they? Have they been tested?

146

Interview several different people if possible and try to get some cross-references between these sources.

Interviewing is a good way to get a general idea of what went on, but it is seldom useful for quantitative or definitive information. If more quantitative information is required, ask to look at plant production records, waste disposal records, spill incident reports, RCRA and Superfund notification documentation, and similar materials. These often serve as good cross-references or clues to where to look for troublesome materials. Try to get a feel also for the kind of housekeeping that went on. Often the "burps and drools" are as important as the plant dump.

While you are in talking with the people who know about the property, get copies of other relevant documents which they may have. These are:

o   Aerial photographs—old and new photos are very useful to visualize past site activities and are helpful in locating spill limits, wastewater treatment areas, abandoned dumps, and similar older features.

o   Topographic maps—same purpose as aerial photographs. These are often useful in showing fill areas and location of new structures on old dumps.

o   Foundation investigation reports—useful in learning local geology and soil conditions and determining depth to ground water.

o   Boring logs from production and monitoring wells—same purpose as foundation reports.

o   Pumping test results from production and monitoring wells—these can give some idea of aquifer yield and water movement.

o   Ground-water modelling studies—these are sometimes available at facilities who rely heavily on ground water for makeup water.

o   Ground-water monitoring results—useful to establish existing baseline ground-water quality.

Using the information gained from these sources you should begin to get an idea of what materials are of concern and where they are likely to be located.

Now a problem arises. How do you know which materials are likely to be a problem? The answer to this requires some knowledge of the various industrial operations,

147

chemicals, and wastes. You are looking primarily for hazardous residues remaining on the site. While there is no comprehensive list of hazardous materials, there are some fairly obvious starting points.

o   Are there residues on the site which are listed by the U. S. Environmental Protection Agency (EPA) as hazardous waste under the Resource Conservation and Recovery Act of 1976 (RCRA) rules?

o   Are there residues on the site which contain chemicals that are listed by EPA as hazardous under RCRA rules?

o   Are there residues on the site which contain chemicals that are listed by EPA as priority pollutants?

o   Are there residues on the site which are listed as hazardous by US DOT?

If any of these are found on the site, then you have a starting point for making a judgement about whether these materials are present in significant quantity to represent a potential problem.

The source evaluation should culminate in a written description of the site which documents the history of the use of the site and shows by means of a map the location of potential problem areas.

The source evaluation is very important for one other reason. It helps you define appropriate personnel safety procedures to be utilized when working on the site, conducting any subsurface exploration, and doing the sampling and analysis. Use the information obtained at this point to develop a site safety program.

HYDROGEOLOGIC BACKGROUND

Almost invariably a central issue in the Termite Inspection is the extent to which the soil, ground water, or surface water is contaminated as a result of activities on the site. An important related issue is the extent to which contaminated ground water is moving off-site. To address these issues you should begin to determine the geologic and hydrologic setting of the site. The important questions here are:

o   What is the local stratigraphy (geologic profile)?
o   What is the depth to ground water?

o   What are the characteristics of the aquifer?
     Confined? Unconfined? Is it used? Where? Who
     uses it? What is the water quality? Where does
     the ground water go?
o   Is any or all of the site flooded? At what
     frequency?
o   What happens to surface runoff?
o   What are the important subsurface features (e.g.,
     mines)?

Get answers to these questions by consulting existing
records available from public agencies. You should procure
and review:

o   Soil conservation service soil maps--useful
     indicators of natural soil present in the area, a
     tool needed to measure whether the site soil is
     contaminated.
o   Results of previous water quality analyses on
     receiving streams--useful indicator of expected
     background conditions.
o   Existing site aerial photography and topographic
     mapping from Corps of Engineers, State Department
     of Transportation, Nuclear Regulatory Commission,
     or other similar agency and local utility
     company--useful supplement and cross-reference to
     materials obtained from site owner.
o   208 Regional Water Quality Study Reports--often
     provides summary information to place the site in
     a regional context.
o   USGS and state geologic survey geology,
     ground-water, and surface water reports--used to
     supplement site specific information and
     understand geologic context.
o   USGS or Corps of Engineers flood mapping for the
     site.

SITE MAP

Having a suitable site map is critical to an effective
integration and interpretation of the information collected
in the Inspection. As a minimum, you should secure a base
map showing the location of existing buildings, structures
and other physical features as well as property boundary
lines. A suitable map can usually be obtained from
existing records. Use the base map as the device for
consolidating all relevant information. Show on the map
significant areas which might be sources of contaminated
soil or ground water--e.g. underground storage tanks,

149

loading/unloading areas, lagoons, and disposal areas. Also show the location of existing bore holes, water quality sampling points, monitoring wells, important geologic features, and observations made from the field reconnaissance. The map should also show the location of sewers, buried pipelines, and buried utilities. These buried items can serve as conduits for migration and can tell you where not to dig if you plan to do any subsurface work.

If possible, obtain a recent topographic map of the site. The topographic map is more useful than a simple plot plan because it provides relative elevations. The topo map should be at a scale of one inch equals 200 feet (or larger if the site is smaller) and show at least five foot contour intervals (two foot contour intervals are preferred at small sites). The topographic map is used for interpretive purposes such as plotting the direction of ground-water flow, evaluating the rate and direction of surface water runoff, and evaluating flooding potential. The map is also a valuable tool in planning and interpreting future work. The map will aid in laying out a soil sampling and drilling program, locating ground-water monitoring wells, constructing geologic sections, predicting ground-water movement, calculating the volume of material in landfills, showing the boundaries of contaminated areas, setting the limits on remedial measures, setting the limits on future land use or land title constraints and explaining to the buyer/seller where things are located.

SITE RECONNAISSANCE

The next step in the inspection is a detailed visual site reconnaissance—a careful and thorough walkover of the property. Bring your map and wear the proper protective gear if that is necessary. Note on the map and describe the things which you see and smell. Use Table 1 as a guide to the things you are looking at and evaluating.

Water features are particularly important since water is a principal conduit for movement of materials off-site. Geologic features provide important clues to the direction and rate of water movement. The presence of waste materials may provide confirmatory evidence of things which people told you about in the source evaluation. Pay close attention to suspected spill areas, old dumps, and old chemical processing areas. Use your knowledge of the site to look for meaningful confirmatory evidence, but be alert

TABLE 1

ENVIRONMENTAL TERMITE INSPECTION
SITE RECONNAISSANCE CHECKLIST

1. Water Features. Note the location of and describe:

   - wells   - "squishy" areas   - direction of runoff
   - springs   - marshy areas   - direction of runon
   - seeps   - ponded areas   - surface erosion
   - swamps   - streams   - evidence of flooding

2. Geologic Features. Note and describe:

   - topography and slope   - mining activity
   - soil characteristics   - quarries or pits
   - rock outcrops   - gas and oil wells
   - sink holes   - diversion ditches
   - excavations   - soil stockpiles
   - spoil piles

3. Waste Evidence. Note the location of and describe:

   - drums, barrels, containers (Condition. What do
     labels say?)
   - waste materials
   - construction debris
   - discolored soil
   - odors
   - leachate seeps
   - "unnatural soil"
   - ash or blackened areas

4. Vegetative Features. Note and describe:

   - type
   - maturity
   - density
   - condition
   - stress

5. Physical Features. Note the location of and describe:

   - roads
   - power lines
   - public buildings
   - dwellings
   - structures (e.g., fencing, water towers,
     buildings)
   - rights-of-way

TABLE 1 (Continued)

6. Security Features. Note the location of and describe:

   - access roads
   - fencing and gates
   - vegetation barriers
   - bike trails
   - campfire remains
   - boat launching areas
   - picnic and beer party residue

7. Adjacent Land Use. Note the location of and describe:

   - surface water
   - roads/utilities
   - housing /industry
   - vacant land
   - vegetative types

for new findings or findings which contravert what you were
told. Evaluate the vegetation on the site. Where are the
bald spots? Where does the vegetation look stressed?
Where are the dead trees?

Noting the physical features (e.g. roads, buildings,
power lines) is useful primarily because these become
reference points for future on-site work. Evaluating the
site security, however, is important because it can give
clues to potential human exposure pathways which are not
obvious. Children often use abandoned property for
entertainment, e.g. mini-bike trails, campouts, campfire
and beer parties, boat launching, hide-and-seek, and mock
war games. Look for evidence of these activities and check
the condition of the fence and gate. Look also at adjacent
land. How is it used? Are people close-by? Who would be
affected by hazardous materials migrating from the site?

Collate all the information developed from the field
reconnaissance and see how it matches up with what else you
know about the site. You should do a preliminary risk
assessment at this point and evaluate what you have. There
may be enough evidence to reach a preliminary conclusion.
However, you may decide you need more information.

NON-INVASIVE EXPLORATION

In some situations you may want to explore the
property further without digging anything up. There are
specialized tools available for very selected purposes.

Metal Scanning - There are hand-held devices used to
scan for concentrations of buried iron--e.g., barrels,
drums, and pipes. This type of detector can locate buried
items to a maximum depth of about 5 feet. The device is
handy but is not useful in areas where the soil is heavily
laced with materials like slag or bottom ash which contain
iron residues.

Ground Surface Radar - This technique is a geophysical
tool which can be of value in determining the depth to
ground water, the location of barrels or tanks, and the
boundaries of a fill. The technique consists of moving a
source of electromagnetic radiation across the ground
surface at regular intervals and evaluating the input of
reflected radiation. The technique is rather highly
specialized and cannot be used without trained, experienced
personnel to interpret the results.

Resistivity – This involves placing paired electrodes into the ground at selected intervals and measuring the electrical resistivity between the electrodes. Since different soils and refuse materials have different resistivity, the technique can be used to locate buried materials and ground water. This technique also requires a trained, experienced specialist to interpret the results.

Radioactivity – Measurement of readioactivity at the ground surface using an appropriate radiation detector is a very specialized application for use when radioactive materials are suspected of being present on the property.

Organic Vapor Detector – Several devices are available to quickly measure the presence of hydrocarbon vapors present in the atmosphere using a photoionization detector. In its simpliest form the machine can simply be carried around on the site making measurements of the open atmosphere. Results from this methodology are seldom useful because of vagaries in the movement of the open air. The device is usually more valuable when it is used as the detector in a head space analysis, where suspected soil or suspected water is "sniffed" in a controlled manner.

INVASIVE SUBSURFACE EXPLORATION

In most Environmental Termite Inspections, suspected soil and/or ground-water contamination are the principal issues. This usually implies some level of soil, waste, ground-water, or surface water sampling and analysis is desired. Here the options for what to do begin to multiply dramatically and the costs of the investigation also increase dramatically. An exploratory program can be of almost any scope and magnitude, ranging from collecting soil samples with a shovel to conducting detailed ground-water modeling using the results of extensive in-situ permeability testing. What to do and how to do it are a matter of judgment which you will have to determine based on site-specific conditions. Remember, however, that this is a Termite Inspection and you do not need to collect any more information than is necessary to reach a decision about the real estate transaction. This means proceeding in small increments of work in a step-wise fashion. It is also very important to determine in advance how you will use information which is collected.

In order to figure out what to do, here are the principal questions which must be answered before you start.

o   What are you trying to find out?

- What information do you want?
- What materials are you looking for?

o Where should you dig/drill/sample?
- What location?
- What depth?
- How many pits/borings/wells/samples?

o How should you sample and analyze?
- Grab or composite samples?
- How should you grab or composite to get representative sample?
- What are the sample conditioning or storage requirements?
- What analysis should be used?
- Should you do field or lab analysis?

o How will you know if you have something bad?
- What yardstick will you use to compare against your results?
- Where and how should you take control samples?
- How will this information be used in the risk assessment?

o How confident will you be that the results are representative and correct?
- How many replicate samples/analyses are required?
- What frequency of sampling/analysis is required?
- What statistical analyses should be applied?
- What control samples should be taken?
- How many control blanks should be used?

o What safety measures must be taken to insure proper personnel protection

Remember, the most important factor is to plan the work carefully so you know what the results will mean and how to interpret them.

Usually some subsurface work is required to define the site-specific stratigraphy (geologic profile), confirm the presence/absence of contaminated soil or wastes, determine the depth to ground water, and provide access to the ground water for sampling. Shovels or hand augers are sufficient to accomplish these objectives where shallow sampling (less than three feet) is required. Larger equipment is needed to go deeper. A backhoe is an effective tool for digging test pits or trenches to a depth of about 15 feet. The backhoe can dig quickly and can move rapidly from one location to the next. The open excavation allows visual confirmation of the stratigraphy and soil samples can be conveniently collected. The major disadvantages of the

155

backhoe are that it can only go to about 15 feet, it disturbs a lot of ground, and it does not allow for ground-water sampling. Where deeper excavation is required, a truck mounted or skid mounted drill rig is most appropriate. There are various types of drilling methods available, each with some important advantages and disadvantages. You should discuss these with an experienced hydrogeologist before deciding on a particular type.

A determination of the number, location, depth, and method of excavation of test pits, borings, monitoring wells, piezometers, and other measuring points are all a function of site-specific conditions. In determining where to look for contamination, be guided by your source evaluation. Look in and downgradient from spill areas, old dumps, and process/manufacturing areas.

What to look for in the soil and water should be determined by what you think is there and how you can conveniently analyze for that constituent. Some sacrifice in detail or accuracy may be warranted for the sake of convenience. Laboratory analyses are very accurate but also are very expensive. It is often appropriate to use field analytical methods (which are usually less quantitative or less specific) for rapid determination of one constituent or parameter and back up this field work with some limited laboratory work. This is particularly true for soil samples. You cannot hope to fully characterize the level of soil contamination in a 25 acre site to a depth of 25 feet without using some effective semi-quantitative screening techniques.

Consult with an experienced field chemist to review alternative methods and to achieve the proper balance between accuracy and convenience. For water samples, there are convenient prepackaged kits (such as that manufactured by HACH) to do measurement of temperature, pH, and specific conductance and many other dissolved species. For soil/sludge/waste samples, consider soil test kits for direct measurement or simple aqueous field extractions followed by analysis of the extract using a HACH kit or similar device. For volatile organics, consider head space analysis of soil/water samples using a total hydrocarbon analyzer or, if you know specifically what compound to look for, consider a portable GC unit.

A most important consideration is knowing what to do with the results of the sampling and analysis when you get them. How will you know if you have found something bad?

For ground water this is relatively easy. Compare the
constituents you have found against the primary drinking
water standards. If you exceed the standards, then you
have a potential problem. But how do you evaluate
hazardous materials, like EPA listed priority pollutants,
for which there are no standards? Similarly, how do you
evaluate contaminated soil? Except in a few instances
there are no standards for things like pesticide residues
or chromium residues in soils. To evaluate the results you
must create a yardstick against which you can compare the
results. The yardstick can be created from the literature
or you can develop one for your property. You simply want
to know, what is the concentration of material X in
uncontaminated soil/water? Either obtain the answer from
values extracted from the scientific literature or, safer
yet, develop some control samples on or adjacent to your
property which would represent uncontaminated soil/water.
In other words, since there are no universally accepted
standards, develop a yardstick which is site specific.

This discussion of the yardstick raises another
important question. How much data do you need in order to
be confident that your results are representative and
correct? Many people will be satisfied with the results of
one round of sampling and will base their real estate
decision on those results. Most scientists and engineers
are aware, however, that these results are a bit tenuous
and several replicates are usually required to instill some
confidence in the investigator that the results are
meaningful. Simple statistical analyses such as a
student's t test or analysis of variance are entirely
appropriate to soil/waste/ground-water sampling. Properly
designed into the exploratory program and properly applied
to the results, these statistical analyses are powerful
tools which greatly enhance your ability to understand and
interpret the results. The desire to use statistical tools
will enhance your desire to use quick field tests which can
be run over and over very quickly and cheaply. Using the
field tests allows you to develop enough information to be
able to use the statistical tests effectively and at a
reasonable cost.

OTHER TYPES OF EXPLORATION

Depending on the situation at your site, you may want
to characterize other media on your site. You may want to
consider sampling and evaluation of bottom sediments,
sludges, surface runoff, nearby surface waters, ambient
air, or local biota. Use the same principles outlined for
soil and ground water to tailor your program to suit the
issues.

RISK ASSESSMENT

Having some information now at hand you are in a position to do a risk assessment. This means answering these questions:

o   To what extent does the property now adversely affect human health and the environment?
o   To what extent could the property potentially represent a threat to human health and the environment in the future?

Two dimensions should be considered: on-site and off-site. Would materials on-site be hazardous to construction workers or future property users? Are hazardous materials migrating from the site at such a rate that they represent a health/environmental threat?

For both the on-site and off-site exposure you need to consider:

o   What materials are present?
o   How are people potentially exposed (e.g., inhalation, skin contact, ingestion)?
o   What is the probability of exposure?
o   What are the effects of exposure (e.g. skin irritation, carcinoma, death)?
o   What population is at risk? (i.e. how many people are likely to be exposed)

This is where you begin to tie all the pieces together. The source evaluation and site exploration work have provided reasonably good information by which you can answer the questions raised in the first three bullet items above. By the field work you have confirmed what hazardous materials are on-site. Using available literature, such as that shown in Exhibit 1, you should look up information relating to the effects of exposure. Examples of the type of summary information you are looking for is shown in Exhibit 2.

Next you should evaluate what population is at risk. How many people are likely to be contacted by the hazardous materials? Would it just be during construction? If the material is in the ground water, where does the ground water go? Who will drink the water? Is there enough in the water to cause a problem? Evaluate each material in each medium and then do a collective evaluation. The output of this exercise is a summary statement that addresses the degree and immediacy of the hazard associated with the property.

The method described here is entirely qualitative. Various workers are developing more quantitative ways to assess the risks from properties containing contaminated soils and ground water. More work along these lines is appropriate, particularly so that the results can be developed quickly.

WHAT DOES AN INSPECTION COST?

Our experience indicates it is appropriate to design a program in each case to fit the needs of the situation. The costs of any Termite Inspection vary greatly. However, we do see a trend emerging whereby three levels of inspection are conducted. In each of these the level of effort and cost varies markedly.

Level 1 - A "quick and dirty" evaluation based primarily on a few discussions with knowledgeable people and a site reconnaissance. The risk assessment is largely subjective and based on the experience of the investigator. Typical cost $2000-10,000 per site.

Level 2 - A modest evaluation involving discussions with individuals familiar with site use; an evaluation of existing hydrogeologic literature; a thorough site reconnaissance perhaps involving non-intrusive sampling; some limited water, soil and/or waste sampling; and a cursory evaluation of pathways and receptors (population at risk). The risk assessment is reasonably conclusive and plans for additional work (if any) are definitive. Typical cost $15,000-30,000 per site.

Level 3 - A detailed site investigation, perhaps involving exploratory borings; sampling and analysis of surface water, ground water, air, biota, soil, waste, and/or sediments; investigation into population and epidemiological considerations; ground water modeling; evaluation of remedial action; evaluation of particular health hazards; or other special areas. It is difficult to generalize on this level because these studies are highly site specific. Typical costs $50,000-$300,000 per site.

In our experience, the Level 1 investigation is often useful for commercial, light industrial operations, and small land parcels. However, it is seldom sufficient to draw meaningful conclusions for large, heavy-industry properties, especially those involving any chemical processing (including chemical cleaning and rinsing) or land disposal operations. Level 2 perhaps followed by a small Level 3 is more typical for these types of

159

properties. Level 3 is never justified without first doing
Level 2. Level 3 should always be broken into the smallest
possible number of increments so that the work can be
discontinued as soon as possible.

SUMMARY AND CONCLUSIONS

The Environmental Termite Inspection is a new term
being applied to site evaluations conducted in relation to
commercial and industrial real estate transactions. The
Inspection involves a look at what hazardous materials are
present on a parcel of land being considered for sale; what
pathways might serve as conduits for off-site migration;
what population is exposed to these materials; and what
adverse conditions could result from this exposure. The
output is a health/environmental risk assessment of future
use of the site.

The Inspection itself will vary depending upon the
level of effort appropriate. However, it usually entails a
source characterization, evaluation of hydrogeologic
conditions, development of a site map, a detailed site
reconnaissance, some subsurface exploration, some
evaluation of the population potentially exposed, and
finally, a summary risk assessment.

Because it is done in the context of a real estate
transaction, the Inspection is used to evaluated the
financial risk or exposure to the buyer or seller and is
used in part by the buyer/seller to determine the
conditions/exceptions/limitations of the sale.

EXHIBIT 1

SOME SOURCES OF INFORMATION ABOUT CHEMICAL AND
TOXICOLOGICAL PROPERTIES OF HAZARDOUS MATERIALS

Drinking Water and Health Part II. A Report of the Safe
Drinking Water committee, National Academy of Science.
1977.

Drinking Water and Health, Volume 3. Safe Drinking Water
Committee. National Academy of Science. 1980.

G.D. Clayton and F.E. Clayton (eds.). Patty's Industrial
Hygiene and Toxicology, 3rd Ed., Vol. 2a, b, c,
Toxicology. John Wiley and Sons. New York. 1978.

N.H. Proctor and J.P. Hughes. Chemical Hazards of the
Workplace. J.B. Lippincott Co. Philadelphia. 1978.

N.I. Sax. Dangerous Properties of Industrial Materials.
Van Nostrand. New York. 1979.

N.I. Sax. Cancer Causing Chemicals. Van Nostrand Reinhold
Co. New York. 1981.

EXHIBIT 2

EXAMPLES OF SUMMARY HEALTH EFFECTS INFORMATION
USED IN ENVIRONMENTAL TERMITE INSPECTIONS

Methylene Chloride

Methylene chloride is a highly volatile solvent (boiling point 40°C) with limited solubility in water (1 part in 50 parts). The primary effects associated with this compound are that of a mild central nervous system depression and irritant of the eye and skin. This chemical has not been shown to cause significant chronic effects. Weak mutagenic effects have been demonstrated and it has been observed to cause some organ damage. Methylene chloride does not appear to be teratogenic nor has it been shown to be carcinogenic. A no-effect level has been shown for rats at 2.3 gm/litre. The NAS (1980)* has recommended a SNARL** for 24 hour of 35 mg/liter or 7-day exposure at 5 mg/liter for drinking water. No chronic exposure level has been established. The EPA has recommended a water criterion level of 12 mg/l for the protection of public health. This criterion is based on noncarcinogenic effects and includes a safety factor of 100,000.

Benzene

The chemical is stable and slightly soluble (0.82 g/l). As a result, this nonpolar material can bioaccumulate and has a high lipid solubility. It is also rather mobile in the environment. Although benzene is associated with acute poisoning, which results in nausea, vomiting, ataxia followed by depression and coma, the major concern is its chronic effects as a hematoxin. The available evidence does not demonstrate carcinogenicity in animals, but there is strong evidence implicating benzene as a cause of leukemia in humans. The NAS (1980) has recommended a SNARL of 12.6 mg/l but this ignores mutagenic and suspected carcinogenic effects. The NAS was not able to recommend a criterion for protection from chronic exposure due to what it considered to be a lack of evidence. The EPA has considered setting ambient water criteria based on cancer risk levels. For a risk of 10^{-5} (one additional cancer for every 100,000 persons exposed) from cancer, a criteria would be 6.6 ug/l. Similarly, a risk of 10^{-6} would be associated with a criteria of 0.66 ug/l.

*Drinking Water and Health. Volume 3. Safe Drinking Water Committee. National Academy of Science. 1980.
**SNARL = Suggested Non-Adverse Reaction Levels

CHAPTER 11

USE OF AERIAL PHOTOGRAPHY AND REMOTE SENSING
IN THE MANAGEMENT OF HAZARDOUS WASTES

John Grimson Lyon
    The Ohio State University

INTRODUCTION

Concern over proper disposal of toxic wastes has resulted in a need for information to facilitate waste management. Use of maps, chemical sampling and company documents can only provide a portion of this information. Other techniques for gathering information are required, and this has fostered an interest in applications of aerial photography and remote sensing for management of hazardous wastes.

Interpretations of aerial photographs and remote sensor data can provide information on historical and current conditions of hazardous waste sites. Historical data extracted from aerial photographs are very valuable for locating presence or absence of landfills, subsequent land use on closed land fills, and historical vegetation, soil and hydrological conditions. Current aerial photographs or remote sensor data can be employed to assist the monitoring of soil and water conditions, characteristics of vegetative cover, and in helping to prioritize abatement activities.

Aerial photographs and remote sensor data are particularly valuable for siting hazardous waste landfills. Compliance with requirements of the Resource Conservation and Recovery Act of 1976 (RCRA) necessitates detailed planning. Interpretation of aerial photos can supply map and resource information, and assist in prioritizing the use of funding and human resources.

These capabilities are discussed in following sections on historical and current applications of aerial photographs and remote sensor data. Several examples are provided to facilitate understanding of the utility of these data for planning and management of hazardous wastes sites.

HISTORICAL INFORMATION AND APPLICATIONS

Historical aerial photographs can supply information for detection of fill areas, and assist in the inventory of waste sites. It is possible to determine the extent, location and possible nature of landfills with interpretation of aerial photographs.[1] The information supplying capabilities of this approach are great, and interpretations along other sources of data can provide details on hydrology, soils, vegetation, and flooding characteristics.[2,3]

Successful interpretation of aerial photography requires an understanding of edaphic, hydrologic and vegetative characteristics.[4] Many "clues" are developed from such an understanding, and these clues can be vital in reconstructing the history of a particular site. The capability of the interpreter to extract information on the site is dependent on the person's skill and is partially dependent on the types of materials in a landfill. Hence, the interpreter must have knowledge of resource characteristics and data on the contents of the landfill.

Hydrology and soils

Aerial photography is valuable for locating several types of soils, and hydrological conditions associated with certain soils. To understand problems of leachate movement, it is important to determine soil type through grey-tones, texture and cross-sectional shape of gullies. For example, steep-sided gullies indicate cohesive materials which prevent leachate movement, while gently sloping gullies indicate sandy soils and the potential of movement of leachate and contaminates to the ground water.[4,5] Particularly wet areas are associated with cohesive soils and they appear as black or dark tones on photography. Dark tones in fields indicate areas of wet, partially organic soils, and they are conduits of runoff. In addition, historical photographs can be used to locate wetter areas for placement of leachate collection systems. Ponded water can be seen on early spring photographs, and it is a good clue of fine textured soils and wet areas.

Vegetation

Vegetation can provide information on presence of environmental conditions[6] and of toxic materials. Abandoned areas rapidly generate woody vegetation and eventually tree cover. This is evident from examining abandoned farmsteads and the growth of woody plant species. On toxic sites, woody vegetation are particularly suspect, because areas with adequate rainfall usually have some plants growing on them. Absence of plants is a source of concern, and should be investigated with soil and water sampling.

Land Cover

The location of inactive sites and the determination of subsequent land use are important concerns. In the case of Love Canal, NY, aerial photographs of 1938 clearly show the canal, while 1958 coverage reveals a school on the site and the presence of dead vegetation.[1] Conversely, residents of the Frayser neighborhood of Memphis, TN, complained of exposure to toxic wastes.[7,8] An intensive sampling program revealed no exposure beyond a background level of chlordane and other pesticides attributed to an insect control program. The public was difficult to convince, but examination of historical aerial photographs reveal no possibility of a landfill for the last fifty years and this proved to be persuasive evidence.

Coverage

Aerial photographs have been acquired of the U.S. on a consistent basis since 1936. These photographs are commonly vertical black and white images in a nine by nine inch format. The nominal scale is 1:20,000, and most coverage was flown during "leaves off" conditions.

Several government agencies catalog aerial photographs. It is common to find ten dates of photography for a given site over the last forty-five years. It is important to contact each group and obtain all available aerial photographs, and to interpret the photos for both change in size and characteristics of the landfill. The determination of soil, water and plant characteristics are vital to determining integrity of a site, and to provide warning of problem areas. By employing all available coverage, a thorough examination of a waste site can be achieved.

When requesting information on available aerial photographs, it is customary to specify the latitude and longitude of the site, and the U.S.G.S. quadrangle. The following agencies will supply information upon request. For pre-World War II photographs (1941), contact the National Archives and Records Service (Cartographic Branch, GSA, Washington, D.C., 20408). Post W.W. II photographs were flown by several groups including the U.S. Geological Survey (User services, EROS Data Center, Sioux Falls, SD. 57198), the Agricultural Stablization and Conservation Service (Aerial Photography Field Office, ASCS, P.O. Box 30010, Salt Lake City, UT., 84130) and the Soil Conservation Service (SCS, Cartographic Division, Federal Center, Hyattesville, MD., 20782). Photographs of coastal and airport areas may be obtained from the N.O.A.A. National Ocean Survey (Coastal Mapping Division, Rockville, MD. 20852). Specific coverage of an environmental nature may be obtained from the U.S. Environmental Protection Agency (EPA, Remote Sensing Branch, P.O. Box 15027, Las Vegas, NV. 89114 or EPA Interpretation Center, P.O. Box 1587, Vint Hill Farms, Warrenton, VA., 22186). Local and regional aerial mapping firms often maintain older coverage. Locate these firms through the Yellow Pages and through professional directories in "Photogrammetric Engineering and Remote Sensing," which is published by the American Society of Photogrammetry, 210 Little Falls St., Falls Church, VA. 22046.

CURRENT INFORMATION AND APPLICATIONS

Siting and Proximal Resources

Interpretation of aerial photographs and remote sensor data supplies information on resources proximal to the site. Analysis of hydrologic and edaphic resources adjacent to the site is vital for locating problem areas. For hydrological purposes, aerial photographs and remote sensor data can be useful in several ways and aid in compliance with the RCRA.[8] Some of the advantages which stem from interpretation of images include:

1. Avoidance of sensitive areas, such as wetlands, coarse textured soils, permafrost areas, and critical habitats for plants and animals.
2. Determining flood plain areas to comply with RCRA.
3. Locating areas with minimal problems of erosion and slumping.
4. Locating diversion structures (e.g., dikes, drainage ditches) to minimize surface water flow into fill areas.
5. Locating sites where hydraulic transport of leachate to an aquifer is unlikely.

In addition, aerial photographs and remote sensor data can be helpful in siting as they can be used:

1. To determine the haul distance or to plan routes.[9]
2. To locate secure landfill sites in clayey areas.
3. To assist in locating gas collection systems.

Hydrology and Soils

Color infrared photography (CIR) is valuable for monitoring water characteristics. Infrared light is extinguished in the first millimeter of water. Hence, very thin layers of water can be distinguished. This is valuable for locating leachate movement, ponding of water, erosion, and condition of diversions.[10] CIR supplies information that is extremely difficult to obtain with ground inspection, and it has been particularly useful in monitoring water characteristics associated with surface coal mining and reclamation.[11] The combination of Color and color infrared film is most useful for monitoring water characteristics which can indicate the integrity of the disposal area.

For example, neighboring the diked, dredge spoil disposal area on Dickenson Island in the St. Clair Flats, MI., wetland resources were mapped with color and CIR aerial photography.[2] These maps have been used in management plans in the delta area. They also can be employed in the future evaluations of the integrity of the disposal area. Vegetation can provide an indicator of movement of toxic leachate (mercury poisoned spoils) included in this site. This "vegetative indicator" approach to monitoring the effects of diked disposal areas can be employed for similar projects in the Lake Erie coastal zone, or similar hazardous waste sites.[12]

Vegetation

In monitoring a site, plants can provide an indicator of site characteristics. Individual species and groups of species respond to the presence of toxic materials. Much work has been done on locating ore bodies from vegetative characteristics.[13] This subject is called geobotany,[14] and results of this work can be generalized and used in management of hazardous wastes.

Plants respond in several ways to environmental stress. In toxic exposure areas, the number of species will decline and only tolerant plants will be present. In high and medium exposure areas, plants will often die or begin senescence (fall colors) much earlier (August versus September). They will appear chlorotic (without chlorophyll), or yellow/orange in color. As an example, work in the Lost River Valley, WV. gas field revealed that Red Maple was tolerant of this exposure and had not been replaced by oak species that "out compete" and exclude Maple on non-exposure sites.*

---

* Barry Rock, Personal communication, Jet Propulsion
  Laboratories, Pasadena, CA., 1981

Plants also respond to increased levels of nutrients, such as high levels of phosphorus and nitrogen compounds. Color infrared aerial photography displays vigorously growing vegetation as bright red in color.[2] For example, CIR has been valuable for locating failed septic tank systems, because the grass grows vigorously and this is evident on CIR as areas which are bright red in color. This technique saved property owners $5000 apiece, as a sewage-line program was deferred when the U.S. EPA funded program revealed that just three land owners with failed septic tanks were causing limited eutrophos in Crystal Lake, Benzie County, Michigan.[15]

Old or dead trees and shrubs can be valuable in dating closure of a site. Woody plants deposit annual rings of woody material, and these rings may be counted by coring the tree. This tree ring information and aerial photographs can provide strong argument for dating a particular fill. Dates of closure are potentially valuable to current land holders, as they can establish the need for remedial efforts using "superfund" monies.

Closure

Environmental monitoring practices are critical to ensuring environmental safeguards for the public. During the post-closure period, aerial photography can be used to monitor soil integrity, slope and vegetative cover, and all diversion and drainage structures.

A well functioning site requires synergistic employment of cover media to achieve longterm integrity of the filled site.[8] Aerial photographs can be valuable for locating proximal sources of daily fill, and monitoring the placement of fill.

Acquisition of Data

A program to collect aerial photographs and remote sensor data on current sites can be helpful in management. This program can be a formal or informal approach. The formal approach involves contracting annual or biennial coverage by an aerial photography firm. This can be expensive but the photography can be used for detailed mapping, and help to reduce human exposure to wastes. Most large industrial plants have a regular aerial photography program, and it is relatively easy and low-cost to add another "leg" to a given flight.

A less formal and very low cost approach involves use of small format photography (35mm) of the sites. A company employee can rent a pilot and aircraft, and take photographs out the window. A better approach is to build or purchase a camera mount to acquire vertical coverage with small format systems. This has been a valuable, low cost and quantitative approach.[16,17] The use of color and color infrared aerial photography and a small format system provides a good monitoring program.[6] The system is very responsive, and coverage can be obtained when a contingency exists.

Monitoring

A chronology of photographs provides a record of position and types of hazardous waste within a landfill. This record can assist in future recovery if it proves economic, or to facilitate understanding of the cause of undesirable chemical reactions.

Preliminary results indicate it is feasible to locate and characterize the contents of liquid chemical waste storage drums.[18] With proper mission design, it should be possible to detect thermal differences among drums filled largely with aqueous solvents, organic solvents, clay packing materials, and other substances. Finer levels of separation among liquid chemicals are not likely to be obtained through airborn thermal sensing. In addition, airborn thermal scanner sensor systems should be valuable for locating landfill fires. In almost all cases, scanner radiometers detected surface burning, and many instances the fires' progress was monitored and mapped in the case of coal mine fires.[18]

DISCUSSION

From historical and current applications, it is evident that much information can be abstracted from aerial photography and remote sensor data. The success of each application results from knowledge of resource characteristics and their expression on imagery. The advantages of using these data sources can be enjoyed by users who develop this understanding of resource characteristics, and use it to assist the interpretation of imagery.

As an additional source of data or an additional element in a sampling design, aerial photography and remote sensor data can save money and human resources, and decrease exposure to toxic chemicals. Also, it can potentially save money by firmly establishing which fills are candidates for reclamation with monies from Superfund.

CONCLUSIONS

1. Aerial photographs and remote sensor data are useful in providing both current and historical information on sites of hazardous waste.

2. Siting, monitoring and c-osure of hazardous waste sites is greatly facilitated by utilizing aerial photography and remote sensor data in planning and management activities.

3. Utilization of these tools can potentially save money, human resources and reduce exposure to toxic wastes.

BIBLIOGRAPHY

1. Erb, T., W. Philipson, W. Teng and T. Liang, "Analysis of landfills with historical airphotos." Photog. Eng. Remote Sensing 47:1363-1369, 1981.
2. Lyon, J., "Remote sensing analyses of coastal wetland characteristics: the St. Clair Flats, Michigan, "Proc. 13th Inter. Symp. Remote Sensing of Environ., Ann Arbor, MI., p. 1117-1129, 1979.
3. Lyon, J., "Data sources for analyses of Great Lake wetlands." Proc. Annual Meeting Amer. Soc. Photogrammertry, St. Louis, MO., p. 516-528, 1980.
4. Way, D., Terrain analysis. McGraw-Hill Book Co., NY, 483 pp.
5. Mintzer, O., "Terrain analysis procedural guide for surface configuration." Eng. Topographic Lab., Corps of Engineers, U.S. Army, Fort Belvoir, VA, 456 pp., 1981.
6. Lyon J., "The influence of Great Lake Water levels on wetland plants and soils in the Straits of Mackinac, Michigan." Ph.D. Dissertation, School of Natural Resources, Univ. Mich., Ann Arbor, 120 pp., 1981.
7. Maugh, T., "The dump that wasn't there." Science 215:645, 1982.
8. Sittig, M., Landfill disposal of hazardous wastes and sludges. Noyes Data Corp., Park Ridge, NJ., 365 pp., 1979.
9. Garofalo, D. and F. Wobber. "Solid waste and remote sensing." Photog. Eng. Remote Sensing 40:45-59, 1974.
10. Sangrey, D. and W. Philipson. "Detecting landfill leachate contamination using remote sensors." Res. Rep. EPA-600/4-7-060, EPA, Las Vegas, NV., 67 pp., 1979.
11. Mintzer, O. and D. Spragg. "Mini-format remote sensing for civil engineering." Trans. Eng. Journal, ASCE:104, 1978.

12. Philipson, W. and D. Sangrey, "Aerial detection tech-
    niques for landfill pollutants." In Proc. 3rd Annual
    EPA Research Symp. on Management of Gas and Leachate in
    Landfills, St. Louis, MO., 11 pp., 1977.
13. Lyon, R. "Correlation between ground metal analysis,
    vegetation reflectance and ERTS brightness over molyb-
    denum skarn deposits, Pine Nut Mountains, Western
    Nevada." Proc. 11th Inter. Symp. Remote Sensing of
    Environ., Ann Arbor, MI., p. 501-512, 1977.
14. Bolviken, B., R. Honey, S. Levine, R. Lyon and A.
    Prelat. "Detection of naturally heavy-metal-poisoned
    areas by Landsat-1 digital data." J. Geochem. Explora-
    tion 8:457-471, 1977.
15. E.P.A., "Environment Midwest". Region V., Chicago, IL.,
    1980.
16. McCarthy, J., C. Olson and J. Witter. "Evaluation of
    Spruce-Fir forests using small-format photographs."
    Photog. Eng. Remote Sensing 48:771-778.
17. Meyer, M. "Operating manual-Montana 35mm aerial photo-
    graphy system." College of Forestry, Univ. Minn.,
    St. Paul, MN., 50 pp., 1973.
18. Philipson, W. "The Cornell remote sensing newsletter."
    Dept. Civil Eng., Cornell Univ., Ithaca, NY, 4 pp., 1982.

CHAPTER 12

CHARACTERIZATION OF ABANDONED
HAZARDOUS WASTE LAND BURIAL
SITES THROUGH FIELD
INVESTIGATIONS

S. P. Maslansky, CPG
Geo-Environmental
Consultants Inc.

INTRODUCTION

Although some state and federal agencies, specialty consultants, and spill cleanup contractors have highly trained and dedicated crews, many individuals associated with inspection, evaluation, and remedial actions at abandoned hazardous waste disposal facilities lack adequate training. Training is often perfunctory for those involved in sample collection or on-site inspection. The employers of such individuals have both a legal and moral obligation to insure that their employees are prepared to undertake such work.

When evaluating a site containing known or suspected hazardous materials, one must proceed cautiously. Investigatory personnel must be screened not only for their technical expertise, but also for their alertness and common sense. Prior planning and proper training should, in most cases, prepare personnel for unexpected conditions and mitigate the potential for serious accidents. Abandoned land burial facilities should be evaluated within a problematical framework which characterizes the component parts of the landfill, its environment, its wastes, and its existing or potential ability to cause an impact, not only to the environment, but also to the investigator.

The following is intended as a general outline which addresses some of the problems encountered when working at hazardous or potentially hazardous sites.

The evaluation of an existing or abandoned disposal site and its potential to impact the environment is a complex task, requiring a multidisciplinary approach. The observations of well trained field personnel cannot be minimized. Team members should attempt to characterize the landfill, its surrounding environment, its wastes, and its impact during the initial screening phase. In this way the undertaking of borings, monitoring wells, tests pits, geophysical surveys and sample collection can proceed in an organized and safe manner.

Characterization of the Site

An important, but basic consideration is how big (depth and area) is the site. It is important that multiple dumping areas be identified and categorized. The investigator must ensure that he has covered a large enough area to demonstrate the lack of or presence of satellite areas. The utilization of maps, photographs (particularly old aerial photographs), and interviews with neighbors, former employees, local fire services, and regulatory officials may yield information on site access, time sequencing, types and sources of wastes as well as potential hazards. Such historical perspectives may also give information on the existence of site control features such as dikes, berms, covered areas, liners, drainage controls or monitoring wells. Field investigators should ascertain the condition of any readily apparent control features.

Characterization of the Environment

The investigatory team must have sufficient expertise so that it can characterize the physical and biological conditions of the site and its environs. The team's earth scientist should attempt to delineate the site's soils and geology. Important considerations include: the type, condition, attitude, and apparent depth to bedrock; the type, and physical-chemical characteristics of the surface soils; the apparent depth and quality of ground water, its local recharge-discharge relationships, and the distance to, use and type of nearby wells. The degree that the topography of the site and its environs has been modified should also be evaluated. The team member versed in the biological sciences should characterize the local habitat and indicate if there is evidence of stress within the aquatic or terrestial systems. Team members should delineate the land use

174

of adjacent properties including existing use and the
evidence for historical land use.

Characterization of the Wastes

The physical, chemical, and biological properties of
the waste can be initially ascertained from historical
records, interviews, visual observations and by utilization
of portable measuring devices. Team members should identify
the form (i.e. powder, liquid) and volume of the wastes.
Identification should include the method of containment such
as sludge or liquid lagoons, bulk piles, drums or other
containers. When present, the condition, color, material
and presence of labels or other coding on containers should
be recorded. It should be noted that wastes are many times
put in a container with an old label on it.

An estimation should be made for the volume (by weight)
of any suspected contaminated vegetation or soil.

Characterization of the Impact

The delineation of the site and the initial characterization of its wastes and environment will give indications
as to the degree of hazard manifested at the site. The
degree of impact or potential for impact must also be
evaluated during the first phases of the field investigations. Although a site may be extremely hazardous its
potential for offsite impact may not be imminent. Field
investigators should look for evidence of surface leachate
and/or well contamination. Where possible the quantity and
characteristics of leachate should be estimated, and changes
in downgradient conditions should be assessed. The sensitivity and distance to nearby receptors should be estimated,
as well as the presence of natural or artificial barriers
already in place or that can be hastily employed for impact
mitigation. It is useful where possible to identify the
condition of upgradient surface and ground waters.

CLASSIFICATION OF INVESTIGATIONS

Investigations should be undertaken based on a classification system of knowns and unknowns. Accordingly, six
classes of investigations can be assigned as shown in Table
I.

175

Table I. Classification of Investigations

Severity	Class	Nature of Wastes	Typical Site Characteristics	Protective Equipment
Greatest Hazard	A/K	Positive identification of significant quantities and/or concentrations presenting grave threat to uninformed or unprotected investigatory personnel	Existing disposal records or monitoring data available	Fully encapsulated suit,* pressure demand SCBA, inner gloves, heavy-duty overgloves, hardhat, safety overboots, safety harness and line.
	A/U	Unknown, but hazardous materials highly likely	Abandoned industrial disposal site with mixed wastes	*a splash suit with SCBA may be sufficient at some sites.
Moderate Hazard	B/K	Positively identified, but with only indirect threat to investigatory personnel	Industrial landfill of known generator	Full face or ½ mask chemical-mechanical respirator, hardhat with face shield, splash suit and hood, (or chem. resistant overalls) inner and outer gloves, safety boots, overboot
	B/U	Unknown, but significant quantity/concentration of hazardous materials not suspected	Municipal landfill, little industrial waste disposal recorded	
Least Hazard	C/K	Positively identified, but not hazardous	Sludges/residues from non-hazardous generator	Boots, coveralls, hardhat, goggles, gloves, particulate filter
	C/U	Unknown, but hazardous materials not suspected	Controlled access municipal landfill	

With each class of investigation the degree of planning, special training, supervision, and protective equipment will vary. However, when dealing with unknowns, a worst-case scenario should be assumed, with personnel prepared and protective equipment readily available.

PERSONNEL CONSIDERATIONS

Investigatory teams should consist of no fewer than two people, although three is an ideal size for safety and control purposes. Team members should represent such diverse fields as environmental engineering, geology, biology, and/or environmental health science. Specialties should be assigned on an as-needed basis.

The supervisor of the team should be thoroughly trained in the nature of hazards involved. He or she must be familiar with precautions necessary for the known or suspected wastes to be inspected or sampled, including toxicological and industrial hygiene data when available. The supervisor must insure that sufficient equipment of the appropriate type is present at the site to cope with any emergency and that all personnel are familiar with the wearing and functioning of each item of protective gear.

Because early detection of potential danger signs at a site is of paramount importance, investigatory personnel should have a superior sense of smell (usually precludes smokers). They should be familiar with the characteristics, including odors, of materials that may be encountered; and, because many danger signs can be detected by reading color-coded labels or noticing the color of fuming wastes, personnel must not be color blind.

Odors, should not be used as the criteria to put on respiratory protection at class A or B sites, since in many instances the odor threshold concentration for a chemical substance may be many times its toxic level. In addition, many substances induce olfactory fatigue, giving a false sense of security.

It is paramount that all team members know the limitations of the measuring devices they are using. For example, some combustible gas meters will not function in an oxygen deficient atmosphere or will give inaccurate readings when exposed to gases at inordinately high concentrations.

Individuals prior to working around suspected or known hazardous wastes should have a complete medical history taken and undergo baseline medical profiling. Tests should include pulmonary function, blood chemistry, urine analysis and liver and kidney functions. These tests also serve as indicators of possible preexisting abnormal conditions or unusual sensitivities to certain types of toxic compounds. Other tests which may be appropriate include chest x-ray, electrocardiogram, and specific tests for exposure to certain chemicals or physical agents. Medical profiling should be repeated (as a minimum) at one year intervals. Those suffering from obesity should not be part of a team due to inordinate susceptibility to accumulate organic solvents or fat-soluble compounds within their tissues, and potential lack of agility. Contact lenses tend to concentrate materials around the eye, while soft plastic lenses can adsorb chemicals directly. In addition, rapid removal of contact lenses may be difficult in an emergency, these contact lenses should not be worn at land fill sites. Eye glasses can prevent a good seal around the temple when wearing goggles or face masks; special spectacle adapters are available. Likewise, the presence of beards or heavy mustaches can cause a poor fit of protective masks.

All personnel should have the phone numbers of local fire and police services and of CHEMTREC (Chemical Transportation Emergency Center - CMA), if responding to a spill, and know the location of the nearest telephone. They should be familiar with basic first aid techniques including, as a minimum, training in artificial respiration and cardiopulmonary resuscitation (CPR), preventing loss of blood, first aid for poisoning, preventing or reducing shock, first aid for burns, and techniques used to move a victim. Team members should be familiar with decontamination procedures and the administration of field-expedient antidotes (only if a life-threatening situation exists).

PROTECTIVE EQUIPMENT

Investigatory teams should be familiar with potential routes (inhalation, skin or mucous membrane contact, and ingestion) by which specific toxic substances can enter the body, and specific measures to prevent exposure. The inhalation of toxic substances is the primary hazard at a disposal site. To this end, breathing equipment, including self-contained breathing apparatus (SCBA) and chemical/mechanical respirators of the appropriate type, must be readily available, correctly fitted and properly maintained. SCBAs should be worn when less than 19.5 percent oxygen is present, still air prevails around unknown wastes, containers of unknown or known toxic materials are being opened or investigated, in enclosed spaces such as pits, or un-ventilated structures containing wastes, and whenever there is any doubt. Only positive pressure (pressure demand) SCBAs should be worn at class A sites.

Cartridge and canister respirators must be used with an approved cartridge, suitable for certain specific chemicals or group of chemicals. Since most cartridges and canisters function primarily as an adsorbent medium they must be changed periodically. If possible the NIOSH respiratory protection factor required should be determined as an aid in identifying the required type of respirator.

Respirators, regardless of type, must be donned im-mediately when experiencing breathing difficulty, dizziness, unusual tastes or smells, and when in doubt. Under these circumstances the individual should remove himself from the suspected source of discomfort.

Recommended levels of protection are summarized in Table I. It is important to note that one size does not fit all, nor is protective clothing necessarily resistant to all chemicals. With regard to the latter it is important to check manufacturers' performance data.

Because even innocuous materials can cause adverse hypersensitivity reactions when in a powdered form, all unknown powdered materials should be considered hazardous and appropriate protective measures taken for lungs and eyes. In addition, many powdered substances and vapors can react with perspiration to produce localized skin irrita-tions and in some instances severe burns.

Many organic solvents, although not in themselves particularly toxic, can enhance the permeability of the skin and facilitate the dermal absorption of toxic substance which are normally unable to penetrate this barrier.

Eye protection is a must when working around sites even where the only danger is from blowing dust. Normal safety goggles (vented type) should not be worn around corrosive fumes or vapors. Eye washing solutions must be available on site in addition to first aid and snake bite kits.

Measuring devices for combustible and toxic gases, oxygen deficient atmospheres, and for radiological measure-ments are requisite pieces of equipment during the initial screening of class A and B sites. Measuring devices must be calibrated and checked prior to field utilization.

Care must be exercised around sites in highly visible areas to balance the possible conflict between the safety of investigatory personnel and undue anxiety among local residents caused by the presence of protective equipment.

OPERATIONAL CONSIDERATIONS

An investigation of a disposal site can be complicated by weather conditions, terrain features, and general site layout. Strong winds prevent adequate evaluation by masking and diluting odor concentrations. Precipitation can cause slippery walking surfaces, splashing of hazardous materials, the formation of reaction products, and generation of toxic fumes.

High ambient temperatures can cause increased volatilization (requiring more mask time) as well as personnel fatigue. Excessive perspiration and high humidity can decrease the resistance of permeable protective clothing and make the wearing of impermeable clothing impractical for more than a few minutes. Low temperatures decrease freedom of movement due to bulky clothing requirements.

Rough terrain conditions such as thick underbrush and steep slopes can make a proper evaluation difficult. Special care is required when working around overhangs and sharp objects as they can tear protective clothing and cause injury to personnel. Even minor lacerations and punctures can be adventitious routes for harmful toxic agents.

A site safety plan should be developed, reviewed and made familiar to all personnel who will work on the site or will act as outside support personnel. Prior to site entry, an operations plan should be reviewed addressing anticipated hazards, equipment to be utilized, safety practices to be followed, emergency procedures and signals, and communications. Entrance and exit routes should be preplanned and emergency escape routes identified. The buddy system should be utilized with visual contact between members at all times. Field personnel should observe each other for any signs of physical or mental anomalies such as confusion, complexion changes, lack of coordination, or changes in speech patterns. Investigators should immediately inform each other if experiencing headache, dizziness, blurred vision, cramps, respiratory distress, or irritation to eyes, nose, or mouth.

Decontamination procedure will vary greatly depending on the size of the site, the nature of the wastes (both toxicity and form), the degree of hazard, and the nature of site activities (inspection, drilling, remedial action). In general at class A sites specific work areas, control lines, barriers, and checkpoints should be established. The contaminated area in which only protected personnel have access is delineated by a marked control line known as the hot line. On the other side of the hot line is the contamination reduction area. In this zone, wash stations and protective equipment drops which are designated personnel decontamination stations are located. The area is separated from clean areas, consisting of support and marshalling areas, and a command and control point, by a marked barrier designated the contamination control line.

If possible the line of decontamination should be in the direction of the wind, with personnel moving into the wind as they drop equipment and protective clothing. The protective mask should be the last item removed. Vehicles and test equipment should remain on site until no longer needed and must be thoroughly decontaminated upon completion of the field work. All disposable protective clothing, contaminated dirt and water must be disposed of properly.

SUMMARY

The topics that should be considered when evaluating a hazardous or potentially hazardous site are diverse and require input from a variety of investigatory personnel. Characterization studies of such sites should be coordinated by an experienced investigator who is responsible for site evaluation, personnel safety and equipment, as well as operational considerations. A thoughtfully planned and organized approach to site evaluation by properly trained, equipped, and supervised personnel will facilitate the evaluation of hazardous waste land burial sites while minimizing the risks involved.

CHAPTER 13

HAZARDOUS WASTE SITE FIELD INVESTIGATIONS:
GEOPHYSICAL TECHNIQUES UTILIZED FOR COST EFFECTIVENESS

G. David Knowles, P.E., L.S.
George W. Lee, Jr., C.P.G.S.
Scott J. Adamowski
    O'Brien & Gere Engineers, Inc.

INTRODUCTION

The techniques available for the subsurface field investigations at a hazardous waste disposal site (HWDS) commonly consist of drilling test borings or test pits at the site and extracting soil or rock samples directly for analysis. The test borings should be strategically located such that their arrangement will provide an adequate representation of existing subsurface conditions. However, due to the possibility of punctures, explosions, gas leaks, and other potential dangers inherent to a site containing hazardous wastes, only the gathering of peripheral data can generally be attempted using these exploratory methods. Clearly, other data-gathering techniques must be employed if one is to properly evaluate the horizontal and vertical limits of the hazardous waste refuse and the geologic environment in which the refuse is in contact.

The purpose of the field investigation is to identify the potential extent of contamination and to aid in the identification of subsurface conditions prior to the development of remedial alternatives. This paper presents the evaluation and practical application of four modern, safe, non-destructive geophysical techniques to the cost-effective investigation of an existing inactive site known to contain hazardous wastes. The investigatory techniques that shall be discussed include:

1. Earth resistivity investigations
2. Magnetometer investigations
3. Ground penetrating radar (GPR) investigations
4. Seismic refraction investigations

DESCRIPTION OF PROBLEM AREA

The hazardous waste disposal site was operated from the mid-1950's through 1970 for the disposal of waste materials including chlorinated solvents, solvents, waste oils, polychlorinated biphenyls (PCBs), scrap materials, sludges, and solids. The site encompasses approximately 11 acres, which consisted of a 6-acre lagoon (used for open dumping) and 1 to 3 acres of drum disposal area. The remaining 2 to 4 acres were used for access roadways and miscellaneous dumping areas.

Countywide soil and geological information revealed that the site was located on stratified glacial deposits of gravel, sand and silt overlying either a non-stratified till or a shale bedrock formation. The depth of till over bedrock was reported to be generally greater than 10 feet. The till was reported to be a dense, heterogenous mixture of sand, silt, clay, and rock fragments. The bedrock is primarily soft shale, approximately 400 feet thick of Cambrian Age (approximately 550 million years ago), often interbedded with sandstone. The migration of the wastes through the preceding strata via groundwater flow, and a surface water course on the westerly side of the site became evident in the late 1960's.

After several years of complaints from local citizens, documented downstream fish and cattle kills and uncontrolled fires at the site, the regulatory agency initiated legal action against the private disposal company, leading to a Supreme Court Order and Judgement against the company to stop the discharge of untreated hazardous wastes and other wastes into classified waters of the State, to rectify the existing water pollution problem at the site, and to remove a portion of accumulated drummed wastes from the site. The remedial activity, however, was not effective and additional contamination began to appear in the late 1970's.

This ultimately led to the present field investigation which was required prior to the further development of a remedial plan. The methodology outlining this field investigation appears in the next section.

METHODOLOGY

The goal of the field investigation was to safely and cost-effectively determine the horizontal and vertical limits of the site and to define the pathways and extent by which hazardous wastes were migrating from the site. Cost to conduct soil borings at a HWDS increases dramatically over normal soil drilling techniques due to problems associated with implementing a safety protocol, prevention of cross contamination of borings and cleaning of all field equipment. Therefore, a more practical approach to the realization of the goal of the field investigation is presented below.

The key to planning a cost-effective field investigation is to thoroughly research all available historical information concerning the particular site. A clear, concise history of the HWDS in question allowed the investigators to choose the most effective tool(s) in proceeding with the investigation. This rational systematic approach was a key factor in the development of a cost-effective geophysical and boring program. The review of historical information also proved very important prior to entering the site. Familiarization with operating records, newspaper clippings and photographs, as well as interviews with residents, former employees, and waste haulers, developed information by which the amounts of waste could be estimated, and the methods of disposal and types of chemical wastes disposed could be determined. Soil borings previously completed in the general vicinity by governmental agencies also proved valuable by providing data regarding expected soil type, depth to water table, and depth to bedrock.

One important type of investigatory technique which proved very useful in the investigation was the use of chronological aerial photographs. Review of historical airphoto coverage by a skilled professional provided the investigatory team with an indication of the manner in which the site had developed and the milestones in the establishment of the HWDS. Changes in traffic patterns, shifts in water courses, variation in vegetative cover, apparent decreases in depressions, quarries, and borrow pits, significant changes in areas of exposed refuse, alterations of typical erosion features, and other physical evidence can prove to be of use in the understanding of the historical development of the HWDS, and as an aid in establishing the most appropriate locations for utilization of geophysical equipment and drilling of soil borings.

GEOPHYSICAL TECHNIQUES EVALUATED

The geophysical techniques of earth resistivity, magnetometer, ground penetrating radar and seismic refraction, were evaluated for their potential effectiveness given the existing conditions at the HWDS in question. Descriptions of the techniques and information regarding the limitations, disadvantages, advantages, and costs of each are presented in the following paragraphs.

Earth Resistivity

Earth resistivity is a geophysical technique that measures the apparent resistance of soil to the passage of electric current. Measured resistivity values are obtained and compared with characteristic ranges of resistivity known to exist for certain earth materials. In this manner, the different soil strata present may be identified.

Resistivity investigations are particularly useful in the identification of groundwater aquifers, saltwater intrusion, and contaminated groundwater. The use to identify leachate plumes at land disposal sites has been documented in other studies[1,2] based on the principle that contaminated groundwater or leachate has a high conductance (the inverse of resistivity) due to dissolved solids.

Earth resistivity measurements are generally made using four electrodes driven into the ground along a straight line, with the electrodes equally distant from one another. An electric current is passed through the two outer electrodes, and an electric potential drop is measured at the inner two. From OHM's law, and knowing the current and voltage drop, the resistivity of the earth is determined.

Sources[3,4] indicate limitations of the technique to be as follows:

1. Technique is not effective on frozen surfaces due to the difference in conductance of ice lenses;

2. Accuracy is limited if complex geological formations are present, and cannot be accounted for in the spacing of the electrodes;

3. Technique is limited to the detection of conductive pollutants and major changes in geologic stratigraphy over large areas;

186

4. Technique not effective in areas where man-made pavement or structures exist; or in areas of electrical interference such as near buried power lines, or pipelines.

Generally, the earth resistivity technique can be a useful tool in the evaluation of subsurface conditions at the HWDS, possessing advantages such as rapid data gathering (approximately 5 acres per day), the ability to adapt to a computer for interpretation and plotting, and a relatively low cost.

Capital costs for hardware associated with the earth resistivity technique are estimated at $2500. Field work consists of a 2-man crew at $400 to $600 per day per crew, and interpretation of results may be $100 to $200 per acre. At HWDS's, these costs may vary if special safety equipment is required, or if driving the electrodes presents a problem.

Magnetometer

The magnetometer utilizes the total magnetic field intensity of a media to determine the presence of metallic objects. The depth at which the metal is located cannot be accurately determined; however, the presence of the metal is indicated by high magnetic field intensities.

The magnetometer, therefore, is a useful tool at HWDS when used in the definition of horizontal limits of fill, especially when the site may contain metallic drums, and the surface of the disposal site is no longer clearly defined. It is recommended that the stationing of the magnetometer grid network be located by established survey baseline control since the need to accurately determine the location of each station is very important.

Limitations[3,4] of the magnetometer technique are listed below:

1. Magnetometer technique can only detect metallic objects, and detects the upper surface of the buried metallic objects only.

2. Technique cannot determine the depth, number, or arrangement of buried objects.

3. Technique is only effective to a maximum depth of 2 to 4 times the largest representative dimension of the buried object.

4. Technique does not differentiate the various geologic strata or groundwater level, but merely identifies a metal object.

Advantages possessed by the technique consist of the ability to be used on any terrain, including overgrown areas and frozen surfaces. Generally, the magnetomer is a rapid technique (one acre per day with grid stationing on 25-foot centers) for the determination of horizontal site limits when the refuse is suspected of containing metallic objects. The technique has a relatively low cost when compared to other techniques yielding comparable data, and can usually detect much smaller objects and objects buried deeper than can be indicated by a metal detector.

Capital costs associated with the magnetometer are estimated at $3500 to $4000. Field costs for a two-man crew are approximately $250 to $400 per crew per day, and interpretation of data is estimated at $500 per 3 acres. If adapted to a computer, however, the costs will generally be in excess of the above.

Ground Penetrating Radar

The geophysical technique of ground penetrating radar (GPR), sometimes referred to as Electromagnetic Subsurface Profiling, "ESP", or Subsurface Interface Radar, "SIR", is utilized to detect and measure depths and discontinuities in subsurface strata by the reflection of short duration electromagnetic pulses radiated into the ground. GPR can provide a continuous profile of subsurface conditions by towing a sled-mounted antenna by hand or behind a vehicle. The minimal hardware generally used consists of a radar set and antenna; however, optional equipment such as a magnetic tape recorder with micro-processor are available, and generally necessary.

The GPR technique is appropriate for use at a HWDS when subsurface profiles and the delineation of buried objects are desired. The technique is relatively new, more sophisticated than most other geophysical tools, and requires a trained technician to operate the equipment and interpret the data. This contributes to the relatively high cost associated with GPR. When site conditions are favorable (open, relatively flat terrain accessible by a vehicle), however, the technique is rapid and the data can be stored for playback and interpretation at a later date. Sources[3,4,5,6] indicate limitations associated with this technique are as follows:

1. Technique has limited ability to determine size and shape of buried objects.

2. Effective depth of GPR varies with the inverse square of the frequency of the electromagnetic pulse. Size of the antenna, moisture conditions, and soil type also affect the depth of penetration and response;

3. Underlying objects may be obscured by those above, if interruption of the electromagnetic pulse occurs;

4. Technique requires open terrain and is not suitable for overgrown areas. The antenna, sending and receiving, must be towed at a constant speed over the surface;

5. Sophisticated field operation and interpretation of data is required.

Some advantages associated with the GPR technique include the ability for use over paved surfaces, a deeper effective penetration and ability to reveal more information than the magnetometer.

Costs incurred through use of GPR are the greatest of the techniques evaluated. Necessary hardware ranges from $21,000 to $35,000. Operational costs vary from $1,000 to $2,000 per day and associated interpretation costs by a qualified professional per day of field data are between $600 and $700.

Seismic Refraction

The seismic refraction method is based on the physical principal that the time of passage of elastic compression, seismic waves through the various subsurface earth formations can be recorded over a measured distance. It has been proven that the shock waves travel through various soil media (sand, clay, bedrock, etc.), at different velocities. The greater the density of the material, the greater the velocity. The shock wave is usually produced with a small explosive charge or, in the case of a HWDS, a dropped weight or sledgehammer.

Seismographs are available as multi-channel units (usually 6 or 12 channel), with one geophone for each channel. The geophones are placed at fixed intervals along an established section line at the HWDS. A shock wave is generated, picked up by the geophones, and transmitted to the seismograph. The operator can then select the time of first arrival of each wave and plot graphs of time versus distance. The time-distance relationship yields a calculated velocity of the wave through the material. This calculated velocity is then compared to characteristic velocities of known materials, thus enabling the definition of various soil and rock strata and other subsurface conditions at the HWDS. The velocity contrasts between the various media are relatively well defined and can usually be easily differentiated.

Seismic refraction poses some limitations, as follows:

1.   Technique is not effective if a more dense material lies above a less dense material; (i.e., most effective when density increases with depth);

2.   Technique is not always effective on frozen surfaces, due to the density of the frozen surface;

3.   Technique is not effective if there is not a sharp velocity contrast between formations;

4.   Interpretation becomes difficult where complex geology or stratigraphy occurs, particularly if lenses of different geologic configurations exist;

5.   Depth of penetration is dependent on strength of energy source.

Advantages of the seismic refraction techniques are evident in its usefulness for the determination of vertical limits of a HWDS, for defining depths to bedrock, and for detailing of subsurface conditions; especially at sites where drilling may be very dangerous. The method may be conducted in overgrown areas and is usually much faster than the drilling of test borings.

Capital costs for instrumentation are approximately:

6 channel - $ 7,500 to $10,000
12 channel - $18,000 to $25,000

Operation in the field is performed on 20 to 100 foot centers with an approximate price per day of $750 to $1,200 for a two-man crew using a dropped weight or hammer. Interpretation of the data per 1,000 feet of seismic line is estimated to be $350 to $500.

GENERAL APPLICABILITY

Two types of disposal areas were found to exist at the HWDS in question: a drum disposal area and a lagoon area. The general applicability of geophysical techniques to these types of areas are discussed below.

Field investigations at areas known to contain waste in metal drums usually require the use of non-destructive field investigatory techniques to determine subsurface site conditions. The geophysical methods that can effectively be employed and provide subsurface information include the use of magnetometer, resistivity, seismic refraction, and ground penetrating radar. Generally, it is not necessary to use all of these techniques at a site since some yield similar results that would not be cost-effective to repeat. Information from test borings will usually substantiate results obtained from the geophysical techniques.

The other type of HWDS considered was the lagoon area, or area of open dumping of contaminants. Field investigations at areas known to have received liquid contaminants or non-containerized contaminants may utilize subsurface investigatory techniques of the destructive and non-destructive nature. The destructive techniques include the use of test borings, trenching, test pits, hand augering, etc. The non-destructive techniques include resistivity and seismic refraction; however, these are not the only techniques that can be used at a lagoon disposal area. A discussion of the techniques actually utilized at the HWDS in question follows.

UTILIZATION

The objective of the field investigation was to determine the vertical and horizontal extent of contaminants within the site, to define the ways in which these contaminants were migrating from the site, and to ultimately develop a remedial program for containment of the waste materials.

The magnetometer, ground penetrating radar, and seismic refraction investigations were conducted at this particular HWDS. The decisions to utilize these techniques were based on thoughtful consideration of technical feasibility, ease of implementation, and cost.

A magnetometer survey was conducted at the drum disposal area to determine the horizontal extent of ferrous metallic materials (drums) buried at the site. Magnetometer work was also done at the lagoon area of the site to insure that there were no buried debris which would pose a potential drilling hazard.

The magnetometer survey was performed using an EG & G Geometrics Model G-816/826 portable proton magnetometer. Hardware included the magnetometer unit, sensor, backpack, and collapsible staff, which enables use by one or two persons. Verification of conclusions from this survey were provided by visual observations, a review of chronological aerial photographs and from the seismic refraction, and GPR investigations performed at a later date. Information obtained from the magnetometer survey was used to locate test borings outside the limits of areas suspected of containing drummed waste materials. This was done to minimize the potential for rupturing a drum during the drilling operation.

Ground penetrating radar equipment was also used at the HWDS. Although this technique possesses the ability to detect the presence of buried metallic objects and provide a continuous profile of subsurface conditions, the desired depth of penetration could not be achieved at the site with the antenna unit that was available. Therefore, use of GPR at the site was terminated and a seismic refraction survey was pursued. Seismic refraction and GPR techniques, used in unison can generate reasonable data; however, the high costs associated with them and the repetition of data obtained due to the similarities in the two techniques discourage their use together in a cost-effective manner.

Seismic refraction profiling conducted at the site served to determine the vertical extent of the fill in the site and to identify the subsurface soil and bedrock profile beneath the site in areas suspected of containing buried metallic objects.

The seismic refraction profiling investigation was conducted using an EG & G Geometrics Model ES1210F Multiple Channel Signal Enhancement Seismograph with 12-element, 240-foot geophone spreads (20-foot intergeophone spacings) and a 50 lb. weight drop. Seismographs from the ES1210F were recorded on chart paper by an electrostatic printer for each of the 12 channel responses over a selected time window (200 millisecond seismographs of 0.002 seconds). Geophone spreads were established along the ground surface. Drop points were located at 80-foot intervals along each 240-foot spread, providing seismographs for each point. Data were reduced according to the standard, "cross-over distance" method, where signal arrivals were plotted as a function of distance from the drop point.

From the data generated by the seismic refraction profiling, and from the subsurface information obtained by the test boring program, cross-sections through the site were developed and utilized in the preliminary design of a remedial program. Essentially, the combination of magnetometer and seismic refraction techniques exemplified an ease of implementation and a quality of portability coupled with an associated cost that was substantially lower than any other combination of geophysical equipment capable of achieving the desired result.

The geophysical technique of earth resistivity was not used in this investigation since the technique could not provide the type of data necessary to satisfy the program goals. Nor could the technique supply enough geological information to enhance the level of understanding of the site to warrant a cost-effective approach.

The results obtained from the geophysical techniques utilized (primarily the magnetometer) were instrumental in determining the locations of the 11 test borings subsequently drilled. These borings provided descriptions of subsurface configurations which were compatible with those obtained by the seismic refraction investigation. For this reason, a lesser number of test borings were drilled, resulting in a cost savings. It is estimated that the number of borings required by the regulatory agency would have been doubled if the non-destructive geophysical techniques had not been utilized at the HWDS. The reduction in the number of borings due to the use of the non-destructive geophysical techniques also reduced the possibility of numerous, costly mobilizations of the drill crew, reduced the amount of time and materials utilized by the rigorous cleaning protocol implemented during drilling, and reduced the amount of earthwork (dozers, loaders, etc.) sometimes necessary to provide accessibility to overgrown or otherwise remote areas.

A comparison of the costs of this investigation utilizing the geophysical techniques of magnetometer and seismic refraction and the cost of the investigation if they were not employed appears in the following section.

DIFFERENTIAL COSTS

The geophysical techniques utilized in the evaluation of the HWDS resulted in an accurate determination of subsurface configurations, which in turn provided a basis for the realization of the goal of the field investigation undertaken at the site.

The geophysical techniques of magnetometer and seismic refraction performed in lieu of numerous test borings provided information at a considerably lower cost than would have been incurred had the conventional test boring program been initiated. The costs incurred through the use of the geophysical non-destructive techniques as compared to those that would have been incurred had the conventional destructive test boring program been initiated are illustrated below.

Comparisons are based on the assumption that cost items common to both types of investigations such as the establishment of ground truth by drilling a minimal number of test borings, the installation of groundwater monitoring wells, and the mobilization of drilling crews would remain unchanged, and that an average boring depth of 45 ft. would be attained. Capital costs for geophysical equipment were not included in the comparison.

Geophysical Method

Magnetometer:
Field: 3 days @ $400/day = $1200
Interpretation: 6 acres @ $500 per 3 acres = $1000

Seismic Refraction:

Field: 2 days @ $1200/day = $2400
Interpretation: 5000 ft @ $500 per 1000 ft = $2500

TOTAL COST OF MAGNETOMETER
AND SEISMIC INVESTIGATION - $7100

Conventional Method

Eleven additional test borings consisting of:

-     Set-up charge = $100 per boring

-     Drilling 45 L.F., 4" I.D. @ $25 per L.F. (includes provisions for safety protocol and continuous sampling of soil) = $1125 per boring

-     Washing augers and drilling equipment 4 hours @ $80/hour = $320 per boring

Total = $1545 per boring

TOTAL COST OF ELEVEN ADDITIONAL
TEST BORINGS - $16,995

The high cost associated with each test boring includes provisions for implementation of a safety protocol, intense cleaning of equipment to prevent potential cross-hole contamination, continuous sampling of soil, and geologic consultation. The safety protocol consists of disposable, chemical resistant boots, suits and gloves, protective goggles, hardhats, and on-site accessibility of dual carbon filter respirators and a self-contained breathing apparatus (SCBA) pressure-demand type, to be used as necessary. All drilling equipment was thoroughly cleaned using soap and water, hexane, and rinsed with distilled water prior to reuse.

Continuous soil samples were taken with a 24" long, 1 5/8" O.D. split spoon sampler driven by a 140 lb. hammer falling 30 inches.

Assuming all other costs common to both investigations remain constant, the use of the geophysical techniques of magnetometer and seismic refraction indicate a differential cost savings of approximately $9,895, as utilized in the investigation of this HWDS.

SUMMARY AND CONCLUSIONS

The cost-effective field investigation conducted at a hazardous waste disposal site should consist of a historical search of all available information concerning the site, a well thought-out and implemented, non-destructive, geophysical evaluation program, and a substantiation of the results of the geophysical program through the drilling of a limited number of soil test borings.

Geophysical evaluations of HWDS's generally utilize any or all of the following techniques: earth resistivity, magnetometer, ground penetrating radar and/or seismic refraction. The geophysical techniques represent non-destructive methods designed to aid in the definition of the vertical and horizontal limits of a HWDS, assist in the determination of subsurface geophysical characteristics, and locate with relative accuracy the probable sources and pathways of contamination within, beneath, and adjacent to the HWDS. The results obtained should be substantiated by a minimal number of soil test borings. The test borings will eventually be utilized in the sampling and monitoring program since their strategic location enables groundwater monitoring wells and/or piezometers to be effectively installed in the borings.

Decisions regarding the choice of geophysical techniques depend upon site variables including type of refuse, topography, geology, climate, size, accessibility, and other variables. However, there is usually a technique or combination of techniques for most circumstances.

The above rationale was employed at the HWDS identified in this paper. A thorough historical evaluation revealed the site to consist of a lagoon area and a drum disposal area, underlain by glacial deposits of gravel, sand and silt, non-stratified till and a shale bedrock. The geophysical techniques of magnetometer and seismic refraction were successfully applied to the site. The magnetometer served to determine the horizontal limits of buried metallic refuse (drums), while seismic refraction established the vertical site limits and provided an accurate account of subsurface geological conditions. It is important to realize that while considerable faith is placed in the accuracy of results obtained by the non-destructive techniques, the drilling of test borings to substantiate these results is essential to a complete investigation.

The costs incurred utilizing the geophysical techniques of magnetometer and seismic refraction, not including the cost of the 11 test borings required to substantiate these techniques was approximately $7,100.

If the non-destructive geophysical techniques had not been utilized, a conventional method consisting of drilling additional test borings would have been required. Estimates from the regulatory agency place the number of borings that would have been required at approximately double the number that were installed during the geophysical program utilizing the non-destructive techniques. Therefore, an additional 11 borings would have to be installed at a cost of $1,545 per boring, or a total of $16,995.

The differential cost realized between the two methods reflects a savings of about $9,895 due to the use of the magnetometer and seismic refraction. The high costs for the borings can be attributed to the safety protocol and cleaning of drilling equipment that is necessary during drilling operations at a HWDS.

REFERENCES

1. Stollar, Robert L., and Paul Roux, "Earth Resistivity Surveys - A Method for Defining Ground-Water Contamination", Ground Water, 13(2): (1975).
2. Kelly, William A., "Groundwater Pollution near a Landfill", Journal of Environmental Engineering, Vol (#102): 1189-1199 (1976).
3. Pease, R. W., Jr., and S. C. James, "Integration of Remote Sensing Techniques with Direct Environmental Sampling for Investigating Abandoned Hazardous Waste Sites", National Conference on Management of Uncontrolled Hazardous Waste Sites, Washington, D.C., (October 28-30, 1981).
4. Blasland, W. V., G. D. Knowles, E. R. Lynch, R. K. Goldman, and G. W. Lee, Jr., "Hazardous Waste Site Field Investigations: Evaluation of Cost-Effective Techniques and Procedures", Fourth Annual Madison Conference of Applied Research and Practice on Municipal and Industrial Waste, (September, 1981).
5. Morey, Rexford M., "Continuous Subsurface Profiling by Impulse Radar", Engineering Foundation Conference on Subsurface Exploration for Underground Excavation and Heavy Construction, American Society of Civil Engineers, (August, 1974).
6. Benson, R. C., and R. A. Glaccum, "Radar Surveys for Geotechnical Site Assessment", Atlanta Convention and Exposition, American Society of Civil Engineers, (October, 1979).

CHAPTER 14

HAZARDOUS WASTE SITES:
ASSESSMENT BY COMMON SENSE AND
PRACTICALITY: SITE ASSESSMENT,
SITE SAFETY, AND SITE SAMPLING

K.M. Harsh
   Ohio Environmental
   Protection Agency

ABSTRACT

   The Comprehensive Environmental Response, Compensation, and Liability Act of 1980 (CERCLA); Love Canal; Valley of the Drums; and Chemical Control in Elizabeth, New Jersey have all served to spotlight toxic wastes as a legacy of the 1980's from previous years.

   A number of different approaches have been taken in attempting to clean up various waste sites. Commonly, an independent firm is called in which prepares to clean up the site by moving all of the drummed waste into neat rows two by two. After a certain time period, the entire site is neatly arranged like an article in a book. The site looks good to the untrained eye. Then the entire site is systematically sampled in very precise technical manner.

   All of this is very nice, and very expensive, but it may not be cost-effective.

   In the State of Ohio, several major abandoned or uncontrolled waste sites have been approached in an entirely different manner.

   Sites with approximately 55,000 total drums have been viewed with both analytical tools and minds, and common sense. This is best exemplified at the former Chem-Dyne site near Cincinnati, Ohio, which had about 30,000 drums at the beginning of 1980. Rather than re-arranging all of the drums, the site has been treated as an archeological find. About 80 percent of the waste has been identified to its original generator. This has been accomplished partly by searching records, but mainly by looking at drum color, type, closures, marking and location.

A visual inspection can normally also determine what is contained in a drum. This merely requires opening the drums. Drum condition is also critical in determining contents.

Sample of wastes likewise can be approached using common sense. The degree of protection should be no greater than the degree of risk. Any greater degree is wasted.

HOW TO APPROACH A SITE

Find out when the site was in operation. Was the waste all pre-1970 or post 1970? Did the site solicit local or national business? Most sites are one of these two types. For example, one of the major sites in Ohio (15,000+ drums) did over 80 percent of its business with firms 50-100 miles away, while Chem-Dyne served most of the major chemical companies. This makes a tremendous difference in the variety of materials on-site. Unless the local site specialized in laboratory waste, small bottles, cans and jars, the variety of material on hand will be much smaller. National sites like Chem-Dyne have waste from all over the country, some of it unique and very toxic.

Figure 1. Portion of Chem-Dyne Site in Hamilton, Ohio Looking south.

200

Figure 2. Chem-Dyne site looking north towards Hamilton
City Power Plant and buildings.

Ask neighbors and ex-employees what customers were on-
site. Ask ex-employees as soon as you locate them, one of
the key witnesses in an Ohio case had a "heart attack" one
week before he was scheduled to "tell all".

One should obtain all of the available company records.
If the site is a newer one there may be invoices and truck
records detailing individual shipments. Over 7,000 invoices
were taken from Chem-Dyne after it was closed. In addition,
many boxes of letters and files were also taken.

The invoices were examined and a master computer log
made of them. If for example only one company has shipped
ten drums of Carbon tetrachloride to a site and you find
ten drums of that material, it probably belongs to company
"X".

On the next page is an example of a typical invoice
from Chem-Dyne.

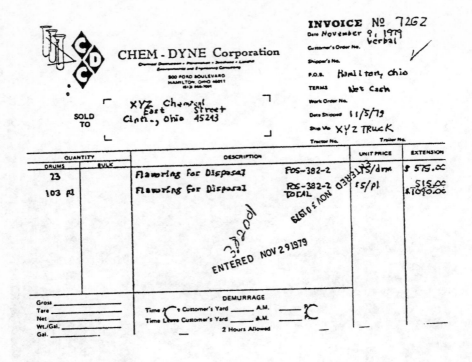

Figure 3. Typical Chem–Dyne Invoice.

A list of all generators was compiled from available
records, and lists made of what each generator shipped.
The files provided lists of materials shipped, dates, and
code numbers. These files may last two to three years after
a company has closed. For other sites a good approach is
to obtain cancelled checks. Many banks maintain five to six
year records and may be able to provide lists, copies of
statements or of who paid the disposal site. If you know
who paid, you may contact the generator and ask him what
type of waste was sent to the site. Under the indemnity
provisions of CERCLA, it may save the generator money in the
long run to help identify the waste rather than risk law
suits.

HOW TO IDENTIFY DRUMS

General Age

Identifying drums is an acquired art, and takes work.
First, the site should not be re-arranged before it is
examined. Normally drums come onto a site in loads of 80
or more. The drums are normally unloaded in one area.

Smaller loads of ten to forty drums are also unloaded into one area. Normally, shipments in the same area on-site are approximately of the same age. If one stack of drums can be dated to August, 1979, then the other drums in the same area are probably of the same age. An investigator will not be able to immediately identify every drum in a stack. However, there may be one "key drum" which contains enough bits of information to identify the whole stack. For example, there was a stack of 120 drums at Chem-Dyne surrounded by two to three other stacks which were dated conclusively between two dates in 1977. The drums were known to contain corrosive material, so all invoices between these dates were pulled and searched, and a match made.

At another site in Ohio, a company was staging the drums in neat rows. However, they were just picking out drums. Instead of matching like drums, they were creating confusion.

Waste sites should be approached as an archeological treasure from which much information can be obtained. If one goes through the drums like a bull in a china shop, all of that information will be lost.

Drum Type

There is an amazing variety of drum sizes and basic types:

1. One gallon pails
2. Five gallon pails
3. 15 to 30 gallon small drums
4. 55 gallon drums
5. 85 gallon "overpack" drums
6. Other odd sizes.

Basic drum construction may be of steel, or plastic, or a combination thereof. The drums may also be coated or lined. Drums containing laboratory chemicals may contain a variety of containers, test tubes, vials, and bottles.

The majority of drums with plastic liners or the all plastic drums themselves, will contain liquid corrosives or highly halogenated materials. Out of 30,000 drums at Chem-Dyne, perhaps 2,000 were either plastic or lined; only 10 percent of these from one generator were not halogens or corrosive.

The steel may be thick or thin, depending on original drum manufacture.

Opening/Closure

Opening configuration and closure type are important considerations.

There are two types of drum tops; either closed head or open head. Closed head means that the top is an integral part of the drum and is not an added on piece. Closed head drums will have various openings or bung holes. Normally, closed head drums contain liquids, or at least the material was liquid when it was placed in the drum.

Figure 4. Closed-head 55 gallon drum with large opening.

Open head drums, on the other hand, have had a top added and attached in place by drum locking rings and various gaskets. These rings can be either fastened by bolts or by a steel lever lock arrangement. Virtually all drums with lever locks that I have observed have contained solid or semi-solid materials.

Figure 5. Open-head drum with lever-lock closure, and with stencilled number on side.

Closure type is very important. Most drums have one two-inch opening and one 3/4 inch opening. A few drums may have two two-inch openings or two openings and a third opening for a stirring rod. The closures can be steel, zinc, or plastic, and may be gasketed. If the drum has a center opening for an agitator, it will be 1 1/2 inch in diameter. The original generator will normally specify one combination. Closures may key 10 to 20 percent of a site's drums. In addition to the basic closures, another 10 percent or so of the drums will have cap seals covering either the large or small plug's opening. At Chem-Dyne, for example, only two to three of the generators used steel or plastic capseals. One and only one, generator used plastic capseals (white) over both opening; these drums all contained hardened polymeric isocyanates. Remember, 80 to 100 industrial companies drums have been identified at Chem-Dyne. Another generator used capseals which had been customized with the company name.

Figure 6.   Drum with capseal over stirring port in middle
            and large closure.

Condition of Drums

Drums <u>do</u> <u>not</u> <u>last</u> <u>forever</u>. They can be dated by their
condition. Metal drums containing acids and highly halo-
genated materials <u>do</u> not last over <u>two</u> years. One generator
shipped approximately 130 thin steel drums with plastic
liners to Chem-Dyne in 1979. The material inside was
two-phase: an aqueous phase with pH of over 10.0 and a
bottom phase consisting of carbon tetrachloride, chloroform
and other heavy chlorinated organic materials. By the
spring of 1981, nearly every drum had leaked and rusted away;
the metal was like cardboard. The plastic liner had been
too thin to stop the material from seeping through.

Figure 7. Deteriorated drums containing organophosphate
solution. Some with only plastic liners

There were approximately fifty plastic-lined steel
drums of organo-phosphates shipped to Chem-Dyne in 1977.
By late 1979, only the liner remained; the metal rusted
away. Polyethylene drum liner inserts of either 40 to 60
mil thickness do not guarantee good drum condition.

Another generator shipped 50 drums of fuming, water
reactive benzoyl chloride to the site. These drums had
thick plastic liners, and were still in relatively good
shape after four years. Neutralizing the material was
another matter, however.

On drums over five to six years old, the first part
to rust through will be the lid. On about 60 drums of Methyl
methacrylate monomer which had been shipped to the site in
late 1975, by 1982 only the lids needed replacing; the
drums were still in shippable condition.

Drums containing solvent or solvent sludges and less
than 10 percent halogenated materials will last five to ten
years outside and maintain their structural integrity.

Figure 8. Fifty-five gallon open-head drum with lid removed showing inner plastic liner.

Color

Many companies seem to have fashion coordinators to design the color of their drums. Among the color patterns at Chem-Dyne were:

1. Black
2. Green
3. White
4. Yellow
5. Red
6. Blue
7. Aquamarine
8. Grey
9. Red with a yellow band
10. Black with blue band
11. Black with yellow band
12. Blue with white band
13. Brown
14. Black and white
15. Red with blue band
16. Grey and white

17. Orange
18. Silver
19. Red and white
20. Fiber drums
21. Many other colors.

Perhaps the most difficult drums to identify are the plain black drums. The more distinctive the color pattern, the easier the drum is to identify. For example, two of the largest generators at Chem-Dyne had very distinctive drums, which together accounted for almost 3,000 drums.

Markings/Codings/Labels

Together, drum types, openings, condition, and color should help the average researcher identify about 25-40 percent of all on-site drums. The other 40 percent of the drums at Chem-Dyne were identified by the markings, drum codes, or any labels attached to the drum. There are two major types of markings: (1) Original, and (2) added on later.

Most generators buy raw materials in drums from other chemical companies. They then use these same drums to dispose of waste materials. Most generators make sure that they do not use their own original drums for waste disposal. Only two generators at Chem-Dyne used their own marked/ painted drums. All of the rest used other people's drums.

Added on markings are generally much more important than original markings. However, if the original lot number is on the drum, you may call them and ask to who they sold that product. If you have a list of all known site refuse contributors and one company matches up, then the generator has been found.

If you do not know who made the material, there are several quick techniques for tracking down companies. The Uniteds States Environmental Protection Agency (USEPA) has published a useful series of reference documents.[1,2] There is a whole volume which lists Trade Names and manufacturers. As part of the implementation of the Resource Recovery Act all hazardous waste sites were required to notify the USEPA. Approximately 56,000 entities are listed in the ten volume series.[2] The listing includes company name, responsible individual, and telephone number. This is normally the company's environmental manager, who will be your best source of information.

In addition, Chemical Marketing Reporter[3] and Chemical Purchasing[4] both contain alphabetical lists of who makes what chemicals. These, together with directory assistance, should enable you to track down 99 percent of all generators who are still operating.

Figure 9. Coding on drum showing label, lot number and other

As an example of this, a certain lot of adhesive compound was traced back to the original manufacturer. I had six guesses as to the original buyer; only one matched and the generator was identified. Companies are usually most anxious to assist in proper waste identification to avoid their own liability. Most companies are extremely helpful.

Original paper shipping labels may still be found on 5 to 10 percent of the drums, and can last outside for two or three years or more. If one label out of 80 drums had an address: XYZ Chemical, 7763 Middleroute, Anywhere U.S.A., then the source has been located.

Another company stencilled purchase order numbers either

on the drum or on paper labels; these were matched up with the original purchase order number on the invoices. Some companies conveniently stencil their name and address or a traceable waste code number on the side of the drum; for example, at Chem-Dyne, Fos 337-10 may mean flammable toluene from 222 chemical, or the drum may be stenciled "ZXY Acid Waste."

Figure 10. Drum with labels, old painted over flammable sticker, and contents of drum labeled "ethyl acetate."

One generator used plastic labels on its drums; the material was a nitrocellulose lacquer. A quick call to the suspect company and the query "do you use labels on your drums" identified the generator.

Another company painted bright yellow "A," "B," "C" on the sides of their drums, with a certain color spray paint

Some others embellish freely with spray paint "resin dump" in a certain style handwriting. The scrawl and the paint clearly identify the drum.

In addition, many waste handling, loading dock personnel mark the waste in their own peculiar way.

Figure 11. Drum with spray-painted "Resin Dump" on side of
            drum.

Perhaps the most embarassing thing for a company is
when the responsible man's name is painted on the drum.
Lets say Mr. John Doe visits the site; he is shown a drum
with the label "Return to Mr. John Doe at Building B,
Powerhouse." Labels and tags must be examined carefully;
they are excellent evidence.

Other

One of the last resorts is to match up a generator
with the material inside the drums. Laboratory chemicals
are very good for identification purposes, but <u>extreme
caution</u> must be exercised when examining laboratory chemical
drums. These drums usually contain the worst possible mixture, and the materials that can explode. However, company
and individual's names are often on the jars, especially if
they are samples of product.

Sometimes combinations of clues will be needed to
positively identify the manufacturer. It may be a black
drum with a green lid with yellow label and a white "A."

Perhaps the best sleuthing on drums that I ever did was with one group of about a dozen drums. A survey was being made at Chem-Dyne with a geiger counter. Two or three drums showed low level radioactivity readings. The drums were black with a stencilled number and no other markings. There were no labels of any kind. As a last resort each of them was opened up-still no labels. The low level stuff was a heavy yellow solid, there were three lab packs, and four drums of bags of a "Iminone" black powder. Worse, the maker of the black powder no longer existed and no one knew what it was for. Finally, all invoices of less than 15 drums were pulled and a series of phone calls made. Still no luck in locating a source. Finally, I guessed that the stuff might be a glass additive. Several dozen more phone calls were made; finally a man knew what Iminone was used for. The yellow solid was a cerium compound. Company representatives visited the site and confirmed that the material was theirs.

Figure 12. Plastic drum open-head drum, and closed-head drum with various closures, open-head has label on side.

Cooperation

Most companies can recognize their own drums if shown the appropriate clues and the material inside the drums.

SAFETY

There are many misconceptions about safety at hazardous waste sites. Overprotection seems to be more common than underprotection. In approximately 500-1000 visits to various waste sites, the worst effects I have had were not from the chemicals, but from heat prostation and over-exertion. In the summer of 1981, I was moving drums around wearing boots, gloves, respirator, and heavy chemical protective rain gear. I started seeing stars and had to lie down for several hours to recover from heat prostation. Levels of protection should be based on common sense and rational ideas rather than strictly following some guidance manual.

In perhaps the best example of blind obedience, a contractor was doing sampling at a hazardous waste site. They had set up safety gear and equipment just outside the site fence. The day they were set up outside the fence, the wind was blowing the chemical odors towards their command post.

They carefully had their personnel put on self-contained breathing apparatus before crossing the fence-line. The area of the site they were sampling was a ten minute walk from the fence; so the walk in and back consumed twenty of the thirty minute air supply. They would exit the site and be carefully decontaminated. Meanwhile, I was wearing boots, respirator mask, and other protective gear. I walked to their safety command post and found that odors were worse there than anyplace else in the area. Overprotection and poor planning cost thousands of dollars in time and effort in this one case. This is not to imply that protection is not vitally important in examining waste sites. There are several levels of protection which may be used.

In eight years of experience with numerous hazardous waste sites and thousands of spills of hazardous chemicals, I have worn a full-encapsulated suit only once.

Protection

Protection is broken down into seven areas:

1. Head protection
2. Eye protection
3. Foot protection
4. Hand protection
5. Respiratory protection
6. Hearing protection
7. Body protection

These basic areas of protection can be further broken down into examples of protective gear for each part of the body:

Part	Gear
1. Head	(a) None (b) Bump hat or (c) Hardhat with face shield
2. Eye	(a) Glasses with safety lenses (b) Safety glasses (c) Protective goggles (d) Faceshield alone or attached to a bump hat.
3. Foot	(a) Leather boots with metatarsal and toe protection (b) Pull-over boots, ankle, knee, or hip length, made of rubber, vinyl, neoprene or other materials (c) Disposable boots made of paper, rubber or plastic material. (d) Rubber boots, ankle, kee, or hip length. (e) Vinyl or neoprene boots or other material.
4. Hand	Gloves are made mostly of rubber, neoprene, vinyl, polyethylene, leather, rayon, nylon, or other materials. They may be: (a) Disposable (b) Wrist length (c) Arm length (d) Shoulder length (e) Various weights and sizes.

Thin surgeon type disposal gloves (vinyl) work
nicely underneath other gloves in the winter, to help keep
hands warm. Good gloves are very important! I have tried
many variety of gloves; some are too thin and tear easily,
some are cumbersome or too hot, and some are not chemically
resistant. It is a matter of personal preference to try
different varieties until you find the ones best for your
purpose.

5. Respiratory Is vitally important; you will get
   Protection   sick and possibly die without the
          right protection. Skin may grow
          back, but lungs heal very slowly
          or not at all. <u>Do Not Use a, b, or
          c.</u> where insufficient oxygen is
          present or when concentrations of
          vapors exceed cartridge capacity

        (a) Dust masks are almost <u>totally
           useless</u>. One "chemist" used
           a dust mask when working with
           hazardous chemicals. Dust
           masks may filter out abestos
           particles, but no other vapors
           or gasses. <u>Do Not Use Them.</u>

        (b) <u>Half-face Cartridge</u> masks
           with appropriate cartridge
           are used almost 75% of the
           time at waste sites, espe-
           cially when making reconnais-
           sance rather than sampling
           inspections. Make sure you
           use the appropriate cartridge.

Figure 13. Half-face respirator, with bump hat, and
     coveralls.

(c) Full-face Cartridge respirators
are excellent for sampling drums,
and much other work, and also offer
eye protection.

Figure 14. Full-face cartridge respirator, personnel also
wearing hardhat with faceshield, boots, gloves,
anc chemical Resistant plastic gear.

(d) <u>Full-face Canister Respirators</u>
can be used in much higher concentration of contaminants. I prefer
to use one when sampling drums,
especially back-mounted canister
units which draw air further from
the actual drum opening.

(e) Oxygen or other gas air supplies
for escape purposes only.

(f) <u>Self-Contained Breathing Apparatus</u>
is a must for any really risky
operation and comes in a variety
of models and brands.

217

Figure 15. Front-mount full-   Figure 16. Back-mount full-
            face canister   face canister
            respirator.   respirator.

   6. Hearing   (a) Not normally used on waste sites,
                              unless there is machinery making
                              excessive noise.

   7. Body   (a) None
                         (b) Fabric coveralls
                         (c) Disposable coveralls (Tyvek,
                              etc.)
                         (d) Plastic chemical resistant
                              coveralls or rain gear.
                         (e) Full encapsulated enclosed
                              "acid suits" which are to be
                              used only with self-contained
                              breathing apparatus.

    Levels of protection can be based upon a number of
technical and common sense factors.

Technical/Non-Technical Evaluation

Perhaps the best field survey meter for the basic detection of organics is either a portable Gas Chromatograph (GC) (like the Century OVA), photoionization detector (HNU), or colorimetric tubes (Mine Safety, Drager, etc.), or carbon tubes which are analyzed at another outside laboratory. The HNU or similar device is good for identifying "hot spots." These devices were required as documentation or proof of substances for scoring of abandoned unmitigated waste sites as part of Superfund submittals in late 1981.

One of the best devices which man has available and which requires no batteries is his own nose. I have used my nose hundreds of times to identify chemicals and "hot spots." The nose is much more sensitive to many chemicals than many field instruments. If you do not have a "good" nose do not use this technique. Many chemists, however, can be cautious amd map site odors. I can walk over the Chem-Dyne site and point out a mercaptan odor, halogenated solvent odors, plastic odors, sulfur waste, and other "hot spots." Remember, this technique can be used only by trained personnel. Do not disregard the technique just because it involves man rather than machine.

Figure 17.   Site evaluation in area with no risk.

Required Levels of Protection

Preliminary Evaluation/Initial Visit

For the first visit to any site, I wear boots, gloves, hard hat, disposable coveralls, and half-face cartridge respirator. This is merely for initial reconnaissance. If I notice any odors or eye irritation, I switch to full-face or cartridge respirators.

If the site is very small and the materials are known to be of low toxicity (from record searching), then only boots and coveralls may be required.

Sampling

If any sampling is required it _must_ be done with boots, gloves, hard hats with face shields, respirator (preferably full-face) coveralls and caution, and a good universal mon-sparking bung wrench.

I use the breathing apparatus only when sampling very toxic or reactive materials, which cause such great harm as to be life-threatening.

One of my most memorable moments was one hot summer day in 1980. I was wearing a self-contained breathing apparatus and trying to overpack drums of water reactive vanadium oxytrichloride and vanadium pentoxide. Fumes from those compounds are toxic, corrosive and water reactive. I was straining under the hot sun and sweating. Unfortunately, the drums were a little too tall for the overpacks. I tried to wedge them in when the drum began to get to hot and give off nasty orange fumes. I backed off and asked for help from the original generator.

Perhaps the most serious experience was with some suspected nitrocellulose. I spotted some drums containing a dried white powder and labeled nitrocellulose. Extra help was called in to handle the waste. When pure nitro-cellulose is dried out and aged, it may become shock sensitive. The material can either be detonated or carefully wetted with water or a solvent. Solvent (acetone) was carefully added to the drums and they were moved to an isolated location. About one-half hour later there were two loud explosions. Unfortunately, the material inside the drums was not nitrocellulose, but calcium hypochlorite an oxidizer, which caused the explosion.

Some other materials requiring extreme protection are cyanides, pesticides, and laboratory chemicals. In all cases the degree of protection required should be no greater than the risk. Worker comfort and stress should be a factor in choosing safety gear.

SITE SAMPLING

Sampling can be divided into areas for any site:

1. Bulk tank waste
2. Drummed waste
3. Groundwater
4. Soil/bulk waste piles
5. Ambient, surface water, runoff and air

1. Perhaps the most difficult of the five to sample, and this may seem unusual, is the bulk tank waste. Some of the most innovative techniques are those utilized to get representative samples of bulk tanked waste. Wastes in bulk tanks are not homogeneous, but rather are normally layered. If the tanks was used as part of a "fuel program" it may contain:

a. A light solvent layer
b. An aqueous/water layer
c. A floating or submerged polymer layer
d. A "heavy" chlorinated organic layer
e. A blanket of sludge

Tanks may be liquid, solid or frozen. At a minimum, the tanks should be sampled at the middle, top, and bottom. If the tank has sample ports, this will be easy; if not I would recommend a Kemmerer sampler or similar device which can be lowered into the tank opened and a sample withdrawn. Another good method is to use a portable peristaltic pump and plastic tubing. A plastic hose is lowered at a fixed rate through the tank while a sample is pumped into a suitable sample container. This works very nicely and rapidly on tanks containing 100% liquids. A third option is to withdraw a thief sample with a plastic or steel tube, or aluminum conduit tube. The fourth and final option is to improvise; whatever works best.

2. Drummed waste is not difficult to sample, but it can be risky. Only a few items are required to sample hundreds of drums daily.

a. A good non-sparking bronze universal bung wrench.

b. One meter (or 4 foot) glass rod at least 10

221

millimeter (mm) outer diameter; or aluminum conduit cut into convenient lengths and available from most hardware stores.

c. Sample jars; for Gas Chromatograph Mass Spectrometer (GC:MS) samples; 40 milliliter Teflon septa vials work very well for organic matrix samples. For other samples, either 500 or 1,000 ml jars with Teflon lined lids. For samples where only flash points or other simple tests are required you may utilize any convenient container.

d. Protective gear

e. Some device for marking the drums sampled. I personally prefer yellow or red Markal[R] Paintstik[R] Type B. These are convenient, easy to use and durable. For a while, I tried spray paint and stencils, but that took much longer. Crayons and other materials are not as good as the Paintstik[R] (Markal Company, Chicago, Illinois).

Sampling is best done in two or three-person teams. One person can select and open drums to be sampled, another can do the actual sampling, and the third can record the appropriate data. Several hundred drums can be sampled and researched daily by each team. I do not like to puncture drums to take samples, but if necessary I will use a bronze pick. I have yet to see a good, non-destructive, efficient mechanical drum opener.

I have personally opened thousands of drums at sites all over the State of Ohio. I have never been injured opening a drum. There are a few simple rules to remember when opening drum.

1. If the big bung is rusted shut, try the smaller closure. It will take less torque and open more easily.

2. Open the closure _slowly_ and listen for any hisses indicating pressure build-up inside the drum. _Do not_ remove the closure until the hissing or liquid outflow has stopped.

3. Drums of liquids (solvents) will _always_ hiss if there is pressure.

222

4. Experienced samplers can determine the viscosity of the material inside a drum by its sound when thumped with a rod or stick. A hollow ring means liquids, a muted ring semi-solids. Solids also have a distinctive sound. Practice this and perfect it.

5. The most dangerous drums are drums of semi-solid polymers. There may be very little head-space and the resin will seal up the threads so that they may not hiss to indicate pressure. The polymer may suddenly spurt out at the sampler. Make sure you have a faceshield. I can recall that one day a company employee wanted to sample one last drum. As he opened the drum, resin glop/snarg flew out of the drum and over his head. He immediately ran to the shower; unfortunately the resin hardened upon contact with water. The stuff had to be literally chipped off his head.

6. Shove the rod/conduit all the way to the bottom of the drum to obtain a representative sample and place that sample in the jar.

7. Reseal the drum.

Lots of drums can be sampled in a short time period if representative drums are sampled in separate site areas.

I normally take composite drum samples; samples of similar materials are more cost-effective if run en masse. I have submitted samples for GC:MS analysis containing up to 100 drums per/sample. Samples are normally analyzed for polychlorinated biphenyls, pesticides, and a general broad spectrum[5] GC:MS analysis.

Every drum does not need to be run individually for GC:MS analysis. If a site has 30,000 drums and GC:MS analysis were run on each drum individually, most of the GC:MS capability in the country would be utilized and the cost would probably exceed 7-10 million dollars (U.S.). Wherever possible, the material should be generally categorized as it is sampled.

ON-SITE CHARACTERIZATION OF MATERIALS

Many materials can be identified by visual inspection. Familiarity with various waste "stream" and industrial processes is very helpful. Among the distinctive materials are:

1. Resins and plastic polymers which are viscous, gooey, and slimy, or may be already hardened.

2. Solvents by the odor, specific gravity or by a quick "will it burn" test. Take a small thimble of the material, and light it to see if it will burn.

3. Chlorinated solvents by their sweetish odor, or by the fact that they may be in a layer near the bottom of the drum or beneath an aqueous layer. Take a glass of water; put a couple of drops of material in it; if it (liquid) sinks, it is probably chlorinated, and may be trichloroethylene, perchloro-ethylene or other similar materials.

4. Aqueous, watery materials are very obvious, but they may have a thin organic layer on top or bottom.

5. Acids and Bases are quickly identified by the corroded nature of the drum, or by pH paper.

6. Oily waste looks like oil; it should be checked for PCBs.

7. Paints, inks and laquers are readily apparent because of their colors, which may be very bright.

8. Laboratory wastes are identifiable once the drum has been opened up. Use extreme caution with lab waste, unpack them carefully and do not drop.

9. Water reactive materials give off fumes so as to become immediately noticeable. I have observed fumes from such water reactives as benzoyl chloride, vanadium compounds, phosphorus trichloride, thionyl chloride, (very nasty), chlorosulfonic acid, calcium carbide (be careful, may explode), titanium tetrachloride, oleum, and nitric acid, among others.

10. Rubber or rubber compounds look like gooey tires.

11. Soaps, drugs and shampoos may be in their original containers, as may pharmaceutical products. Expect the unexpected at waste sites.

12. Asphalt waste is very obvious.

13. Chrome compounds are discernible by the yellow (distinctive) color.

14. Heavy metal wastes containing lots of lead (Pb), Cadmium (Cd), Cerium (Ce), Gold (Au), are obvious because of weight of the drums. Drums containing chlorinated materials are also very heavy.

15. Mercury is obvious.

16. Other materials, which may look mondescript, may be some of the most toxic on the average site. For example there was a lot of black material on the ground at Chem-Dyne that looked like black glass or obsidian. This material contained polychlorinated diphenyl dioxins. Another black tarry material was actually arsenic sludge bottoms. Some little round heavy balls were actually balls of pure cadmium.

    Again I would like to warn you that laboratory package drums can contain pyrophorics, explosives, reactive materials, cyanides, corrosives and carcinogens, all in the same drum. Some may also contain radioactives and precious metals. Be cautious of empty and partially empty gas cylinders.

17. Radioactives can be quickly detected with a geiger counter.

SUMMARY

Historical searches of records and checks, plus a common sense approach to hazardous waste safety and site evaluations can quickly give an investigator a good indication of the immediate scope of the problem at that site. You may be able to go to a site and count the number of drums and determine in a few hours by non-destructive testing techniques whether or not the material is paints, oils or polymers. The first impression of a site may be the best. I have been to some smaller sites (less than 500 drums) only once, yet I know what wastes are there. Some larger sites have taken half a dozen to dozens of trips to characterize the site. The larger sites (over 500 drums) will take longer to evaluate.

Hopefully, some of these practical considerations will help you make a better site evaluation in an easier manner, in more comfort, with less risk, and at lower cost.

REFERENCES

1. United States Environmental Protection Agency Office of Toxic Substances. "Toxic Substances Control Act Chemical Substance Inventory Volume III (Washington, D.C., U.S. Government Printing Office, 1979).

2. United States Environmental Protection Agency, Office of Water & Waste Management. "Notification to EPA of Hazardous Waste Activities, "Region 1-10 (Washington, D.C., U.S. Government Printing Office, 1980).

3. Kaualer, A.P., Ed. "1980-1981 OPD Chemical Buyers Directory," Chem. Marketing Reporter (1980).

4. Bane, M., Ed. "Chemicals Directory 1980-1981," Chemical Purchasing 16 (11) (October, 1980).

5. Harsh, K.M. In: 1982 Hazardous Material Spills Conference Proceedings (Rockville, Maryland, Government Institutes, Inc., April 1982) 199-204.

CHAPTER 15

RAPID SCREENING OF
CHLORINATED HAZARDOUS WASTE

P. P. Dymerski
    O.H. Materials Co.
    Technical Services Division

J. R. Dahlgran
    O.H. Materials Co.
    Technical Services Division

INTRODUCTION

The increasing occurrence of environmental contamination
due to chemical spills and hazardous chemical waste sites has
prompted major efforts to deal with these problems. To support these efforts, the analysis of environmental contaminants
has developed rapidly in recent years to the point where a few
reliable approved methods are now available.[1,2,3] Both the
Environmental Protection Agency (USEPA) and the American
Society of Testing Materials (ASTM) have recommended or approved sophisticated analytical methods for extraction, separation, identification, and quantification to parts-per-
billion levels of contamination in potable water and wastewater. Corresponding methodologies for other environmental
samples such as oils, soils, and sludges are under close
scrutiny and will soon be approved. These methods are designed for a high degree of accuracy, precision, and sensitivity in identifying and quantitating specific compounds. For
organic contaminants, many of these methods were designed for
low resolution, packed column gas chromatographic separation
and quantitation, and involve extensive sample preparation
schemes. This approach was necessary to ensure maximum sensitivity for trace components, and to reduce interferences
leading to false positive identifications.

While chemical spills require rapid confirmation of
specific suspected materials to support the safety of emergency response personnel and local residents, chemical waste
sites require characterization of totally unknown matrices.
The extreme case includes waste drum sites where thousands
of separate unknown matrices are present. In neither of these

cases is a <u>high</u> degree of accuracy and precision in quantitation required. Rather, accuracy in the identification of specific compounds is required since this indicates the type of hazard involved. Speed and cost are also significant considerations in these cases.

An analytical laboratory scheme has been designed specifically to support the rapid and cost effective characterization of complex hazardous waste materials. This scheme has been devised to answer two specific questions for project management during chemical handling at chemical spill and other chemical sites:

1) Is an immediate chemical hazard present which endangers personnel and residents?

2) Is a long term chemical hazard present in sufficient amounts to warrant additional worker protection?

For chemical spill and other one component matrices, categorization of the gross characteristics of a sample is often sufficient to answer these questions. Characterizing a strong acid, strong base, or explosive, for example, is sufficient to support proper protective measures. For multicomponent spills and abandoned waste sites, however, the degree of hazard depends not only on the major component present, but also on low level components of high toxicity or carcinogenicity. (All of the above categories will be considered hazardous in the following presentation.) This is especially true for complex sample matrices of unknown history. Abandoned waste site samples have been screened in our laboratory which contain over <u>three hundred</u> compounds at various concentrations.

Rapid analysis of hazardous environmental compounds then, includes "compatability" testing for categorization of the gross characteristics of the material, identification and quantitation of metals, and the identification and "semi-quantitation" of organic hazardous compounds in water, soil, sludge, and organic matrices. This paper will discuss a portion of this novel scheme involving the rapid characterization of organic hazardous compounds in various matrices at levels from ten percent to part-per-million. Levels above ten percent are characterized by "compatability" testing and are covered adequately elsewhere.[4] The scheme presented here has been developed over the past two years in our laboratory.

Environmental organic hazardous compounds can be separated into two categories: halogenated compounds, specifically chlorinated and brominated compounds of which most are hazardous to some degree, comprise the first category, and non-halogenated compounds, (most are not particularly hazardous), comprise the second category. After sample extraction, specimens can be screened for these two categories using gas chromatography equipped with a high resolution fused silica column (which resolves a large number of compounds better than the packed column) a halogen specific detector, such as Hall electroconductivity, and a general organic detector. This screening process is outlined in Figure 1. Halogen containing specimens can be categorized for content of polychlorinated biphenyls (PCB's) or non-PCB halogens by comparison to proper standards.

O.H. MATERIALS SCREENING PROCEDURE

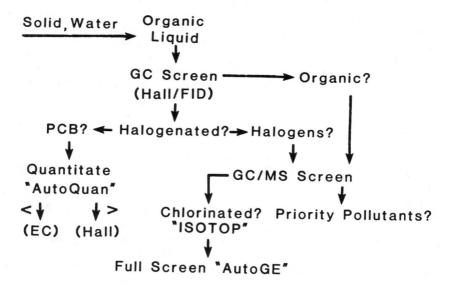

Figure 1.   Separation, identification, and quantitation of environmental compounds are enhanced by use of high efficiency fused silica columns (GC screen) and automated computer programs ("Autoquan," "ISOTOP," and "AutoGE").

Halogens and organics not characterized by GC alone can be analyzed by gas chromatography/mass spectrometry (GC/MS), which enables the chemist to make identifications based on chemical structure information. Halogen containing compounds produce characteristic mass spectra while other organics are often difficult to identify specifically. Target compound analysis, such as the 113 priority pollutants, ensure that the most hazardous species are found, while the remaining unknowns must be searched against comprehensive mass spectral libraries.

Finally, the hazardous rating of all compounds identified must be assessed by a qualified chemist. The nature of the compound, its concentration, length of possible exposure, and other factors must be taken into consideration before recommendations can be made.

SAMPLE PREPARATION

Organic liquids (oils, waste solvents, etc.) are screened directly, and viscous oils are diluted with solvent by a factor of 2 to 10. Water samples are extracted following approved methods.[3] Soil samples and other solid residues present a major problem to a rapid screening scheme. These matrices not only contain a concentrated complex matrix of possible contaminants, but also require extended sample extraction times of 16 hours if approved methods are followed.[1] Several extraction methods are currently being evaluated which will significantly reduce specimen preparation time.

SCREENING

Prepared specimens are characterized for total organic and total halogenated organic compound content utilizing a flame ionization detector (FID: general organics detector) and a micro Hall electroconductivity detector (Hall) respectively. For maximum sensitivity, resolution, and speed, a Hewlett-Packard 5880 gas chromatograph (GC) was equipped with an autosampler, capillary column injector, SE-54 fused silica capillary column, and both FID and Hall detectors.[5] The resultant FID trace for total hydrocarbons versus the Hall trace for halogenated organics is plotted in Figure 2. This performance specimen is an oil mixture diluted 1:10 in hexane with a one part-per-million (ppm) spike of polychlorinated biphenyls. Total screening time per extract is 35 minutes.

This screening procedure produces several important bits of information concerning the nature of the sample (refer to Figure 1).

Simultaneous Detection of Total and Halogenated Hydrocarbons in Complex Environmental Samples

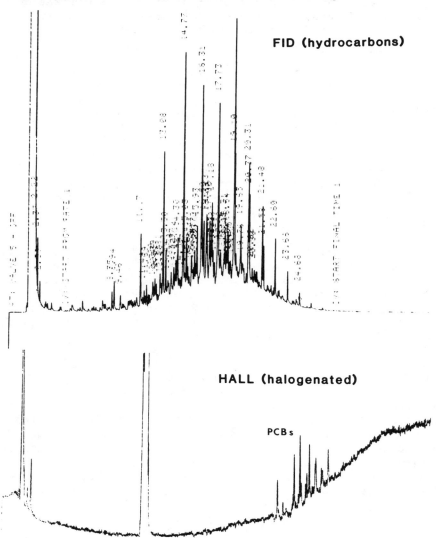

Figure 2. A fused silica SE-54 capillary column split to FID and Hall detectors produce directly comparable chromatograms. The specimen contains diesel fuel (top) spiked with 1 ppm Arochlor 1254 (bottom).

1) Major organic contaminants can be qualitatively characterized from the FID trace. Fuel oil "finger-prints" support mixture identification. Concentration ranges indicate proper dilution factors for further GC and gas chromatograph/mass spectrometry (GC/MS) characterization.

2) A positive Hall trace indicates halogenated organics are present with concentration ranges relative to the total organics present. GC/MS characterization is used for specific compound analysis if necessary.

3) Retention time and GC profile information (Figure 2) indicates the value of the Hall detector in identifying PCB contaminated extracts to 1 ppm (sample concentration).

PCB QUANTITATION

If PCB quantitation is to be performed directly from the Hall detector (for example, RCRA waste oil quantitation to 50 ppm) then proper quality control blanks and bracketing standards must be analyzed together with the specimen batch. Blanks and PCB standards are routinely included in the present scheme both to speed up quantitation if PCB's are found in the sample and to support instrumental quality control.

Samples requiring PCB identification or quantitation below the Hall limit of detection (LOD) or quantitation (LOQ) require analysis by electron capture detection (ECD).[6] The previous Hall screening results are used to determine proper dilution factors not only for quantitation of PCB, but also to eliminate ECD detector overload due to other halogenated compounds (Hall trace) or total organics (FID trace). Correlation of screening traces to proper dilution factors is quickly learned by the chemist. With this scheme, quantitative results can be obtained on samples of unknown origin in the linear of the ECD range on the first pass and detector maintenance is reduced significantly.

PCB manual quantitation is slow and laborious; it is prone to calculation errors, especially if the EPA recommended method[7] is followed for the complex sample matrices discussed here. This method recommends summing the areas of six to eight PCB peak areas in the standard and matching these to the sum of the same peaks in the specimen. Conversion from area to concentration is performed on this one value. In extracts from complex samples, this method often over quantitates PCB content by adding in unresolved interference peaks and does not account for weathering of the sample.

Performed manually, this procedure is cumbersome and time consuming. With a microprocessor/integrator, however, this procedure can be performed automatically and routinely. In the present analysis scheme, a basic program (AUTOQUAN) has been written to quantitate PCB's. AUTOQUAN calibrates on the six largest peaks of up to five PCB standards at known concentrations. For each specimen, a straight line segment bracketing the specimen with a high and low standard quantitates the six corresponding specimen peaks. Interferences are flagged and removed from the average quantitation value reported. This results in a more accurate quantitative value being reported and eliminates the need for "manual calculations."

Table I is a "Quantitation Report" generated by AUTOQUAN showing the average value quantitated, standard deviation of the averaged values, and the number of peaks used. Warning messages indicate whether sample points are extrapolated above or below the standard range, split peaks are outside the retention windows of the quantitation peaks, or that retention time changes have occurred.

Comparison of this individual peak quantitation (IPQ) method with the summation method is identical for clean PCB standards. The standard deviation of the average IPQ value is typically 3-5 percent. With samples spiked with known interferences, both methods give comparable results if all peaks are used in the IPQ method. Standard deviations of the IPQ average value range from 7-20 percent, however, due to deviations introduced by interferences. When outliers are removed from the IPQ method, resultant values diverge from the summation method but the standard deviation is reduced.

HALOGEN COMPOUND CHARACTERIZATION

Halogenated compounds cannot be quickly identified by GC alone. These compounds can be detected and identified by GC/MS using their distinctive mass spectra data. Chlorine and chlorine/bromine isotope clusters are particularly unique in GC/MS spectra. These are examplified in Figure 3 for one through five chlorines per molecule. Despite this unique information, specifically detecting halogenated compounds manually at low levels in complex sample extracts containing a hundred or more chromatographic peaks has hindered the rapid identification of this class of contaminants. (Figure 3 shows one such specimen.)

233

PR: 10:38 DEC 15, 1981

PEAK BY PEAK LINE SEGMENT QUANTITATION

SAMPLE#	[RETENTION TIMES]						AVE.0	ST.D.	#PKS
	13.91	14.37	15.04	15.54	15.96	16.44			
1PPM 1254									
	2835	2414	3344	2820	4185	3884	1.000		6
5PPM 1254									
	9453	9356	13681	17586	17462	16501	5.000		6
347+77 1;200									
** WARNING #1 **									
	-4567	8244	12043	16958	15587	-1478	4.498	.224	4
5PPM 1254	9453	9356	13681	17586	17462	16501	5.000	0.000	6
347-79 ** WARNING #2 ** WARNING #3 **									
	-1064	1223	1654	-1478	2043	1926	.346	.027	4
REG BLK *ND*									
M S 1 ** WARNING #1 ** WARNING #3 **									
	2544	2249	3074	1246	3894	3635	.839	.134	6
M S 2	2693	2373	3272	2709	4174	3900	.972	.032	6

WARNING #1 = SMALL WINDOW QUANT. BUT 1 OR 2 LARGER PEAKS ARE NEARBY
WARNING #2 = 3 OR MORE LARGER PEAKS OUTSIDE SMALL WINDOW
             THEREFORE LARGE WINDOW QUANT.
WARNING #3 = BELOW LIMIT OF QUANT.
WARNING #4 = ABOVE LIMIT OF QUANT.
WARNING #5 = STD'S ARE NOT USING THE SAME PEAKS.

Table I.   Autoquan generated PCB quantitation report.
           All data is made available to the chemist
           through warning messages. Outlier peaks
           excluded from quantitation are indicated
           as negative areas.

Chlorine Isotope Ratios

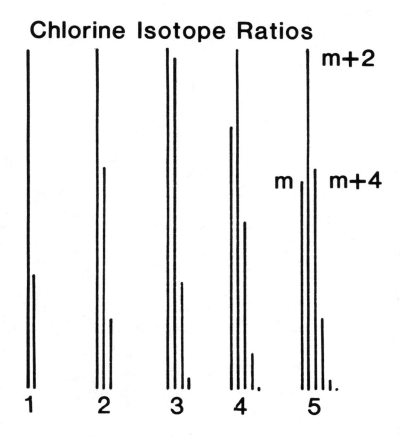

Figure 3. Halogenated organic compounds produce characteristic isotop clusters in their mass spectra. Cl^{35} and Cl^{37} produce mass ion (M), mass ion plus two (M+2), and (M+4) clusters shown here for one through five chlorines.

To enhance the detection of halogenated compounds in complex matrices, a computer program (ISOTOP) has been written to rapidly survey the mass spectral data for halogen fragment clusters. ISOTOP has been used in our laboratory for several months and compared with an experienced GC/MS operator.

With this program, the detection of halogenated compounds in specimens containing 300 peaks is reduced from several hours to ten minutes or less. Figure 4 shows a "halogen reconstructed chromatogram" containing only mass spectra flagged by ISOTOP and plotted on the same scale as the total sample chromatogram.

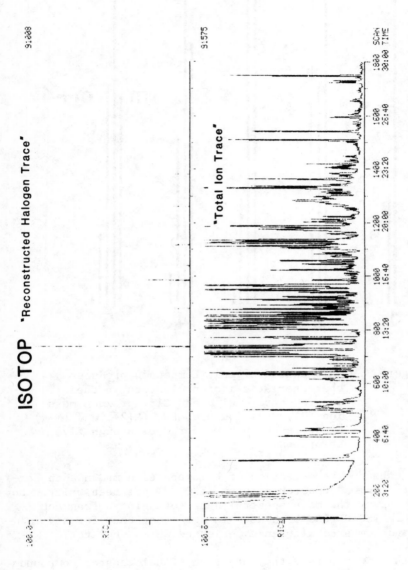

Figure 4. A hazardous waste sample spiked with 20 ppb halogenated contaminants contains 100 discernible peaks (bottom). Screening with ISOTOP indicates only six possible chlorinated compounds (top). Peak highs are comparable on both traces.

Table II compares ISOTOP against library searches and manual identification by an experienced operator in locating halogenated compounds spiked at 20 ppb in the specimen diagrammed in Figure 4. Two hundred thirty-four peaks were found by "Biller-Biemann" autoprocessing which were subsequently searched for halogen containing compounds. Of the eight spectra flagged by ISOTOP, four were confirmed manually to contain halogen. Non-halogenated compounds were missed by ISOTOP. Using full library search, only two of the four halogenated compounds were found due to overwhelming interferences, even with background subtracted out. Forcing the library to search only high mass fragments or chlorinated library entries was still not sufficient to find all the halogenated compounds.

Each library search through the 235 entries required 90 minutes while ISOTOP, followed by library searching of the eight flagged spectra, required only ten minutes. To evaluate the "Biller-Biemann" algorithm, ISOTOP was used to survey all 2,000 spectra in the data file for halogen isotope patterns. An additional halogenated compound was located by this approach which was missed by the "Biller-Biemann" program. Total ISOTOP search time was 10 minutes. The processing times are concluded in Table II.

Compound	ISOTOP	Manual	Auto Processing Library Search
Hexachloroethane	√	√	√
Hexachlorobutadiene(C46)	√	√	√
Hexachlorocyclopentadiene(C56)	√	√	
Bromophenoxybenzene	√	√	√
Hexachlorocyclohexane	√	√	
Time to search	10 min	150 min	90 min

√ Indicates identified component

Table II. Comparison of present methods of identification of halogenated components to ISOTOP identification.

NON-HALOGENATED COMPOUND CHARACTERIZATION

This class of compounds presents the most difficulty to a rapid screening scheme. While some elements have natural isotope patterns, these are not as unique as the halogens. Other classes, such as polynuclear aromatics (PNA), are characteristic in other ways which enable the analyst to quickly pick them out of a chromatogram.

The use of "target" compounds such as the priority pollutants can aid in screening compounds of most concern. This method uses a reduced library of suspect or target compounds along with their GC retention times. Searching narrow retention time windows for these compounds has been highly accepted. In many cases, however, potentially hazardous organic compounds still require a manual library search.

LIMITATIONS

Identification of unknown compounds can be performed adequately only by GC/MS. Several limitations in the characterization of complex sample matrices should be noted:

1) Identification of the major components in a specimen does not correlate with the most toxic or hazardous components, often present at low levels.

2) Even extensive spectral enhancement and comprehensive library searching does not identify all the compounds in a complex waste sample. The remaining 10-15% unidentified components may be significant.

3) GC column overload and subsequent loss of separation reduces GC/MS efficiency. Specimen dilution is of limited value since trace components are diluted below the detection level.

4) Quantitation of priority pollutants or other complex mixtures using accepted methods of external standardization can cause major errors. Signal loss can be suppressed by compounds which elute at nearly identical times by as much as a factor of ten.

SUMMARY

With the present scheme, reliable analytical data can be produced within several hours for characterizing the possible hazardous nature of environmental matrices, even for complex samples. The full capabilities of high resolution inert fused silica capillary column chromatography, and mass spectrometry equipped with computer based laboratory data acquisition systems are only now being realized.

While accuracy in identification can be maintained in the present scheme, accurate quantitation by current standards cannot be. Non-halogenated compounds and the 10-15% unidentifiable compounds present in extremely complex samples still present limitations to the complete characterization of hazardous materials.

ACKNOWLEDGEMENTS

The authors wish to thank the laboratory staff whose efforts have assured quality results from this screening program.

REFERENCES

1. a) Manual of Methods for Chemical Analysis of Water and Wastes, EMSL, USEPA, Cincinnati, OH, 1979. b) Federal Register, 40 CFR 136, Vol. 44, No. 233, Dec. 3, 1979.
2. Annual Book of ASTM Standards, Water, Part 31, ASTM, Philadelphia, PA, 1980.
3. Standard Methods for the Examination of Water and Wastewater, 15th ed., American Public Health Assoc., Washington, DC, 1980.
4. Dymerski, P.P., in progress.
5. Dahlgran, J.R., Journal of High Resolution Chromatography and Chromatography Communications, Vol. 4, Aug. 1981, pp. 393-397.
6. MacDougall, D. et al., Analytical Chemistry, Vol. 52, No. 14, Dec. 1980, pp. 2242-2249.
7. Annual Book of ASTM Standards, Water, Part 31, (D3534076T), p. 700, ASTM, Philadelphia, PA, 1980.

CHAPTER 16

REMEDIAL ACTION AT UNCONTROLLED
HAZARDOUS WASTE SITES:
PROBLEMS AND SOLUTIONS

Jeffrey A. Cassis
Edward P. Kunce
Tom A. Pedersen
        Camp Dresser & McKee

INTRODUCTION

Indiscriminate disposal of hazardous waste in the U.S.
has clearly created a situation requiring extensive remedial
action at a very high cost. This paper describes the remedial action program for the eastern U.S., under the direction
of the U.S. Environmental Protection Agency (U.S.E.P.A.).
The remedial action plan developed for the Pollution Abatement Services (PAS) site is a good case in point because
it involved field investigation, feasibility studies, preliminary and final design, services during construction,
start-up services, and proposed approach for site restoration. This remedial action plan is presented herein as one
example of an implementable clean-up strategy.

The federal Comprehensive Environmental Responses, Compensation, and Liability Act of 1980--known as Superfund--
makes industry responsible for abandoned hazardous waste
sites and waste spills. The Act provides for liability,
compensation, cleanup, and emergency response for hazardous
substances released to the environment and the cleanup of
inactive hazardous waste disposal sites. The Act provides
funds from industry, states, and the federal government
where responsible parties cannot be determined or cannot
afford to pay for cleanup. Unlike the federal Resource
Conservation and Recovery Act of 1976 (RCRA), which regulates
present and future hazardous waste management activities of
industry, Superfund specifically looks to the past.

As part of the Superfund Act, the National Contingency
Plan provides the procedures and standards for remedial
action at uncontrolled hazardous waste sites. The plan was

originally prepared and published pursuant to Section 311 of the federal Water Pollution Control Act. The plan was to be republished within 180 days after enactment of the Act.

States share responsibility under Superfund to contribute at least 10% of the actual long-term costs of cleanup per site, unless the site is publicly owned. On publicly owned sites, states are required to pay or assure at least 50% of the costs. Detailed plans for cleanup of abandoned waste sites will be worked out in conjunction with the states. Cleanup plans can occur through three mechanisms: direct federal contracts, cooperative agreements under which the state takes the lead in directing cleanup, and private cleanup through voluntary or court ordered action.

U.S.E.P.A. implements remedial action through federal term contracts. This simply means that U.S.E.P.A. assigns a number of specific work assignments under the contract. For example, each uncontrolled hazardous waste site assigned to CDM for remedial action is a separate work assignment. A separate scope of work is prepared with an estimate for labor hours and costs for each.

States have to take the lead in developing remedial action plans for a hazardous waste site through cooperative agreements. Cooperative agreements are site specific and do not apply to all sites within the state. An example of this relationship is the Sylvester site in Nashua, NH, where New Hampshire and the U.S.E.P.A. reached an agreement for remedial action.

A private party can implement a remedial action plan through voluntary or court-ordered action. One example is the Woburn, MA, hazardous waste site, where a previous owner is currently negotiating with the U.S.E.P.A. and Massachusetts to assign responsibilities for investigation and remedial action.

On October 23, 1981, U.S.E.P.A. announced that 115 hazardous waste disposal sites were among the worst in the country, that they threaten public health and the environment. The sites were ranked on the basis of their potential danger to health, primarily to drinking water supplies, with threats to the environment also taken into consideration.

The list is used to allocate federal funds under a program established by Congress last year to cleanup hazardous waste dumps. Thirty-nine states placed hazardous waste sites on the list, which was compiled from a group of

282 sites jointly nominated last summer by the states and regional offices of the U.S.E.P.A. The 11 states that did not apply were Alaska, Hawaii, Idaho, Louisiana, Montana, Nebraska, Nevada, Vermont, Oregon, Wisconsin, and Wyoming. Florida led the states with 16 sites on the list, followed by New Jersey with 12, and New York and Pennsylvania with eight each. The 115 sites belong to a group of 400 sites to be listed as candidates for remedial action by U.S.E.P.A. under the Superfund law.

The list is expected to be published in the summer, and it will be made final only after public scrutiny and participation, and after the results of further study and data collection are incorporated into the hazard-scoring.

The four sites in Ohio are:

1. Chem-Dyne Corp., Hamilton:
    900 drums from inactive recycling firm are contaminating river with endrin wastes.

2. Field Brook, Ashtabula:
    Stream contaminated by toxic organic chemicals from multiple, industrial point discharges and runoff from industrial lagoons and land disposal area.

3. Chemicals and Minerals Reclamation, Cleveland:
    Several hundred drums of solvents, acids, and sludges at inactive recycling facility on Cuyahoga River.

4. Summit National Liquid Disposal Services, Deerfield:
    Inactive storage/incineration facility containing 10,000 drums and 13 bulk storage tanks of toxic and flammable wastes.

The U.S.E.P.A. selected CDM to lead engineering studies for immediate remedial action in Superfund Zone 1 (U.S.E.P.A. Regions I, II, IV), which covers 15 states east of the Mississippi River. Two additional zones cover the midwestern and western states.

CDM, in association with several subcontractors, will investigate the extent and degree of contamination at various sites, sources of contamination, and possible remedies. The site specific project plans may involve conducting field investigations, feasibility studies, design, services during construction, start-up-services, and post-closure monitoring.

Several sites have already been assigned to CDM for remedial action. These include Love Canal, NY; Olean Well Field, NY; Pollution Abatement Services, Oswego, NY; Winthrop, ME; Nashua, NH; Price Landfill, NJ; Biscayne Aquifer, FL; and the Bridgeport Rental and Oil Services, Bridgeport, NJ.

PAS OSWEGO SITE

The PAS Oswego site is situated just easterly of the City of Oswego and approximately one-third of a mile southerly of Lake Ontario. The site is generally bounded on the south by E. Seneca Street, to the east by White Creek, to the north by Mitchell Street and the west by Wine Creek. A liquid waste disposal company, Pollution Abatement Services (PAS), operated a high temperature liquid waste disposal incinerator on this site during the period from 1970 through 1977. During the peak period of operation of this facility, it reportedly received approximately one million gallons of waste chemicals per month. Previous cleanup efforts have resulted in the treatment and disposal of waste contained in two on-site lagoons, the filling of these lagoons, and the removal and disposal of PCB contaminated wastes contained in two vertical above ground tanks and one underground tank. In addition, there have been various other activities conducted on-site associated with drum staging and overpacking, removal of approximately 3000 drums from the site by New York State during 1980, installation of groundwater monitoring wells, and significant waste sampling and characterization.

The objective of this PAS Oswego remedial action plan is the removal and satisfactory disposal of the surface drum wastes, drums and the bulk wastes. Field investigations conducted to date under this program have identified 7500 surface drums on-site containing an estimated waste volume of about 100,000 gallons. Also identified are ten bulk storage tanks containing an estimated 80,500 gallons of waste liquid and 1,170 cubic feet of sediment material. The existing site conditions at PAS are shown on Figure 1. There is reason to suspect that there exists one and possibly two additional underground bulk storage tanks on-site. This project is presently under litigation and further information regarding site specific contamination can be obtained from the U.S.E.P.A. (Region II).

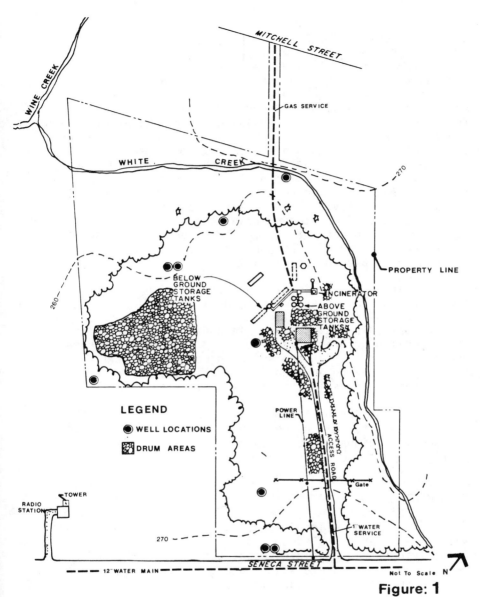

Figure: 1

EXISTING CONDITIONS

P.A.S. OSWEGO, NEW YORK

CHARACTERIZATION OF THE SITE

In order to recommend appropriate treatment and disposal methods for the abandoned chemical wastes at the site, it was necessary to develop a representative list of waste types and quantities. A total of 200 drums were carefully selected from the estimated 7,500 drums by CDM for laboratory analysis. The ten bulk storage tanks were also analyzed and the content volume estimated. These analytical data were used to verify the compatibility of waste types with the consolidation program developed by CDM. The basic concept of the consolidation protocol is to segregate wastes based on water content, acidity, solubility and reactivity in anticipation of ultimate disposal. The data were also used to develop contract documents and specifications for the cleanup and disposal of the chemical wastes which disposal contractors could respond to on a competitive basis.

An air, soil, sediment and water sampling program was conducted to provide an accurate data base of existing site conditions. This information was used to define "clean" and "contaminated" areas on-site. In addition, the results of this sampling program will be used to establish the baseline conditions under which the selected contractor will be required to operate and monitor the site to determine any significant changes in the site's environmental quality.

An aerial survey firm was retained to provide a 40-scale plan of the site for the development of the contract documents and specifications and to obtain false-color infra-red photography of the site.

TREATMENT AND DISPOSAL OPTIONS

At an uncontrolled hazardous waste site, determining waste characteristics is the most important step toward identifying a reasonable disposal option. Unlike industrial waste generation, where wastes are characterized, the wastes at an uncontrolled site are mostly unidentified.

At the PAS Oswego site, the wastes remaining on-site were generally characterized to delineate specific volumes for treatment and disposal. The treatment of wastes on-site was not considered a viable alternative, principally because of the expense of on-site treatment options. The permits required to construct and operate an on-site hazardous waste treatment facility, and the short time schedule established for the cleanup operation, were also prohibitive.

Some on-site preparation activities are not actually considered treatment processes. These activities include bulking wastes, drum-crushing, limited dewatering, and possible neutralization. These activities are designated to achieve two basic objectives.

o Assure a safe, manageable waste type for transport to a permitted treatment and/or disposal site, and

o maintain a waste in a physical/chemical state suitable for a specific treatment process.

Other on-site preparation activities are necessary but do not result in the processing or pretreating of wastes on-site. These non-waste oriented preparation activities include the designation of "clean" and "contaminated" areas, construction of waste bulking and staging areas, implementation of spill control and containment measures, development of site security provisions, and construction of specially designed waste segregation facilities. Along with proper waste handling and disposal, the safe and efficient operation of any remedial action at an uncontrolled waste site hinges on the successful application of these non-waste specific site preparation activities. The recommended layout for the PAS Oswego site is presented in Figure 2.

The hazardous waste types at PAS Oswego site probably encompass the full range of major hazardous waste categories. Results of the sampling program and information obtained from existing waste characterization studies indicate that the following waste types are expected in the drum or tank facilities:

1. Low Level Radioactive Material
2. Strong Oxidizing and Reducing Waste
3. Organic Liquid with Low Halogen Content
4. Organic Liquid with High Halogen Content
5. Strong Aqueous Acid
6. Strong Aqueous Base
7. Aqueous Base Contaminated with Cyanide or Sulfide
8. PCB Wastes
9. Solid Material
10. Empty Drums

Each of these waste types have specific treatment and disposal methods available for both on-site and off-site implementation. It was recommended that on-site treatment should not occur if off-site treatment/disposal capacity was available at a permitted facility. This general policy may,

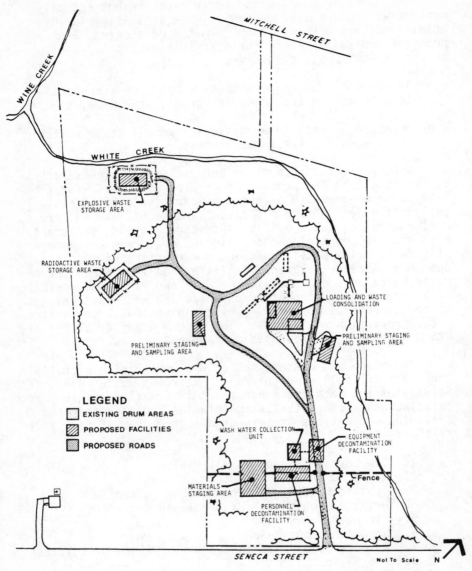

MITCHELL STREET

WINE CREEK

WHITE CREEK

EXPLOSIVE WASTE
STORAGE AREA

RADIOACTIVE WASTE
STORAGE AREA

LOADING AND WASTE
CONSOLIDATION

PRELIMINARY STAGING
AND SAMPLING AREA

PRELIMINARY STAGING
AND SAMPLING AREA

LEGEND

☐ EXISTING DRUM AREAS

▨ PROPOSED FACILITIES

▩ PROPOSED ROADS

WASH WATER COLLECTION
UNIT

EQUIPMENT
DECONTAMINATION
FACILITY

MATERIALS
STAGING AREA

Fence

PERSONNEL
DECONTAMINATION
FACILITY

SENECA STREET

Not To Scale   N

Figure: 2

P.A.S.   OSWEGO, NEW YORK

PROPOSED SITE LAYOUT

248

however, exclude wastes too sensitive to transport, such as some explosives. It may also exclude pretransport activities such as waste-bulking and neutralization. The following sections summarize treatment and disposal methods recommended for the major waste types expected at the PAS Oswego site. Special handling and segregation procedures have been introduced where appropriate. References to costs should be considered approximate. While these costs were based on surveys of treatment and disposal industries, the actual range of costs may vary because of unforeseen factors. These factors include special on-site handling and transport requirements and various contaminants which may limit or require special treatment and disposal processes.

LOW LEVEL RADIOACTIVE MATERIAL

Radioactive materials at the PAS Oswego site should be separated immediately from the bulking and staging areas. They should be placed in a secure, covered storage building.

The storage building should be isolated from any other drum wastes or explosives. Personal protection requirements for handling radioactive material will be determined by the site health and safety officer based on the risk to individual and population. Ultimate disposal would be at a licensed facility in South Carolina.

STRONG OXIDIZING AND REDUCING WASTES

Deteriorated drums and the overall age of the wastes in them have probably reduced the full oxidation or reduction potential these wastes may initially have had.

Once a strong oxidizing or reducing waste is found on-site, it will be segregated in a separate staging area from other wastes. In general, it is extremely dangerous to store flammables (low flash point) with oxidizing agents. If the wastes are stable enough to be transported in re-packed drums, the wastes will be transported off-site for treatment/neutralization. If the waste cannot be made stable for transport, pretreatment to neutralize some of the full oxidation or reduction potential will be performed on-site, away from other waste materials.

Costs will vary for disposal of strong oxidizing or reducing wastes. Characteristics of various contaminants and disposal requirements for sidestreams resulting from treatment will affect total cost for disposal. Typical costs for disposal of strong oxidizing or reducing wastes is about $100 per drum.

ORGANIC LIQUIDS WITH LOW HALOGEN CONTENT

This major waste category includes solvents, oils, and other liquids. Due to the low halogen content, the potential disposal options are typically more diverse and less restrictive than organic liquids with a high halogen content.

The typical disposal modes, by order of environmental and possibly economic preference, for low nonhalogenated organic liquid wastes are product recovery, burning (via fuels blending or as an auxiliary fuel), and destruction through incineration. Land disposal within a secure landfill in a free liquid form, while viable, is not institutionally or environmentally acceptable. In a recent fuling by the U.S. EPA on March 1, 1982, liquid waste can be disposed in landfills for at least three months. The following subsections describe the most preferred and probable methods of waste disposal for the various categories of low-halogenated organic liquids.

Solvents

The finest grade of waste solvent would probably be recovered through a distillation process. Based on the past history of the site, however, it is assumed that any high quality solvents are probably no longer on the site.

Fuels-blending and off-site incineration would probably be the primary disposal options for this category of solvents found on-site. Fuel-blending is practiced by many waste disposal firms. Within New York State, blended fuels, including waste solvents, are used in cement kilns. Disposal costs for a drum of solvent (accepted at site) going to a fuels-blending process range from $25-35 per drum.

Low halogenated solvents found at the PAS Oswego site, which have undesirable characteristics for recovery or fuels blending, would be destroyed at a high temperature waste incinerator. Typical disposal costs at a waste incinerator range from $40-65 per drum.

Oils

As with solvents, the typical disposal options, in order of preference, are product recovery, fuels-blending, and incineration. Like solvents, expected quantities of waste oils found on-site that may be suitable for recovery are probably very low. Most of the low nonhalogenated waste oils would be used in a fuels-blending process or would be incinerated in a waste incinerator. While the range in disposal costs is comparable to that of solvents, quantities of waste oils and related products could be much lower.

Other Organic Liquids

Unlike waste solvents and oils, few other categories of organic liquids have recovery potential. Therefore, other organic liquids will be disposed through fuel-blending or by destruction in a waster incinerator. Waste organic liquids with a BTU value of 90,000-100,000 BTU per gallon--and relatively free of metals, monomers, other gelling agents, and water--could be suitable for fuels-blending. If the wastes are not suitable for fuels-blending, they would have to be destroyed in a waste incinerator. The disposal costs would probably be comparable to those presented for solvents and oils.

Solidification of liquid organics with industrial absorbants, followed by land burial, is a potentially viable alternative. This, however, is the least preferred of all the alternatives discussed above.

ORGANIC LIQUIDS WITH HIGH HALOGEN CONTENT

Because of the many associated institutional, environmental, and technical concerns, it is usually more difficult and more expensive to process and dispose of halogenated liquid wastes. CDM uses 2-percent halogen content to delineate between nonhalogenated and halogenated wastes.

Many halogenated organic compounds are known carcinogens, with strict regulations controlling their disposition. In addition, halogens, especially chlorine compounds, create very corrosive emissions, such as HCl, during incineration. As a result of strict institutional standards and the complex technical issues associated with the processing of these wastes disposal costs for halogenated organic liquids are generally 50-100 percent higher than for nonhalogenated organic liquids.

251

Many issues concerning the various subcategories of nonhalogenated organic liquids apply to the disposal of halogenated organic liquids. However, because of institutional and environmental restrictions, most halogenated liquids would probably be destroyed in high temperature incinerators. Fuel-blending will be restricted by the final halogen content of the blended fuels. When bulking wastes for either fuel-blending or incineration, care must be taken to ensure that the batch of mixed wastes are not contaminated with (a) materials such as metals, which would limit fuel use due to potential emission problems, or (b) monomers or other gelling agents, which could create unique handling problems.

All pesticides and residue from pesticide production found at the PAS Oswego site will be destroyed by high temperature waste incinerators. Typical cost for disposal is about $2.50 per gallon.

AQUEOUS ACID

Aqueous acids are wastes with a greater than 10 percent water content and pH less than 7.0. Aqueous acids are either inorganic or organic. Drums containing acids must be carefully handled. If a drum has a bung, it must be unscrewed slowly to gradually release any internal pressure. A number of materials reach with acid and acid fumes to create heat, hydrogen, and flammable and explosive gases. Removal of acids during any bulking operations must be done with full knowledge of the surrounding materials. Once bulked, acids should be stored away from the above materials.

Acid waste with pH 2.0 should not be bulked with other acids because of the possibility of an incompatible reaction, namely fire, explosion, and release of acid fumes. These acids, once characterized, will be re-packed, or placed in over-packs, on-site for transport to an approved treatment facility. Neutralization with an alkaline material is the basic treatment for almost all acids.

Acid waste with pH between 2.0 and 7.0 will be handled according to the waste consolidation protocol for bulking. Neutralization on-site with alkaline materials, if available, is feasible. Any precipitate will be removed from the site for off-site treatment and disposal at a permitted facility.

Methods of off-site treatment of aqueous acids will vary, depending on whether they are inorganic or organic. Powdered activated carbon can be added in most treatment processes to absorb any toxic organic compounds. The sludge from this process could be incinerated or land buried. Typical costs for disposal are $0.45 to $0.90 per gallon.

AQUEOUS BASE

Aqueous bases, typically defined, are those wastes with greater than 10 percent water content and a pH greater than 7.0. Similar to aqueous acid treatment, neutralization of alkalies with acids is the basic treatment process, provided waste acids are found in suitable quantity and composition at the site. Alkalies in solution are generally referred to as the hydroxides and carbonates of the alkali metals, alkaline earth metals, and the bicarbonate and hydroxide of ammonium. A potentially hazardous situation exists when sodium hydroxide comes in contact with water and produces extreme heat. It is important to wear the appropriate protective clothing and to adhere to established safety and health protocols.

Alkalies in solution at PAS Oswego should be bulked and stored separately from those materials that will react. These wastes will then be transported to a permitted facility for ultimate treatment and disposal, if appropriate, on-site neutralization occurs prior to transport.

Typical costs for disposal are $0.45 to $0.90 per gallon.

AQUEOUS BASE CONTAMINATED WITH CYANIDE AND SULFIDE

Wastes containing cyanide and sulfide will be isolated in a separate staging area on-site. There are two viable disposal options for these waste materials:

(a) on-site pre-treatment,

(b) removal from site and disposal and treatment at a permitted facility.

Under most circumstances, option (a) would be the most cost effective. Provisions can be made to detoxify these wastes on-site either as separate drums or in lined pits on consolidated lots.

Option (b) would be more costly because of the transportation costs. It is expected that drums will be available on-site for staging and storage prior to transport to a permitted off-site treatment facility. Although it could be costly, off-site treatment and disposal is recommended for these wastes. Costs for off-site disposal are $100 to $150 per drum.

PCB WASTE

It is anticipated that most of the PCB-contaminated waste found at PAS Oswego will require removal and disposal from the larger containers and tanks. Although some drummed wastes are expected to contain materials contaminated with PCBs and to require special handling, the greatest volume of PCB-contamination wastes appear to be associated with the on-site storage tanks.

There are three general regulated disposal option categories, based on the actual concentration of PCB in the waste. The first category applies to wastes containing PCBs at concentrations of less than 50 ppm, another is for wastes with 50 to 500 ppm, and third category is designated for wastes with greater than 500 ppm.

Liquid wastes contaminated with PCBs at 50 ppm or less should be burned (via fuels blending or as an auxiliary/supplemental fuel), disposed of in a landfill after solidification (when no other method is practical), or incinerated at a high temperature facility. Disposal/treatment is recommended in the above order of preference. The presence and concentration of other hazardous materials in the waste will largely determine the preferred option.

Wastes containing 50 to 500 ppm of PCBs must be managed and disposed in a manner consistent with federal and state regulations. Two general subcategories of wastes with 50 to 500 ppm PCBs are considered: those with potential fuel value, and those without.

Waste oils and solvents contaminated with PCBs at 50 to 500 ppm with potential fuel value may be disposed by combustion in cement kilns. Use of this waste disposal option is also contingent upon other waste characteristics, such as thermal properties, ash content, chlorine content, heavy metals, and other constituents, which could impact air emission or cement product problems.

If this option is not available for PCB-contaminated wastes (50-500 ppm) with potential fuel value, then the wastes should be bulked on-site and disposed by combustion in

high temperature incinerators, or in a secure landfill after
solidification. Incineration is the preferred disposal option. If other waste characteristics prohibit or limit
viable destruction in a high-temperature incinertor, then
the landfill disposal option should be pursued.

Any waste contaminated with PCBs greater than 500 ppm
must be destroyed at one of two existing U.S.E.P.A.-permitted
incinerators in the United States. This category of PCB
waste would be bulked, if possible, and shipped to either the
Enesco or Rollins location for incineration.

The cost for removal and disposal of PCB-contaminated
wastes will vary depending on many factors, including PCB
content, percentage or water, material phase (liquid, sludge
or solid), condition of container, BTU value, and ash content.

SOLID MATERIAL

This waste type includes a wide variety of wastes:
inorganic sludges, organic sludges, tar and residues, and
contaminated soil. These wastes types are expected to be
found in varying amounts at the site. Many of the wastes
may be found in a physical and chemical state suitable for
direct landfilling. Still, others may require some form of
pretreatment prior to secure land burial or incineration.

Inorganic Sludges (Acid or Base)

These sludges are not appropriate for incineration,
leaving secure landfill as the only disposal option, with
pretreatment and/or special landfilling required in some
cases. Depending upon specific characteristics, direct
reuse may be possible, but it remains very unlikely.

Any pretreatment will basically depend on four characteristics: the pH of the wastes, the free liquid content,
the basic stability of the materials, and the solubility of
toxic components. Except for slightly acidic or basic
sludges, neutralization will occur for conversion either on
or off-site. Following neutralization, conversion of contaminants to insoluble or less soluble forms may be required.
For secure landfill disposal within New York State, solidification and stabilization will be required if the sludge
either yields greater than 5 percent free liquid, or flows.
(Landfills in other states may not have this requirement

255

or it may be less stringent). Solidification may be performed with industrial absorbents.

Alternately, select sludges may be solidified/stabilized
by the more sophisticated cement-based, pozzolanic or self-
cementing processes. These processes may perform some
neutralization, and they will greatly stabilize the wastes.
If successful, a solid or more solid product will result
and greatly reduce the solubility of toxic components,
particularly metals. However, these processes are relatively
costly.

Following neutralization and solidification, the material may be disposed in a secure landfill provided that solubility limitations are met. Costs will vary widely depending
on the level of treatment prior to disposal. Estimated costs
range from $100 to $180 per drum.

Organic Sludges (Acid or Base)

Most of the considerations for inorganic sludges also
apply to organic sludges, with a few differences. The limitations on free liquids, instability, and solubility of
contaminants apply for secure landfill disposal. For treatment prior to landfilling, the use of more sophisticated
solidification/fixation methods is questionable. Organic
materials are usually not compatible with these processes.
Solidification, or the take-up of free liquid with absorbents, remains a viable alternative.

Incineration, in addition to secure land burial, may be
viable for ultimate disposal. The most important factors
regarding incineration are: water content, organic content,
BTU value, halogen content, and the presence of contaminants
(heavy metals) which may be released through flue gases upon
combustion.

Tar/Residues

This subcategory of wastes includes still bottoms, tank
bottoms, filter cakes, spent catalysts, plastic, and polymer
wastes. It is, by definition, a heterogenous mix of hazardous substances. Depending on the organic content of the
residue, ultimate disposal may be by secure land burial or
incineration. It may be possible to incinerate only some
types of these wastes.

This category contains substances which may be viscous. However, they are so contaminated, thickened, multiphased, or full of suspended particles that it is impossible to perform injection incineration as with liquid fuels. These same mixtures may not possess the stability to forego containerization. A specific example may be the mixed residues remaining in the large storage tanks after liquids removal.

Contaminated Soil

Soil contaminated from activities at the new staging pads, around berms and bulking tanks, and soil that cannot be separated from deteriorated drums or equipment, is included in this category. These contaminated soils will be removed from the site for disposal.

Methods to decontaminate soils do exist; however, they are largely impractical for surface deposits. Other methods are state-of-the-art. By far, the most practiced method of disposal is secure landfill. Soils and debris are either placed in drums, sealed, and landfilled, or loaded into leakproof, lined, and covered dump trucks. They are then hauled to a secure land burial facility for disposal.

Unless quantities of soil contaminated to a high percent by weight with extremely toxic or uniquely hazardous substances (such as radioactives or pesticides) are encountered, most secure landfill capacity in New York and the surrounding region is available for disposal.

Costs for disposal may vary widely depending on the mode of handling. For bulk disposal, cost ranges from $30 to $50 per cubic yard. For drum disposal, cost ranges from $40 to $60 per unit.

EMPTY DRUMS

The empty drum classification includes all movable containers found at the PAS Oswego site. We anticipate that a significant number of 55-gallon steel barrels on-site will have little or no contents. The classification also includes those drums emptied in order to bulk the contents.

Although some newer drums may be salvageable, site investigation indicates that the majority of containers on-site are deteriorating. Therefore, the only reasonable alternative for empty drums is disposal in a secure landfill, although many handling and processing options for empty drums exist.

Drums, crushed and compacted in some manner, may be disposed in bulk without categorization or within large containers. Sometimes, empty drums are segregated by wastes which they once contained, or by the inseparable waste residues within them. The crushed units are then deposited in the appropriate subcell.

Costs may be calculated by volume (typically $40 to $50 per cubic yard) or as individual units (as much as $20 and as little as $3 per drum have been quoted).

PACKAGED LABORATORY CHEMICALS

During the preliminary staging of drums at the PAS Oswego site, we anticipate finding laboratory chemicals in drums above ground. These laboratory chemicals may be found separately packaged in vials, bottles, plastic bags, and other containers with and without labels. We recommended that these drums be segregated from other drums for separate chemical characterization. The segregation of laboratory type chemicals will ensure that wastes will be appropriately classified for ultimate disposal.

Once the laboratory chemicals have been classified in accordance with established protocol, they will be repackaged in conformance to appropriate federal and state regulations regarding shipment. Classification and repackaging of laboratory materials is extremely labor-intensive, depending on the chemicals specifically found in each drum. Excluding re-packaging, costs range from $150 to $200 per drum. We anticipate transporting the laboratory chemicals to a secure chemical land burial facility for ultimate disposal.

CONTRACT DOCUMENT

The results of the field investigation, treatment, and disposal analyses were incorporated into contract document for the clean-up activities at the PAS site. The contract documents comprise the removal and ultimate disposal of waste chemicals currently stored in above-ground and in-ground bulk tanks, and in drums stored on the ground surface at the site. The work will include the installation of site health, safety and emergency facilities, administration, decontamination and laboratory facilities and security fencing. The contract is a unit cost type where the site contractor is paid for each unit (gallon) of waste type removed in accordance with his bid price.

Award of the clean-up contract has not been determined but it is anticipated that clean-up activities will begin in the Spring of 1982. After the site clean-up activities are completed, it will be necessary to conduct a detailed site investigation for final site restoration. The investigation is designed to provide information on the extent of site contamination and on mechanisms by which waste constituents may migrate from the site. These data then serve as the basis for further restoration techniques.

SITE RESTORATION INVESTIGATION

Once the drums and bulk tanks have been removed from the site for treatment and disposal, final restoration of the site can take place. Site restoration techniques may include installation of a barrier/drain system, site capping with slowly permeable soil materials or impermeable membranes, and revegetation to minimize erosion. The barrier/drain system would be placed in the shallow sand layer to intercept surface groundwater. This section describes our recommended approach for site investigation for identifying and evaluating site restoration techniques and methods.

Approach

In developing the site investigation approach, consideration was given to existing data on site conditions. The PAS Oswego site is located in the Erie-Ontario Plain geomorphic province of New York State, and it is underlain by the Oswego Sandstone. Overlying the Oswego Sandstone are various glacial deposits including lodgement and ablation till and lacustrine sands, silts, and clays. Within these unconsolidated deposits, the migration of wastes is expected to be most significant. Movement into consolidated bedrock deposits may occur, but it is not expected to be a major factor in contaminant transport. Actual field investigations will be designed to verify this assumption.

Hazardous waste constituents can migrate from an improperly sited or managed facility. They can be dispersed as volatile gases to the atmosphere, or as particulates through soils to the ground or surface water, and they can be taken into the food chain by crops or animals. The rate of migration is related to the characteristics and quantities of the waste, the physical conditions of the site, the climate, and, more specifically, the weather conditions during and following disposal. Each one of these potential migration routes

259

is being assessed as part of the site investigation. However, emphasis is placed on investigative approaches to address the contamination of groundwaters and surface waters, specifically, the movement of contaminants through soil.

Review Available Information

In order to supplement the existing data base, additional desktop studies will be undertaken before initiating field work. These basic studies include evaluation reports on the site and vicinity from the U.S. Geologic Survey, USDA Soil Conservation Service, universities, regional and local planning agencies, and state regulatory agencies.

In addition, the U.S. EPA is reviewing aerial photographs of the site. These aerial photographs can provide clues as to which areas have been disturbed. They can provide information on vegetative changes resulting from waste disposal, and on disposal operations and groundwater characterization. This is necessary to ensure groundwater supplies for the area. The data indicate a predominant groundwater flow to the north with some anomalies on site related to fill activities.

White and Wine Creeks have been contaminated from waste disposal on the site. The characterization of these creeks will provide a reference point for evaluating surface waters, as well as a means of correlating groundwater flow at the site.

Subsurface borings have been excavated at the site, and they provide basic information on subsurface conditions and groundwater quality. Groundwater samples obtained in August 1980 contained elevated levels of lead, zinc, toluene, chloroform, trichloroethylene, and methylene chloride. U.S. EPA lagoon sediment data indicate that other contaminants such as cadmium, copper, chromium, benzene, xylenes, and others exist on site. In addition, surface water samples obtained from the White and Wine Creeks exhibited elevated levels of these contaminants.

The depth and velocity of groundwater flow are crucial factors in assessing contaminant migration. The unconsolidated glacial deposits which occur within Oswego County, particularly outwash deposits, serve as major sources of groundwater.

Soil Survey

Soil investigations will be undertaken using recent aerial photographs: blank and white stereo pairs (Scale 1" = 40'), high resolution color photographs, and false color infrared photographs. These remotely sensed data will be examined to determine areas of previous excavation or surface disturbance, changes in vegetation, areas of vegetative stress, and general site configuration drainage and landforms. The aerial photographs with superimposed topographic contours will serve as base maps for field surveillance information during site investigations.

The initial soil survey will be undertaken to familiarize site layout and configuration. Prominent surficial features will be noted as well as vegetative conditions. During site surveys, the survey team will be equipped with the safety and health equipment.

A detailed survey of existing soil conditions will be undertaken for information on potential sources of contamination. The survey will identify areas in which wastes have been deposited, based on examination of surface and subsurface soil conditions. This will involve examination of vegetation and use of hand soil augers. Photoionization equipment will be used concurrently to identify contaminated areas.

Soil morphological features as well as observed vegetative and other characteristics, will be recorded on the base map. They can provide a general picture of areas in which detailed sampling and subsurface investigations should be undertaken.

Soil test pits will be excavated to a depth of approximately 5 feet in the vicinity of major identified waste deposits. This will help determine the degree of contamination and depth to which waste constituents have migrated. A detailed morphological description and photographic record of these pits will be obtained. Soil samples will be obtained from the major soil horizons in each test pit and retained for analyses.

Subsoil Remote Sensing

Subsoil conditions at the site will be investigated by resistivity survey techniques and ground penetrating radar. Resistivity techniques will provide for general mapping of

261

contaminant plumes and groundwater. The information will
also be used as a guide in placing monitoring wells. Ground
penetrating radar will be used to map near surface conditions
in disposal areas only. These include lagoons, barrel storage areas, and staging areas. These data will provide additional information on the extent of buried wastes.

Soil Borings and Groundwater Monitoring Well Installation

The soil boring and groundwater monitoring locations
will be established to provide quantitative data on site
conditions at the PAS Oswego site. This will include identifying areas that are currently contaminated, defining
groundwater hydrology, determining the extent to which the
landfill across Seneca Street contributes to site contamination, and determining whether White Creek and Wine Creek are
hydraulic boundaries that preclude the spreading of contaminants. From this data it will be possible to assess the
extent of the contamination in the upper zone, and to determine how far contaminants may have traveled into the underlying soil stratum and bedrock.

In order to minimize program costs, detailed boring
logs and soil samples will be obtained and monitoring wells
installed in common boreholes. These soil samples will be
obtained by continuous split spoon sampling techniques, and
they will be retained for subsequent analyses.

Wells will be installed in boreholes, backfilled with
appropriate filter sand, and sealed. Casings will be tele-
scoped through the upper, highly contaminated zones to
minimize risk of cross contamination. Borehole permeability
tests will also be undertaken to help assess strata permeability. Pump tests can be run to determine the permeability
of the subsurface strate.

Sampling/Data Compilation, and Analysis

The constituents to be quantified from soil, surface
water, and groundwater samples will be selected after a more
careful scrutiny of existing site data, and the data currently
collected as part of site clean-up operations. Among the
constituents expected to be analyzed are cadmium, copper,
lead, zinc, chromium, toluene, chloroform, trichloroethylene,
xylenes, and methylene chloride.

Chemical analyses will be used to ascertain degree and potential risk due to contaminant transport. The extent of contamination will be defined, and the local groundwater hydrology characteristics will be evaluated. This information will be used to predict the future migration of contaminants, and to identify potential remedial action.

Field investigation results will be mapped to display the contamination and the site characteristics. The displays will include all water quality constituents evaluated and will identify all contaminated subsurface strata and the degree of contamination.

Upon completing the site investigations program, enough data will be available to develop the best engineering techniques to contain, treat, or dispose of contaminated materials, and to minimize any further degradation of the environment.

SUMMARY

The PAS Oswego site is an imminent public health threat. It requires immediate remedial action to remove the sources of contamination. The approach outlined here was designed to meet this objective in a safe and environmentally sound manner, using sound engineering and scientific techniques. Standard operating procedures were developed for the handling of the on-site waste chemicals and containers. An evaluation of the potential treatment and disposal options for major waste categories was performed, and a scope was prepared for the final site restoration investigations. In addition, contract documents were prepared for the removal and disposal of the drum and tank wastes, and the demolition of the existing building on-site. The remedial action program developed for PAS Oswego is an example of a feasible strategy.

When handling wastes at the site, workers will come into direct contact with hazardous wastes. Inadvertent mixing of some wastes can result in release of toxic fumes and gases, and fire or explosion. Often chemicals can be spilled during bulking and staging activities which, if not properly contained, can leak into the environment. Information concerning wastes on-site has shown that many different types of wastes will require specialized treatment and disposal.

Our experience at the PAS Oswego site has shown us that a well developed and easily implemented waste classification protocol is necessary for safe and efficient remedial actions.

Initial sampling data help determine what wastes should be segregated because of extreme hazard risks, such as shock sensitivity or radioactive features. Further sampling activities will determine waste grouping based on bulking compatibility, and eventual treatment and disposal methods for each major waste category. On-site testing of wastes should be comprehensive. This will ensure that adequate information is generated to complete waste characterization forms required by treatment, storage, and disposal facilities.

Characteristics of wastes will primarily dictate how wastes can be handled, processed, and disposed off-site. However, other factors such as recycling/reuse potential, environmental regulations, and transport distances to treatment and disposal sites, will influence the ultimate method selected for each waste type. Ideally, the most preferred method of handling wastes found on-site would be to directly recycle or reuse waste material. However, in the analyses of treatment options for various waste categories, direct reuse of wastes expected to be found on site would account for only a small percentage of the total waste quantity. The clean-up of the PAS Oswego site is a complex, multi-phase project which demonstrates the capabilities of the public and private sector in rectifying the problems at uncontrolled hazardous waste sites.

The U.S. EPA remedial action program is the first of many programs, under the Superfund law, designed to clean up abandoned hazardous waste disposal sites. It will reduce the impending hazard to public health and the environment by removing the waste chemicals that threaten to contaminate our drinking water supplies, atmosphere, and soil. This program's action, coupled with the regulations and guidelines for the transport, treatment, storage, and disposal of present and future hazardous wastes under the RCRA laws, will ensure the health, safety, and general welfare of future generations.

ACKNOWLEDGEMENTS

The authors express their appreciation to David A. Tedone and Theresa D. Perryman of CDM, for their editorial and technical assistance. We also express our appreciation to Robert Raab and Kenneth Stoller of U.S.E.P.A. (Region II) for their insightful comments and timely review.

CHAPTER 17

REMEDIAL ACTION ALTERNATIVES FOR
ABANDONED HAZARDOUS WASTE SITES

P. B. Lederman, Ph.D., P.E.
   Roy F. Weston, Inc.

Engineering and Design for Remedial Action at Hazardous
Waste Sites presents a number of interesting trade-off options.
No two sites are alike, but they can be categorized into three
generic types:

- Pits, ponds and lagoons
- Barrel and/or equipment repositories
- Burial grounds

For each of these types of sites, remedial action plans
will vary from "no action" to "off-site removal and treatment."
These options must be examined in the light of regulatory requirements, public perceptions and the availability of funds
to carry out the remedial action.

Removal and containment at Hazardous Waste Sites has been
categorized into three convenient, but sometimes overlapping,
steps:

- Emergency response
- Planned removal
- Remedial action

Emergency response is that response provided immediately
to prevent and/or minimize injuries to the public health as a
result of sudden releases. Planned removal is that removal
planned over a 6-months period to remove material from a site,
and/or provide containment, which does not require extensive
engineering analysis. This will usually consist of source
removal. Remedial action is a longer term program to insure
that the site will not be a public health problem in the future.
Generally speaking, emergency response and planned removal are
handled directly by the US EPA through their Emergency Response
Team and do not involve engineering studies to optimize the

removal scenarios. This paper will deal with the longer term approach for remedial action.

Remedial action plans are the result of a phased study. Generally, this begins with an analysis of data already assembled to whittle down the options to be developed more fully and the data which is required to support engineering for those options. This is done to minimize the extent of field investigation and, hopefully, result in a plan which can be implemented quickly. Where source removal is important, such as a site containing PCB transformers, it may be best from an institutional and cost effectiveness point of view to phase the operation into one which quickly removes the source of contamination and a second which examines the ground and ground water to determine what, if any, remedial action is necessary.

Once a decision has been made to go ahead with remedial action, based on examination of preliminary data, a field investigation will probably be required to provide additional data on:

- Extent of contamination
- Type of contamination
- Sources of material (inventory)

These results will lead to an examination of alternatives for remedial action.

An evaluation of alternatives starts with a general analysis of remedial action alternatives. These are outlined in Table I and go from simple in-situ groundwater management to removal and off-site disposal. These must all be evaluated against the "no action" option.

Generally, in-situ site management in the simplest form will include management of groundwater by providing recovery wells to insure that contamination from the site does not contaminate downgradient wells. A typical site management, shown in Figure 1, includes dike improvement, a light cover and drainage control. It does not include stabilization nor does it prevent infiltration into the source. Generally speaking, site management is a long-term commitment and has had limited acceptability.

The next alternative is in-situ stabilization, as shown in Figure 2. This provides for addition of material to the pit, pond or lagoon, or land disposal area, to insure that it can support a multilayer cap. A cover system insures that

TABLE I

REMEDIAL ACTION ALTERNATIVES

- NO ACTION

- SITE MANAGEMENT

- STABILIZATION AND CLOSURE

- ENCAPSULATION

- FIXATION

- ON-SITE TREATMENT
  - Chemical Detoxification
  - Incineration
  - Secure Landfill

- OFF-SITE SECURE LANDFILL

- OFF-SITE TREATMENT
  - Chemical Detoxification
  - Incineration
  - Recycle or Reuse

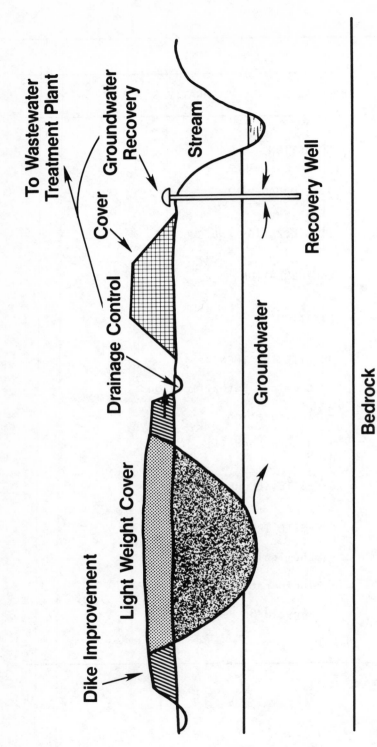

Figure 1. Site Management

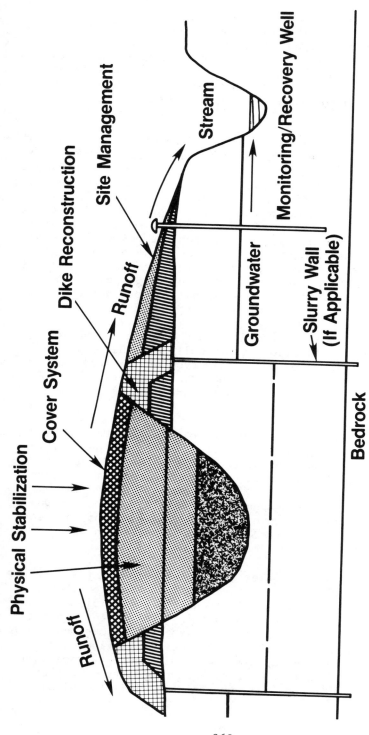

Figure 2. In Situ Stabilization

rain water runs off rather into a landfill. It may be necessary, at times, to include a slurry wall to insure that
groundwater does not penetrate through the landfill. Generally speaking, this type of system is more expensive, reaches
equilibrium in 2 to 5 years and then, requires little or no
attention. Technologies for these systems are well-demonstrated
and have been shown to be environmentally effective and acceptable.

The next alternative is on-site processing and disposal.
Encapsulation, by removing the material, putting down a liner
and replacing the material is one option, as shown in Figure
3. Developing a secure landfill on-site, physically removing
it to the secure landfill, is another. This is followed by
backfilling the previously used site, regrading and revegetation. This alternative is considerably more expensive than
in-situ management because it requires the physical removal
of all contaminated material to a previously prepared site or
the physical removal of material while the site where it was
is lined. The time for actual work on this type of process
is somewhat longer than in-situ stabilization, but the time
to reach equilibrium is the same or somewhat less.

The second on-site treatment and disposal technology
involves chemical fixation and landfilling. Generally
speaking, this involves pozzolanic type of chemical reaction
to produce a chemically stable material. Material handling
is equal to or greater than that in developing an on-site
secure landfill.

Other on-site processing technologies include incineration and reprocessing for reuse. The latter generally
tend to be much more expensive at this point and, depending
on the availability of equipment, can take many more years.

Finally, and most expensive generally, is the removal
of all the material to an off-site facility, either a secure
landfill or an incinerator. These options, again, are
extremely expensive and time consuming.

When these technologies are evaluated against each other
as is done in Table II and Figure 4, it can be seen that the
"no action" alternative, while technically feasible and low
in cost, is generally unacceptable in terms of environmental
effectiveness and the acceptability to the various institutions
concerned. On the other hand, waste stabilization, waste
fixation, and waste encapsulation tend to be environmentally
acceptable, but are more expensive and do carry with them a
greater degree of technical uncertainty. This technical

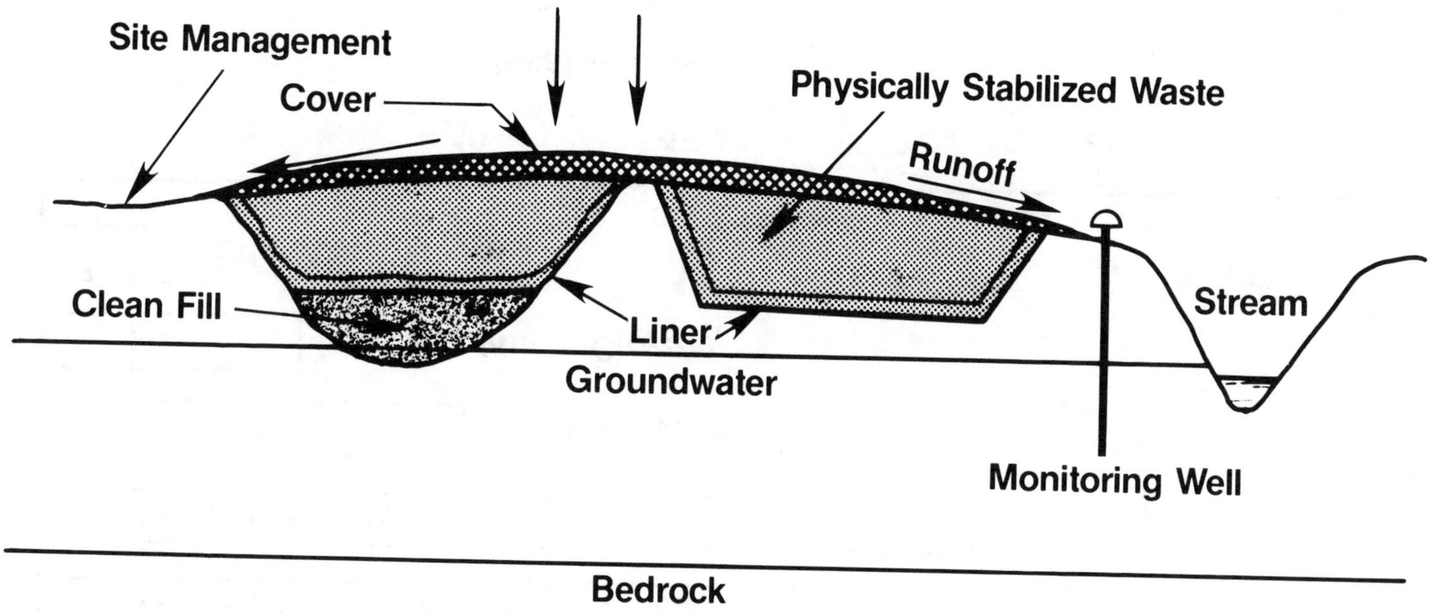

Figure 3. On—site Encapsulation

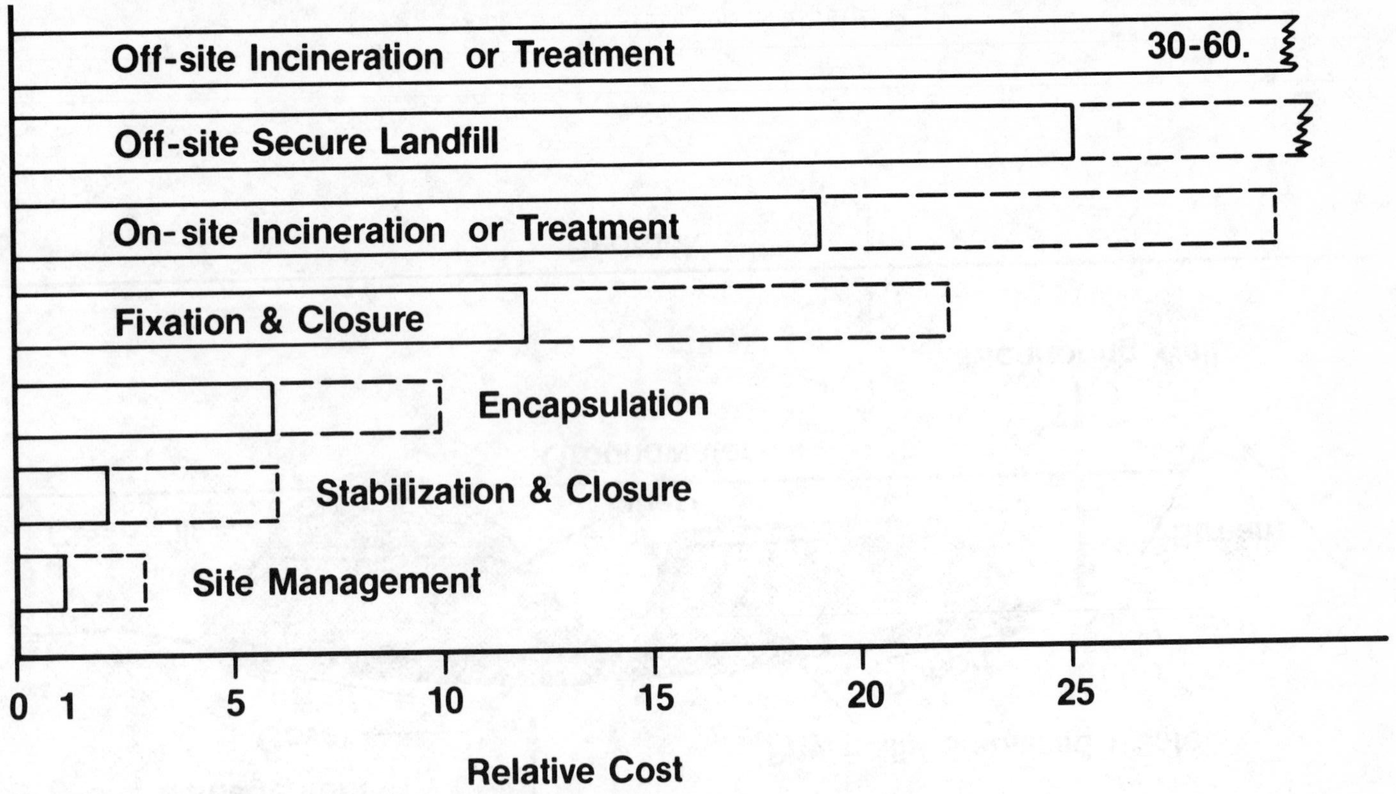

Figure 4. Remedial Action

30-60.

Off-site Incineration or Treatment

Off-site Secure Landfill

On-site Incineration or Treatment

Fixation & Closure

Encapsulation

Stabilization & Closure

Site Management

Relative Cost

0   1   5   10   15   20   25

272

TABLE II

EVALUATION OF REMEDIAL ACTIONS

	Initial Cost	Tech. Feasib.	Environ. Accept.	Time to Stabil. Site
NO ACTION	1	1	4	5
IN-SITU STABILIZATION	2	3	3	3
ENCAPSULATION	3	3	3	4
FIXATION	4	7	3	4
ON-SITE INCINERATION	5	3	2	2
ON-SITE SECURE LANDFILL	6	3	3	4
OFF-SITE SECURE LANDFILL	7	2	1	1
OFF-SITE TREATMENT	8	8	1	1

uncertainty, in many cases, can be resolved utilizing small pilot studies or field tests.

The most desirable remedial action plan from the point of view of the local institutions, that is the microcosm of the site and its immediate surroundings, is usually one of off-site processing and/or disposal. These are certainly technically feasible in that they have low technical uncertainty, but they have high costs. They are no less risky than many of the on-site options.

At this point, it is almost impossible to do a definitive risk analysis as has been traditional in the aerospace industry. Both data and time are insufficient. Yet a risk assessment is important to insure that the remedial action alternative does meet the objectives of not leaving a problem for the future. Table III illustrates such a qualitative risk assessment. It should be noted that for the site in question, which was a lagoon, the "no action" alternative presented a number of medium and high risks for failure. These can be translated into population at-risk. On the other hand, if one compares the waste stabilization, encapsulation, and fixation scenarios, one sees no high risks, 3 to 4 medium risks and several low risks. This would indicate that they are probably all equal in risk. It should be noted that there is at least one additional risk for the waste fixation and encapsulation scenarios because of the materials handling that is involved in removing the material from the initial site. Thus, if one weighs total risk by giving it a numerical value, one would select the waste stabilization option as being the lowest risk of those three. If one then looks at off-site management, one notices that there are several additional risks which involve the new site and, in particular, transportation-associated risks. The risk, then, of off-site disposal to society tends to be higher than on-site stabilization for the site in question. Naturally, if one looks at another site, particularly one which has a large number of above-ground drums, the risk assessment will change dramatically and will probably indicate that off-site removal, at least of the drum sources, presents a lower risk than leaving them and managing them on-site.

There are a number of remedial action alternatives which are suitable for implementation at abandoned hazardous waste sites. These have been discussed in their generic form, but it must be remembered that for each site, careful analysis is required to determine the most effective scenario considering the criteria of cost effectiveness, environmental acceptability, implementability & institutional acceptance.

TABLE III

RISK OF FAILURE

	NO ACTION	STABILI- ZATION	ENCAP- SULATION	FIXATION	OFF-SITE LANDFILL	OFF-SITE INCINERATION
DIKE FAILURE	H	M	M	M	L	L
OVERFLOW	H	L	L	L	L	L
LEACHING TO GROUND WATER	H	M-L	L	L	L	L
AIR EMISSION	M-H	L	L	L	L	L
MATERIAL SPILL ON-SITE	M	L	M	M	M	M
TRANSPORT SPILL	-	-	-	-	M	M

Once a scenario has been selected and approved by all parties concerned, it then becomes a very straight-forward engineering assignment to write a set of specifications for site clean-up and actually carrying out the site clean-up.

CHAPTER 18

HAZARDOUS WASTE SITE CLEANUP
ELIZABETH, NEW JERSEY

R.M. Graziano
   O.H. Materials Co.

BACKGROUND

The Chemical Control Corporation's toxic waste disposal site, in Elizabeth, New Jersey, was one of the country's most notorious. The corporation accepted for storage and disposal a wide array of drums, laboratory samples, and bulk hazardous wastes, and was shut down early in 1979 by the New Jersey Department of Environmental Protection as a result of numerous violations. After the shutdown, large quantities of PCB's, cyanides, and nitric acids were found in the loft area of one building. These extremely dangerous materials posed an immediate threat, and the State was forced to take emergency action to remove them. Both before and after it was finally shut down, the Chemical Control site was virtually a case study in abuse and mismanagement. The 3 1/2 acre dump was repeatedly cited as an example of the country's toxic waste problems; indeed, it had repeatedly been described by reporters as a disaster waiting to happen.

As all of us in this business know, disasters like this never keep you waiting long. Just over a year later, in the early hours of April 22, 1980, a warehouse containing more than 24,000 barrels of chemicals exploded on the site. Drums of toxic wastes and chemicals were blasted sky-high. Fire tore through the site and raged uncontrollably for more than 10 hours before it could be contained. Thick black smoke and ash from the fire blanketed a 15-square-mile area-including New York City--and water contaminated by the fire-fighting spilled into the Arthur Kill and Elizabeth Rivers bordering the site. Five firemen were injured in the blaze, which took five days to extinguish completely. All in all, the Chemical Control blast was one of the worst in history. The combination of its magnitude and its location posed an unprecedented threat--and an unprecedented challenge to our industry.

O.H. Materials was called in by the New Jersey Department of Environmental Protection the morning after the explosion. By that night, personnel and equipment were on-site to begin a $26 million cleanup operation that would take 14 months and more than 250,000 man-hours to complete.

I'd like to take you step by step through the cleanup, and to share with you both what we accomplished and what we learned. The Chemical Control operation posed a number of complex problems, both technical and logistical. The sheer size of the job made many conventional methods inadequate; beyond that, other more specific factors required innovations in management and technology.

I believe it was Archimedes who said in describing the principle of leverage that given a place to stand he could move the world. In the case of the Chemical Control site, we didn't really have a place to stand. Our first priority, then had to be to clear the site for our operations, to make room for equipment and on-site facilities. Based on preliminary testing of air, soil, and water, three zones were determined and established. A clean zone was delineated for housing the contractor's office and D.E.P. command post. The zone also housed lab services, eating and restroom facilities for project personnel, and was used for storage of equipment and supplies, and for refilling breathing equipment. A transition zone was set up for decontamination of clothing, changing of breathing air, and replacement of clothing. The "hot" zone, of course, was the actual working area, and was used as well for safety procedures and decontamination of equipment.

It was clear from the start that the scope and hazard of the cleanup would require extensive safety procedures. In many ways, we were working in the dark. Since much of the storage had been illegal, nobody knew exactly what materials had been stored on the site, much less where they were or what was left of them. Complicating matters further, virtually all drum labels had been destroyed in the fire, making identification impossible. Many drums were dangerously bulged from extreme buildup of pressure during the blaze. In addition, over 200 gas cylinders were found on the site, most of them unidentifiable because of fire damage to both labels and valves.

Safety

Obviously, we could not be too careful. As in any cleanup, we had to assume the worst, and use maximum precautions until further testing could be done. As a result, the sophistication of safety procedures reflected the risk and complexity of the task at hand, as well as our commitment to personal and environmental safety.

First aid stations were set up at various points throughout the site. Personnel in protective gear were also on call at all time with access to first aid kits and emergency oxygen. Constant radio contact was maintained between field personnel and staff supervisors, and each group of six or fewer workers was assigned to a foreman. A team, or "buddy," system was in effect throughout all phases of the operation.

Air monitoring systems were installed both on-site and off; mobile infra-red analyzers, portable air samplers, ionization detectors, and personal air monitors were all utilized on an ongoing basis. Infra-red analyzers were equipped with alarm systems to alert the site to the presence of cyanides, and a warning system was set up in the event of emergency site shutdown. In anticipation of such an emergency, exit lanes--both by land and by water--were cleared to ensure rapid and efficient evacuation of the site.

All of these precautions would have been useless, though, without the strict adherence of all personnel to a rigorous schedule of daily safety meetings and daily activities review. Obviously, systems and equipment can only be as effective as the people who use them; by keeping all personnel informed of and actively involved in the ongoing task of maintaining health and safety standards, we were able to conduct the entire project without a single significant injury.

We found that in addition to their value in maintaining safety standards, daily meetings also enabled management to keep abreast of all phases of the cleanup. At all times

throughout the project we knew who was doing what, when and
how, and thus were able to maintain accurate records of all
project activity. This might seem like a relatively simple
undertaking; bear in mind, however, that at any given time
as many as 88 workers were on site, around the clock.

And in addition to monitoring the work they were doing,
we were also responsible for keeping them supplied with
equipment and protective gear, providing them with sanitary
and food service facilities, and ensuring their continuing
health and safety on the job. I'm proud to say that we were
able to meet these logistical challenges creatively and
effectively by drawing on our experience in prompt deployment
of personnel and equipment, in maintaining extensive documentation
and especially on our dual emphasis on emergency response as
well as long-range organization. Our logistics management is
noted for its military precision, and the scale of the
Chemical Control cleanup required us to make use of it as
never before. I can think of no comment more appropriate,
or praise more valued, than the remark by Captain Fleishel,
then Captain of the Port of New York, who said in a
televised interview after the Chemical Control cleanup that
"if he was going to war, he would want O.H. Materials and
its logistical support behind him."

Equipment

Once work zones and procedures were established, our
first task was to restage the drums left by the explosion so
that sorting and lab analysis could begin. You can see that
one of the problems we encountered was the pileup of drums,
often 4 and 5 deep, throughout the site, making movement by
conventional means difficult and hazardous. The solution to
the problem was the development of specialized equipment.
The barrel grappler is a Caterpillar 225 backhoe whose 32-
foot boom has a gripping device lined with a rubber shield
to prevent sparking during drum removal. Because the
contents of the drums were at this point still unknown,
worker safety was a particular concern. If there is a
machine to do a job to minimize worker contact with
dangerous materials, we will build it. Thus, the grappler's
cab is equipped with a one-inch Plexiglas shield, as well as
its own air-supply and air-conditioning systems; operators
wear protective clothing and carry both fire extinguishers
and emergency escape air packs. The barrel grappler proved
to be a significant improvement, and a major technological
innovation. Not only did it prove to be a safety way to do
the job, it was also a more efficient one: an average of
600 barrels per day were safely moved by each machine.

280

Recovery technicians first segregated and labelled the drums by gross classification as empties, solids, liquids, and laboratory-packed materials, and restaged to an adjacent site which had been ordered closed by the Governor of New Jersey. Empties were crushed and removed. As the others were restaged, chemists worked around the clock in an on-site mobile laboratory to identify the contents of each drum. The lab itself, designed by O.H. Materials, is housed in a 42-foot trailer and contains a full complement of state-of-the-art equipment for fast and accurate analysis of simple and complex samples. The advantages of on-site laboratory facilities in a cleanup the size of Chemical Control are self-evident over 30,000 samples were analyzed in the course of the project, at minimum expense and with minimum turnaround time.

Bulking

Sophisticated analytical capability was crucial to the development compatibility protocols for the bulking of compatible wastes. This is an innovation that I'm sure will be of interest to all of you, and I plan to go into it in some detail shortly. Before compatibility testing could be undertaken, though, samples from each drum were initially tested for hazardous materials. Despite the emergency removal by the state a year before of some 10,000 barrels of toxic materials from the site, in the course of the cleanup we discovered and removed over 500 pounds of explosives and other health and environmental poisons. These include peroxides, crystallized picric acids, azides, and dinitro-tuolenes. As hazardous materials were being removed, extensive qualitative analyses of contents of the remaining drums were taking place. In addition to identifying the substances, complex testing was done to ascertain their compatibility with one another. The goal of this compatibility series was the on-site bulking of compatible wastes. As we all know, the removal of waste materials is a costly business. The time-honored method of overpacking damaged drums with new ones, and then loading them onto a truck for off-site burial could end costing your client around $200 per drum, when you figure in all the costs of all the steps involved. In addition, many of the overpacked drums on your truck are very likely to be less than full. You might be paying to remove a drum that's more than half empty.

Once again, the magnitude of the Chemical Control operation and the strong need for cost efficiency convinced us that there had to be a better way. On-site bulking, as implemented at Chemical Control, is, like many of the best

ideas, a very simple one. It's a process that is equally
applicable to solid, liquid, and lab-packed materials, and
one which dramatically cuts the costs of waste disposal.
Put very simply, instead of overpacking damaged drums,
whenever possible we combined their contents and repacked
them in bulk.

For example, liquid wastes were initially separated
into neutrals, acids and bases. Neutrals, those liquids
whose compatibility was favorable were emptied into a
bulking chamber on-site, mixed together, and siphoned into a
tank truck for removal to an incinerator off-site. This way
we eliminated the need for repacking and loading which are
both time consuming and expensive. Furthermore, we were
assured that liquid waste removal was being accomplished
efficiently instead of a truckload of 50 drums (not all of
which are full to capacity), we had a tank truck carrying a
capacity load of 5 or 6 thousand gallons. By this means we
greatly reduced transportation and disposal costs, as well
as expenses for handling.

The process for bulking of compatible solid wastes
proved to be similarly cost-effective. Once compatibility
has been ascertained, solid and semi-solid wastes were
emptied into a concrete mixing pad, and mixed with fly ash
to neutralize pH. Once neutralized, the materials were
loaded into trucks by means of backhoes. The trucks have
been lined with visqueen and sand for absorbency. Tailgates
were sealed, the beds were tarped, and the waste trucked
away to an EPA-approved landfill. Once again, repacking was
eliminated, and a great deal of time and money were saved.
We estimate that the cost of solid waste disposal by this
method was about $125 per ton--as compared with the $200
per barrel I mentioned before.

Lab-packed Materials

Lab-packed materials found at Chemical Control were
also processed in terms of compatibility. We set up a
portable building on-site to house the processing,
establishing specific areas for segregation of contents.
Chemists identified the contents of the lab-packed drums,
and tested them for compatibility. Once this was
accomplished, compatible materials were repacked in 55-
gallon D.O.T.-approved drums, and transported to a landfill
off-site.

We removed over 100,000 pounds of waste materials from
the Chemical Control site during the cleanup. Thus, it's
based on pretty substantial experience that I can tell you

that the bulking techniques we developed and utilized throughout the project were consistently safe, efficient, and cost-effective. Let me stress once more that safety is of paramount importance to us, as it is to all of you. Efficiency and cost-effectiveness are also vital objectives. Through creative use of compatibility analysis, we were able to meet these objectives with no sacrifice to our stringent standards of health and safety.

By mid-October, all drums had been removed from the Chemical Control site--nearly 60,000 of them. At this point the buildings left on-site were demolished. Rubble was sorted and thoroughly tested for contaminants and other hazardous materials. The rubble that turned up clean was hauled to a landfill; some had to be decontaminated prior to disposal. Debris that chemists found to contain toxic substances was removed to an approved hazardous waste disposal site.

Then the last visible remnants of what had been the country's most dangerous toxic waste disposal site were a number of large storage tanks. Their contents were unknown. Technicians analyzed samples, and found residual oils and sludges, as well as PCB's, which crew members removed and disposed of safely. The tanks themselves were then dismantled, decontaminated, and cut up for scrap. By late fall of 1980, cleanup operations were complete.

The Company was retained by the New Jersey Department of Environmental Protection for treatment and recovery of ground water on the Chemical Control site. Before going on to this second phase of the project, though, I'd like to summarize the operations we've covered so far, and to take a look at the big picture. Obviously the operations we've seen here could only have been undertaken in the context of an overall management strategy. Once again, I have to stress magnitude of the situation, and the complexity of the problems we were faced with. If I sound like a broken record, it's because the Chemical Control site disaster broke so darn many records. Not only had we never seen anything like it--there had never been anything like it.

When we arrived on-site, the rubble left by the explosion and fire contained close to 60,000 drums, whose contents were completely unknown. To describe the situation as high-rise would be putting it very mildly: many of the drums were crushed, or leaking, or heat expanded; their contents

could at any time explode or spew out toxic fumes. And the
damage was extensive as it was intensive--the problems were
as big as they were bad. We were able to solve them with a
game plan that called for highly sophisticated logistics
management, and innovative use of state-of-the-art equipment
and accountability.

To give you a concrete idea of some of the personnel-
related logistics we were dealing with, let me repeat that
the project was one in which as many as 88 people were on
site per day, clocking a total of over a quarter of a million
manhours. No other cleanup in history has ever involved so
many people, doing so much, for so long, and under such
hazardous conditions. Perhaps the numbers become more
meaningful when we consider that the average working American
spends about 2,000 hours a year on the job. Employees at
the Chemical Control site spent 4,000 hours per year at
work. I might add, parenthetically, that a great many of
those hours were clocked in 30-minute shifts, by technicians
wearing heavy protective gear, at times in temperatures
above 90°.

I'd now like to reflect briefly on the technological
innovations we've discussed, and their impact on the
effectiveness of the Chemical Control cleanup. The problems
presented by the Chemical Control site provided a unique
opportunity for growth and advancement in the field of
hazardous waste removal. The problems required solutions,
and the solutions proved to be major breakthroughs for O.H.
Materials and for industry as a whole. Let me recap for you
three particularly significant innovations. First, the
barrel grappler was an important step forward in the
development of mechanized handling systems. Designed to
solve the problem of restaging piled-up drums on the site,
the grappler also provides maximum safety for its operator.
Incidentally, during the course of the project, a further
refinement was added to the grappler's design: a quarter-
inch Plexiglass shield was installed to cover the cab's one-
inch shield. In the event of an explosion,the heavy shield
protects the driver from injury, while the thinner shield
protects the heavy one from damage. By this means, replacement of the far more expensive heavy Plexiglass is kept to a
minimum.

Second, the mobile laboratory unit, adapted to site
specifications greatly enhanced the effectiveness of our
efforts. It helped solve two of the biggest problems we
faced at Chemical Control: time and accuracy. Speedy and
highly accurate on-site analysis of over 30,000 samples
enabled us to reduce significantly the time spent in materials

identification.

Third, the methods developed for bulking of compatible wastes represent a technological quantum leap. Originally designed to address problems of large-scale disposal and cost control, these techniques were exploited extensively in the field, and as I've told you, they proved consistently efficient and cost effective.

Let's move along now to the final stage of the Chemical Control cleanup, which called for groundwater recovery and treatment. Damage from the explosion had resulted in extensive contamination throughout the site. In a couple of instances, soil samples retrieved were found to have contamination levels as high as half a million ppm. To give you an idea of the extent of the problem, throughout much of the Chemical Control operation a lighted match dropped almost anywhere on-site would have caused a ground explosion. In addition, the surrounding rivers had been contaminated by water used in fighting the fire.

Naturally, the variety of contaminants stored on the Chemical Control site meant that we faced an equally complex range of dangerous substances in the groundwater. To cope with the range and degree of the damage, on-site treatment strategy was developed which included an injection and recovery system and a multi-phase treatment series. The groundwater recovery system ran close to 2,000 feet--95% of it underground. Winter was fast approaching by this time, so a 40 x 120 foot heated portable building was set up to contain water treatment equipment, and piping was heated with ductwork. Construction of the entire recovery and treatment system including extensive banking of the site with gravel to prevent erosion, took roughly one month.

While construction was underway, complex soil and water samples were flown to our headquarter's laboratory for analysis. The highly sophisticated GC/MS technology used had proven extremely effective, particularly in terms of high resolution identification of complex samples, so we were confident of obtaining quick and accurate results.

We used three-inch PVC pipe at recovery points throughout the site, through which groundwater was vacuumed into the six-inch leader pipe which carried it into the treatment system. Vacuum receivers shut off automatically when they reached capacity, and pump the collected water into a chemical mixing tank, where liquid caustics were added to neutralize pH. Then, the water moved to a clarification unit, and flocculent was added to precipitate suspended solids and heavy metals. These solids were collected at the bottom of the unit for collection and disposal later. The final non-filtration treatment returned the water to a closed unit for sparging. Groundwater then was pumped to a volatile stripping chamber, where volatile organics could safely be drawn off in a closed system. In this process, organics were vaporized by means of heat induction, and then recondensed into drums for disposal.

Carbon Filtration

Only at this point did we use carbon filtration. An important goal of the treatment system we developed was to minimize the use of this expensive process; by using a series of other methods first, we felt we could realize substantial savings. Thus, carbon filtration was only used on water which had already undergone extensive refinements by other means. The activated carbon in the 2-cell system we used lasted far longer as a result, and we were able to keep costs to a minimum.

The treated water was then sampled, tested, and discharged. Wastes gathered during treatment of the water were collected and safely disposed of. The Chemical Control cleanup was finished.

Like the disposal phase of the Chemical Control cleanup, the water-treatment of the site is noteworthy for two important reasons: The extent of the problems, and the ways in which they were solved. As far as extent is concerned, as I said before, contamination levels were extremely high when treatment began. We didn't consider that our job was done until you could walk safely anywhere on the site without protective clothing, and dropping as many lighted matches as you wanted to while you did. And you couldn't do this until more than six and a half million gallons of water had been effectively treated. Solving the problems once again required us to develop new techniques. Two of them are of particular interest; both of them helped us greatly. One of the problems we were faced with in treating the Chemical Control groundwater was the fact that the site was bordered by water--the Elizabeth river and the Arthur Kill river. We needed to control water on the site to prevent leaching. So engineers devised a means of counteracting the tidal influences, creating a hydrodynamic surge which caused the contaminated groundwater to rise with the tide, and thus minimize the risk of leaching onto the site.

The other innovation was a significant breakthrough in on-site sample analysis: for the first time, GC/MS technology was applied on site. The mobile laboratory was outfitted with a specially adapted GC/MS with full library search capabilities, and in the latter days of the water treatment phase was used to monitor decontamination in each stage of the process. As you can imagine, having GC/MS analytic capability available on-site was an invaluable tool for the treatment effort. It was also completely without precedent. In fact, we were told it couldn't be done. And then we did it. And now that we've proven that it can be done, we plan to do it again. And we plan to do it even better.

The same can be said of all innovations developed for the Chemical Control cleanup. As we've seen the project pushed us to the limits of our ingenuity. We did things which had never before been done, on a scale which had barely been imagined--because that's what it took. It proved to us that our ingenuity and resources were not so limited as we might have thought. And it reminded us that, as New Jersey Commissioner, Jerry English, told our president after it was all over, "impossible jobs can be accomplished with commitment and dedication.

I know that all people in this field share with me a sense of commitment to the job we do, as well as a strong sense of pride on our accomplishments. It's a dirty job, and certainly a dangerous one. It's frequently a frustrating one, too-ideally, funds should be available to the extent that they're needed for doing the job properly. All too often, they simply are not. So we have to do the best possible with the resources that are available to us. Sometimes it seems like a juggling act, balancing high standards with low dollars, cost-efficiency with personal and environmental safety. As we've seen in this case study, though, it can be done. And we'll do it again. And we'll do it even better.

CHAPTER 19

SLURRY TRENCH CUTOFF WALLS
FOR HAZARDOUS WASTE ISOLATION

David J. D'Appolonia
  Engineered Construction International, Inc.

INTRODUCTION

Slurry trench cutoff walls have been used extensively in the past decade as foundation and embankment cutoffs for water-retaining structures and to control seepage into excavations for dewatering purposes. More recently, slurry trench cutoffs have been successfully employed to prevent underground seepage of liquid wastes and to isolate leachate from solid wastes from contaminating groundwater. Advances in technology have enabled construction of exceptionally low permeability cutoffs which remain stable in a contaminated environment and resistant to chemical attack by the waste being contained.

Enactment of the Resource Conservation and Recovery Act resulted in stringent requirements for new disposal areas and imposed new requirements for upgrading isolation of existing and abandoned landfills. In many cases, slurry trench construction is the only practical means of isolating existing waste disposal areas short of excavating and transporting in-place waste materials to new disposal sites.

This paper reviews the various systems for isolating hazardous waste materials and points out the geological conditions where slurry trench construction is most applicable. The important elements of slurry trench design and investigation necessary to insure an effective waste isolation barrier are reviewed. These include mainly the connection between the slurry trench cutoff and the underlying aquaclude and the design of the slurry trench backfill material itself. Important considerations include not only the permeability of the isolation barrier, but also the long-term effects on the engineering properties of the

barrier resulting from contact with or leaching by the
contained hazardous waste material.

WASTE ISOLATION SYSTEMS

The purpose of a waste isolation system is to create
an impervious barrier between the waste and the surrounding
ground, groundwater, and atmosphere. The general procedure
is to form a containment vessel having impervious sides and
bottom within which the waste is stored. Ultimately, when
the disposal area is abandoned, the top is closed, thereby
completing the containment vessel.

Two basic schemes are possible as illustrated in
Figure 1. One involves constructing a lining on the bottom
and sides of the waste disposal area. The lining may consist of a compacted clay blanket, a synthetic membrane,
concrete, or any other material that is impervious and will
not be degraded by the waste being contained. The second
method makes use of a naturally occurring impervious material within the geologic profile to serve as the bottom of
the containment vessel. In this system, the perimeter walls
of the containment structure extend from the surface and tie
into the underlying impervious strata. The walls can be
constructed using any appropriate impervious material
including compacted clay cutoff trenches, synthetic
membranes, etc., or slurry trench cutoff walls.

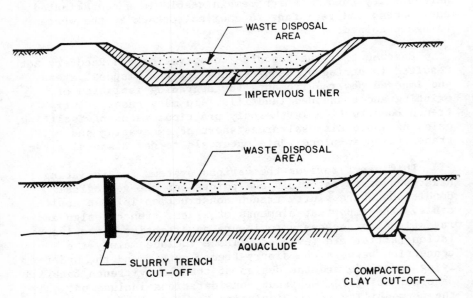

Figure 1. Waste Isolation Systems.

Depending on the waste material being stored, aquifer dilution characteristics, and water quality requirements, a very small amount of seepage from the containment vessel may be permitted. In other cases, no discharge from the waste disposal area can be tolerated. In this latter case, a leachate collection system or a pumping system must be used. A leachate collection system involves a double lining with a pervious material placed between the two liners. Leachate passing the primary liner is prevented from escaping the containment by the second liner and is collected in the pervious material between the two liners and removed. A pumping system can be employed when the waste disposal area is below groundwater level and does not require a double liner. In this system, the fluid level inside the containment is maintained slightly below the groundwater level outside the containment. Therefore, the flow is into the containment and no contaminated fluid escapes.

There are generally very few options available for upgrading existing landfills, sludge ponds, etc., that are not adequately isolated and are causing groundwater contamination. One option is, of course, to excavate and remove the waste and the contaminated ground, placing the excavated material in a new disposal facility. However, a much more attractive alternate is available if an impervious natural barrier exists at a reasonable depth below the ground surface. This alternate involves constructing a vertical barrier around the existing waste disposal area as depicted in Figure 2. A slurry trench cutoff is ideal for this purpose because the vertical barrier can be constructed from the ground surface without uncovering or disturbing the waste.

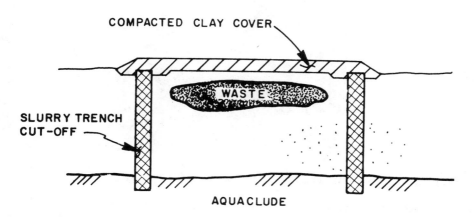

Figure 2. Isolation of Existing Buried Waste.

SLURRY TRENCH CUTOFF CONSTRUCTION

A slurry trench cutoff wall is constructed by excavating a narrow trench or slot (usually 3 ft wide) through the pervious deposits and keying into an underlying aquaclude. Trenches are normally excavated with a backhoe or, at greater depths, with a clamshell. The sides of the trench are prevented from collapsing by keeping the trench filled with a bentonite slurry during excavation and prior to backfilling. Provided that the slurry level is maintained near the ground surface (Figure 3) and at least 2 or 3 ft above the groundwater level, trench depths of 100 ft or more can be excavated and will remain stable. The only significant function of the bentonite slurry is to maintain the stability of the trench walls during excavation until the trench can be backfilled.

Figure 3.   Slurry Trench Excavating Equipment.

After excavation is complete, the trench is backfilled with an engineered material which forms the cutoff wall. The backfill material is designed to have the required permeability characteristics and to be resistant to attack and degradation by the waste materials being contained. A more complete discussion of slurry trench construction procedures is provided elsewhere.[1,2]

The engineered backfill material may consist of concrete, various cement based grout-type materials or a blend of select soil and bentonite clay. By far the least expensive and most effective type of backfill material for cutoff wall applications consists of a blend of sand, clay, and bentonite worked into a paste-like consistency before being placed in the trench. This material is referred to as soil/bentonite (SB) backfill. An SB backfill can be readily designed to have a permeability of 10^{-8} cm/sec (an order of magnitude less than current EPA requirements for impervious cutoff barriers) and can be designed to be virtually unaffected by contact with or leaching by most chemical or industrial wastes.

An SB backfill material is usually mixed on the ground surface adjacent to the trench by tracking and blading with a bulldozer. In special circumstances where space does not permit or where materials are especially difficult to blend, mechanical batchers are used. The backfill material is prepared by blending the soil ingredients with bentonite and slurry until the mix is homogeneous and the material has the proper consistency. The material is then placed in the trench at the point where previously placed backfill rises to the ground surface (Figure 4). The slope of the backfill in the trench advances by a combination of mud waving and sliding down the face of the previously placed backfill, displacing the slurry used to stabilize the trench during excavation. The ideal consistency for backfill placement is a paste having a water content slightly above the liquid limit of the sand-clay-bentonite backfill mix.

ELEMENTS OF SLURRY TRENCH CUTOFF DESIGN

The principal elements of slurry trench cutoff design involve the connection between the cutoff wall and the underlying aquaclude and the SB backfill material itself.

Slurry Trench Connections

In the few instances where slurry trench cutoff walls have failed to perform to design expectations, experience has shown that the failures have been due either to imperfect connection between the slurry trench cutoff and the underlying aquaclude or failure to completely excavate the slurry trench, thereby leaving zones of unexcavated pervious

Figure 4. Mixing and Placing Backfill.

material above the aquaclude. Providing an adequate key
into the aquaclude is of paramount importance in both design
and field quality control.

When keying a cutoff trench into clay materials, a
minimum key depth of at least 3 ft is required to achieve an
adequate connection. As shown in Figure 5a, it is essential that the key be deep enough to penetrate any pervious
lenses, weathered zones, desiccation cracks, or other geological features that might permit seepage under the cutoff.

An effective connection between a cutoff trench and
rock is often difficult. The key must obviously be sufficient to penetrate broken zones and the weathered rock
surface. Often these materials can be excavated with trench
digging equipment. However, excavation into rock, even if
the rock is jointed and fractured, is often impossible
except by using percussion tools or other equipment which
fractures and breaks the rock. Therefore, it is more
appropriate, in some cases, to grout the contact between the
slurry trench cutoff and the rock (Figure 5b) than attempting to advance the cutoff into rock. Prior to backfilling a
trench keyed to rock, positive means such as airlifting
should be employed to assure that all loose material and
cuttings have been removed from the excavation.

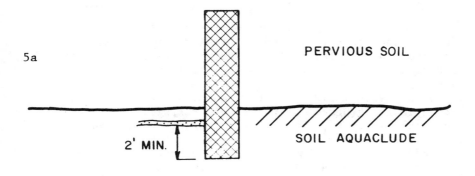

5a

PERVIOUS SOIL

2' MIN.

SOIL AQUACLUDE

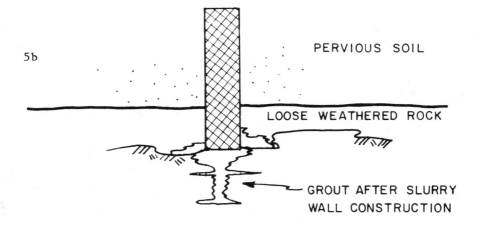

5b

PERVIOUS SOIL

LOOSE WEATHERED ROCK

GROUT AFTER SLURRY
WALL CONSTRUCTION

Figure 5. Slurry Trench Bottom Keys.

Sometimes slurry trench cutoffs are used in connection
with dikes constructed above grade. Depending on available
dike construction materials, it may be most economical to
first construct the dike and then construct the slurry
trench cutoff through both the dike and the foundation, as
shown in Figure 6a. When impervious dike construction
materials are available, it is usually more economical to
construct the slurry trench through the foundation and then
construct the impervious dike over the slurry wall (Figure
6b). In this latter case, the connection between the slurry
trench and dike is an extremely important detail.

A compacted clay fill cannot be connected to an SB slurry trench cutoff; instead, the slurry trench must be connected to the clay fill. An appropriate detail is shown in Figure 6. First a clay blanket (minimum 2 to 3 ft thickness) is placed. The blanket should extend at least 10 ft on the downstream sides of the trench. Then the slurry trench is constructed. Finally, the clay fill is raised above the slurry trench, with the connection assured between the clay blanket and the overlying clay fill. A transition zone of wet plastic clay should be provided directly over the slurry trench for strain compatibility.

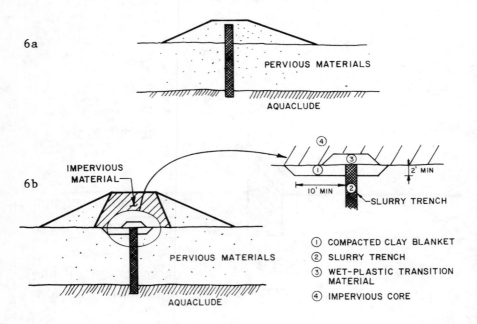

Figure 6.   Slurry Trench Top Connections.

Soil Bentonite Backfill Design

A schematic cross section of a completed SB slurry trench is shown in Figure 7. The bentonite slurry which penetrates the ground and the bentonite filter cake that is deposited on the walls of the trench during excavation contribute to the effectiveness of the cutoff. However, for the low permeability cutoffs required for hazardous waste isolation (10^{-7} to 10^{-8} cm/sec), the beneficial effect of the bentonite filter cake on the overall permeability of the cutoff is basically negligible.[1] Thus, for practical purposes, the permeability of an SB cutoff is equal to the permeability of the backfill material.

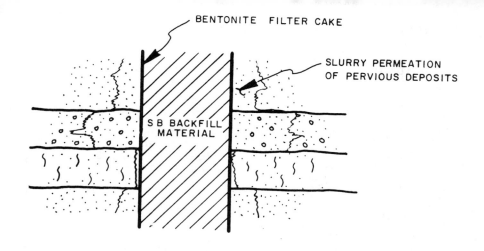

Figure 7. Schematic Cross Section of SB Slurry Trench
          Cutoff.

The permeability of an SB backfill material is dependent on both the soil gradation and the quantity of bentonite
used in blending. Both factors are important. Figure 8
plots the relationship between SB backfill permeability and
the quantity of bentonite in the blend for several different
types of soil. The amount of bentonite normally added to
the SB backfill by way of the slurry used to sluice the soil
during blending is about 0.75 to 1.25 percent, depending on
the initial water content of the soil. Therefore, if higher
bentonite contents are required, dry bentonite must be added
to the mix. Figure 8 demonstrates that low permeability can
be achieved either by adding large amounts of bentonite to
relatively coarse grained soils (silts, sands, and gravels)
or by utilizing significant amounts of clay material in the
blend and relatively small amounts of bentonite. For example, a silty sand backfill material will generally have a
permeability one to two orders of magnitude greater than the
permeability of a clayey sand backfill if both blends have
the same bentonite content. As discussed subsequently, use
of ordinary clay materials is technically preferable to
using large quantities of bentonite, because ordinary clay
is significantly less affected by permeation with most
chemical and industrial wastes than is bentonite.

The dominant importance of soil gradation on SB backfill permeability is demonstrated in Figure 9 which plots
permeabilty of a large number of SB backfill materials as a
function of fines content (percentage of material passing a
No. 200 mesh sieve). Most data points represented in the
figure are for samples prepared in the field using field

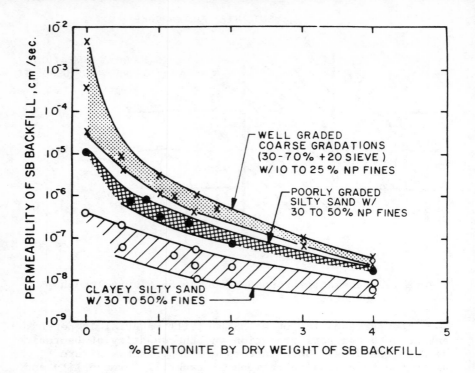

Figure 8. Relationship Between Permeability and Quantity
of Bentonite Added to SB Backfill.

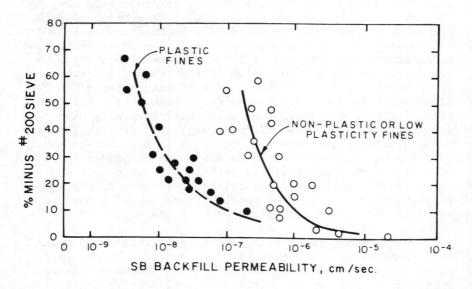

Figure 9. Permeability of Soil-Bentonite Backfill Related
to Fines Content.

blending techniques. The bentonite content of the samples is approximately 1 percent. The data show distinct families of curves, corresponding to the plasticity of the fines used in the backfill blend. Very low permeability is achieved by using blends containing more than 30 plastic fines.

Permeation of an SB backfill material with contaminated water generally leads to an increase in permeability. Thus, even though a low permeability backfill is used, special design considerations must be employed to assure that the cutoff wall will be permanent and not adversely affected by the contained waste material or leachate from the waste.

Two mechanisms can contribute to permeability increase of an SB material leached with a contaminant liquid. First, the soil minerals themselves may be soluble in the permeant, which can lead to a loss of solids and a corresponding increase in pore volume and soil permeability. Second, pore fluid substitution may lead to a smaller double layer of the partially bound water surrounding the hydrated bentonite or other clay particles, reducing the effective size of the clay particles that clog the pore space between soil grains, and thereby increasing the size of the effective flow channels in the soil skeleton. Two independent factors associated with the pore fluid substitution can contribute to this second mechanism: (1) the nature of the pore fluid affects the difference in electrical potential between the clay particle and the free-pore fluid which controls how tightly the double-layer water is held; and (2) the sodium ions associated with the sodium montmorillonite (bentonite) readily exchange with multivalent cations carried in the pollutant which also leads to a smaller double layer.

The permeation time required for the changes associated with pore fluid substitution to be completed is relatively short. Once a sample has been permeated by a volume of pollutant equal to about twice the volume of the pore fluid in the sample, the initial pore fluid has been, for the most part, leached out and the new pore fluid is essentially the pollutant. The time required to complete the cation exchange is related to the sodium cation exchange potential of the bentonite and the concentration of exchangeable cations carried by pollutant. Since sodium readily exchanges with multivalent cations such as calcium, magnesium, and heavy metals, the exchange is typically complete once an equivalent number of ions are supplied by the permeant to satisfy the total cation exchange capacity of the bentonite.

Once both the pore fluid is substituted and the cation exchange occurs, steady state conditions prevail and the permeability remains constant at a higher value associated with the new pore fluid and the new cation montmorillonite (or clay). These points are illustreated by Figure 10 which plots permeability of sand-bentonite mixtures (hydrated with fresh water) as a function of permeation time with a sodium salt and a calcium salt solution. The time scale in the plots is expressed as either the ratio of the leachate volume to the pore fluid volume, or the ratio of exchangeable cations supplied by the permeant to the cations required to satisfy the total cation exchange capacity of the bentonite. In both cases, permeability increases until the pore fluid substitution and/or the cation exchange is complete and then remains constant at a higher value.

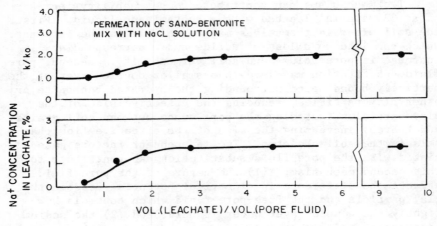

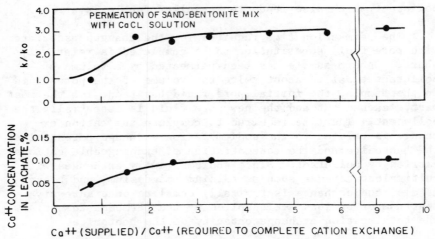

Figure 10. Increase in Permeability of Sand-Bentonite Mix Due to Leaching with Salt Solutions.

The type of bentonite used in preparing an SB backfill does not appear to have a signficiant effect on the permeability increase due to leaching with many contaminants. Table I presents results of leaching tests on bentonite filter cakes using a variety of contaminants. In these tests, various types of bentonite were hydrated in fresh water and filter cakes were formed from the slurry under identical conditions. The filter cakes were then leached until steady state conditions were reached to determine the permeability increase due to leaching. For the several cases tested where comparisons have been made, the bentonites tested all performed similarly.

Table I. Increase in Permeability of Bentonite Filter Cakes Caused by Leaching with Various Pollutants

Permeant	Final Permeability/ Initial Permeability			
	SB 125	NPB	SS 100	M 179
Lignin in Ca^{++} solution	1.9	1.5	2.5	1.4
NaCl based salt solution (conductivity 170,000)	2.7	1.8	2.7	
Ammonium Nitrate (10,000 PPM)	1.8		2.8	
Phenol and salt solution (conductivity 30,000)	1.4	1.5		1.4
Acid mine drainage (pH ~ 3)		1.5	1.3	
Calcium and magnesium salt solution (10,000 PPM)	2.9	3.2	3.2	

Bentonites: SB 125 Slurry Ben 125
NPB National Premium Brand
SS 100 Saline Seal 100
M 179 Dowell M 179

If a small amount of bentonite clay is used in an otherwise sand matrix, a piping failure may occur due to leaching because the shrunken bentonite particles, having smaller double layers, are physically washed out of the matrix. Piping failures do not occur even under high gradients (300 or more) provided that the backfill mix contains about 20 percent plastic fines.

The magnitude of the permeability increase associated with pore fluid substitution and cation exchange depends on two factors: the difference in chemistry between the initial and final pore fluid; and the sensitivity of the clay to the pore fluid chemistry change. Sodium bentonite is the most sensitive clay and hence undergoes the largest change in properties when permeated with a contaminant. Alluvial or lacustrine clays apt to be found near most sites are much less sensitive than bentonite and, therefore, undergo less change with contaminant leaching. Thus, a properly designed SB backfill that is contaminated with the pollutant during preparation is equal to or preferable to an uncontaminated backfill because a smaller change will be induced in the SB material during subsequent permeation with the contaminant.

Figure 11 presents data for two SB samples which have the same gradation (sand with 10 percent clay and 1 percent bentonite slurry). The only variable is that 15 percent by weight of a 3 percent calcium-salt solution was added to one air-dry clay-sand sample and 15 percent by weight fresh water was added to the other clay-sand sample prior to blending with the bentonite slurry. Both samples were then subjected to long-term permeation with the salt solution.

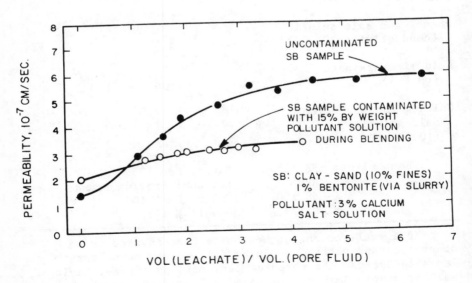

Figure 11.   Effect of Precontamination of SB Material on Permeability Increase Due to Leaching.

The initial permeability of the two samples was nearly the same but the change in permeability due to leaching with the pollutant was somewhat greater for the initially uncontaminated sample. Similar tests using samples having a higher clay content and a higher level of contamination have shown similar results. When the clay content of the SB backfill is more than about 20 to 30 percent, the difference in permeability characteristics associated with using contaminated or uncontaminated samples is usually negligible. The importance of this result is that contaminated soil excavated from a slurry trench constructed around an existing landfill can be safely incorporated in the SB backfill and does not necessarily need to be wasted.

Table II gives a qualitative summary of the effect of permeation by various pollutants on the permeability of SB backfill mixes containing at least 30 percent fines. The table provides only a qualitative approximation of the likely effect of leaching. Long-term permeability tests should always be conducted using the specific SB materials from the site and the actual pollutant in designing a cutoff. Nevertheless, the results of a large number of experiments on a variety of materials using a range of representative pollutants indicate that a well-graded SB material containing more than 30 percent plastic fines and about 1 percent bentonite exhibit only a small increase in permeability when leached with many common contaminants. A permeability increase of a factor of 2 to 4, if considered undesirable, can be reduced by addition of more clay or by increasing the plasticity of the fines contained in the backfill blend. Increasing the clay content will both reduce the initial permeability and the magnitude of the increase.

Larger permeability increases may be associated with leaching by strong acids and strong bases. Figure 12 shows results of long-term permeation tests using 1 percent solutions of hydrochloric acid and sodium hydroxide. Permeability increases due to leaching with the solutions are probably the result of dissolution and alteration of the bentonite and the soil portion of the SB material. The strong base is more detrimental than the strong acid (possibly because amorphus silica is soluble in highly caustic solutions). In the unusual case where the pH of the waste material to be contained by an SB cutoff is less than 12 or greater than 11, a quantitative analysis of the effect of prolonged permeation should certainly be conducted. Permeation tests should be extended for a long time period and the leachate analyzed to determine the effect of solutioning.

Table II. SB Permeability Increase Due to Leaching with Various Pollutants

Pollutant	SB Backfill (Silty or Clayey Sand) 30 to 40 Percent Fines
CA^{++} or Mg^{++} @ 1,000 PPM	N
CA^{++} or Mg^{++} @ 10,000 PPM	M
NH_4NO_3 @ 10,000 PPM	M
Acid (pH>1)	N
Strong Acid (pH<1)	M/H*
Base (pH<11)	N/M
Strong Base (pH>11)	M/H*
Benzene	N
Ethylene Dichloride	N
Phenol Solution	N
Sea Water	N/M
Brine (SG=1.2)	M
Acid Mine Drainage ($FeSO_4$ pH ~ 3)	N
Lignin (in Ca^{++} solution)	N
Organic residues from pesticide manufacture	N
Alcohol	M/H

N – No significant effect; permeability increase by about a factor of 2 or less at steady rate.

M – Moderate effect; permeability increase by factor of 2 to 5 at steady rate.

H – Permeability increase by factor of 5 to 10.

* – Significant dissolution likely.

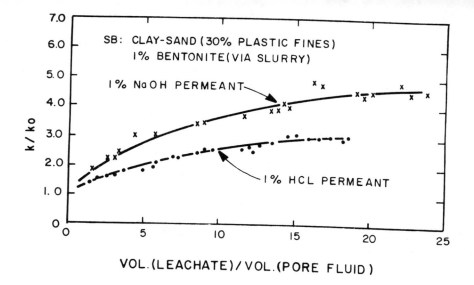

Figure 12. Effect of Long-Term Leaching with Strong Base and Strong Acid on Permeability of SB Backfill.

SUMMARY

Soil-bentonite slurry trench cutoff walls are an effective tool for isolating chemical and industrial waste materials that are disposed in landfills or ponds. Using current state of the art, permanent cutoffs that are resistant to degradation by waste materials and having a permeability of 10^{-8} cm/sec can be achieved and used with confidence.

Important design considerations for SB cutoff walls include the connection details between the cutoff and the underlying aquaclude and the blend of soils and bentonite used for the backfill. In order to obtain a low permeability, containment-resistant backfill, a high percentage of fine plastic material must be used. A clayey sand or sandy clay containing 30 to 60 percent fines blended with bentonite slurry is usually satisfactory for most waste isolation applications.

REFERENCES

1. D'Appolonia, D. J., "Soil-Bentonite Slurry Trench Cutoffs," Journal of the Geotechnical Engineering Division, ASCE, April 1980.
2. Boyes, R. G. H., Structural and Cutoff Diaphragm Walls, John Wiley and Sons, N.Y., 1975.

305

CHAPTER 20

CLAY LINER PERMEABILITY –
A PROBABILISTIC DESCRIPTION

Kingsley O. Harrop-Williams
  Carnegie-Mellon University

INTRODUCTION

The containment of hazardous wastes is certainly one of the most urgent problems facing civil engineers today. The urgency of this problem is highlighted in many recent publications.[1,2,3] This concern is primarily due to the fact that, until just recently, most hazardous wastes have been disposed of at poorly engineered landfill sites. However, although a number of alternatives like ocean dumping or incineration might be available to safely dispose of hazardous wastes, landfill disposal will continue as a popular disposal method. In addition, clay-lined landfills, as compared with synthetic (concrete, asphalt, etc.) landfills, have been proven to offer the most economical solution to the waste disposal problem.[4]

Permeability is the primary criterion used in evaluating the suitability of clay liners for containing hazardous wastes. This characteristic controls the velocity of fluid flow through soil, and thus the volume of flow. This has led many regulatory agencies to adopt regulations requiring clay-lined hazardous waste landfills to have a coefficient of permeability no greater than some fixed value, typically 1×10^{-7} cm/sec. However, like most soil properties, permeability is a random variable,[5] and the difficulty with establishing specific limiting values for it is the uncertainty associated with its measurement and the resultant assurance of desired performance. This is not a unique situation in engineering. Many problems, including earthquake design, off-shore design and design for wind loads, are similar. These problems have been approached successfully using probability theory. In this chapter, a model is developed to predict the uncertainty associated with specified permeability values. Using this model, it is possible to establish confidence levels associated with possible ranges of the coefficients of permeability of clay-lined hazardous waste landfills.

Permeability is generally recognized as one of the most variable of engineering properties associated with construction materials. The range of permeability is over ten billion times from gravel to clay and a thousand times in clay alone. [6] Figure 1 shows the variability of permeability in comparison to that of other engineering properties. In addition to the variation due to the natural homogeneity of soils, the reliability of measured permeability values is highly dependent on the method used for measurement. Laboratory results in some cases bear little resemblance to permeability values obtained from field tests. A list of the possible errors associated with obtaining laboratory permeability values was compiled by Olson and Daniel[7] and is shown in figure 2. These data illustrate the difficulty in placing complete confidence on a laboratory predicted coefficient of permeability for even the most homogeneous of soils.

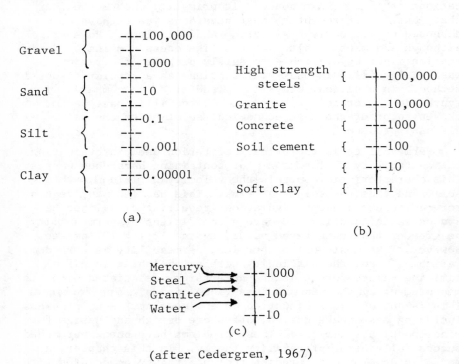

(after Cedergren, 1967)

Figure 1. Variability of permeability compared with other engineering properties (log scale). (a) Permeability, ft/day. (b) Strength, lb/in^2. (c) Unit weight, lb/ft^3.

| | Published Data on Typical |
| | (Measured k)/(Correct k) |

Source of Error

Source of Error	Published Data on Typical (Measured k)/(Correct k)
1. Voids Formed in Sample Preparation	> 1
2. Smear Zone Formed During Trimming	< 1
3. Use of Distilled Water as a Permeant	5/1000 to 1/10
4. Air in Sample	1/10 to 1/2
5. Growth of Micro-organisms	1/100 to 1/10
6. Use of Excessive Hydraulic Gradient	< 1 to 5
7. Use of Wrong Temperature	1/2 to 1 1/2
8. Ignoring Volume Change Due to Stress Change	1 to 20
9. Flowing Water in a Direction Other Than the One of Highest Permeability	1 to 40
10. Performing Laboratory Rather than In-Situ Tests	< 1/10000 to 3

(after Olson and Daniel, 1979)

Figure 2. Summary of Published Data on Potential Errors
in Laboratory Permeability Tests on Saturated
Soils.

Laboratory and particularly field permeability tests are costly and time consuming, allowing for only a limited number of these tests to be made. For these reasons, the permeability of a clay liner is generally characterized by the density to which it is compacted and more importantly by the water content during compaction.[4] The uncertainty associated with a predicted value of the coefficient of permeability of the clay liner is therefore increased by the uncertainty of relating permeability to compaction levels.

Finally, a major problem with clay liners used for waste disposal is that its permeability is usually determined using water as the permeant. However, the true permeability may either decrease or increase depending on the nature of the chemicals that it is expected to contain.[8] Mesri and Olson[9] showed that some of the chemical characteristics that affect the permeability of clay liners containing hazardous wastes are the surface charge density and distribution in the clays; and the pH, valency and concentration of cations, ionic strength, dielectric constant, dipole moment, polarity and viscosity of the permeating fluid.

PROBABILITY DISTRIBUTION OF PERMEABILITY

The permeability of clay soils to chemical wastes depends on many factors, the most important of which are the availability of pore spaces within the soil, the surface charge density of the soil and the chemical nature of the waste. The latter two, however, are chemical rather than physical in nature and their significance may not be as important as the former if the permeant is a neutral agent (e.g. water). As is common in clay liner design, the permeability of the liner is determined based on its permeability to water and modifications are made later for the chemical effects.[8,10,11] In this regard, it would be advantageous to isolate the physical factors from the chemical factors and thereby develop a model for the variation of the permeability of the soil due to physical attributes alone. This model would be more applicable if the permeant was a neutral agent; however, it can later be modified to include the chemical effects of the permeants on the clay.

A randomly chosen point in the clay is either a void or a solid, where the probability of it being a void is equal to its porosity.[5] If for convenience the pore areas in a clay are assumed to be fairly equal in size, then the number of pore areas, m, in a unit cross-sectional area of clay capable of having N pore areas is binomially distributed. That is, the probability distribution of m can be written as

310

$$P_m(m) = \frac{N!}{m!(N-m)!} \, n^m (1-n)^{N-m} \quad ; \quad m = 0,1,2 \tag{1}$$

where n is the porosity of the clay. Further, if m_k is the number of pore areas that are required per unit cross-sectional area in order for the clay to have a coefficient of permeability of value k, then because N is large, m_k can be shown to have a Poisson distribution[5] with parameter $\nu k = Nn$; or

$$P_{m_k}(m_k) = \frac{(\nu k)^{m_k} e^{-\nu k}}{m_k!} \quad ; \quad m_k = 0,1,2 \tag{2}$$

Conversely, if m_s denotes the average number of pore areas in a unit cross-section, then the probability that the permeability of the sample, k_s, is greater than any value k is equal to the probability that m_k is less than m_s; or

$$P[k_s > k] = P[m_k < m_s] \tag{3}$$

As m_s is an integer, this can also be written as

$$P[k_s > k] = P[m_k \leq m_s - 1] \tag{4}$$

Further, in terms of the cumulative distribution functions of k and m_k, equation 4 becomes

$$1 - F_k(k_s) = F_{m_k}(m_s - 1) \tag{5}$$

where [12]

$$F_X(x) = P(X \leq x) = \sum_{X=0}^{x} P(X=x)$$

The cumulative distribution function of k may then be obtained by substituting equation 2 into equation 5, and is found to be

$$F_k(k_s) = 1 - \sum_{m_k=0}^{m_s-1} \frac{e^{-\nu k}}{m_k!} (\nu k)^{m_k} \tag{6}$$

The probability density function of k can be obtained by differentiating equation 6 with respect to k. That is

$$f_k(k_s) = -e^{-\nu k} \left[\sum_{m_k=0}^{m_s-1} \frac{\nu(\nu k)^{m_k-1}}{m_k!(m_k-1)} - \nu \sum_{m_k=0}^{m_s-1} \frac{(\nu k)^{m_k}}{m_k!} \right] \tag{7}$$

By cancelling out all terms until the last term, equation 7 reduces to[13]

$$f_k(k_s) = \nu e^{-\nu k_s} \frac{(\nu k_s)^{m_s-1}}{(m_s-1)!} \tag{8}$$

311

This implies that the permeability is gamma distributed with parameters m_s and ν. In a continuous, rather than discrete, form equation 8 becomes

$$f_k(k_s) = \frac{\nu^{m_s}}{\Gamma(m_s)} \, k_s^{m_s-1} \, e^{-\nu k_s} \qquad , \ 0 \le k_s \le \infty \qquad (9)$$

where

$\Gamma(\ )$ is the gamma function

The mean value and variance of the permeability, evaluated from equation 9, are[12]

$$\bar{k} = \frac{m_s}{\nu} \qquad\qquad\qquad (10)$$

and

$$\text{var }(k) = \frac{m_s}{\nu^2} \qquad\qquad\qquad (11)$$

respectively. Equivalently, the parameters m_s and ν can be written in terms of the mean value and variance of k as

$$m_s = \frac{\bar{k}^2}{\text{var }(k)} \qquad\qquad\qquad (12)$$

$$\nu = \frac{\bar{k}}{\text{var }(k)} \qquad\qquad\qquad (13)$$

Equations 12 and 13 indicate that although the probability density function of permeability (equation 9) was derived from a microscopic point of view, its parameters m_s and ν can be determined from macroscopically measured values of the mean and variance.

Equation 9 ignores the chemical effects of the permeant on the clay. However, if a functional relationship between the clay's permeability to the permeant, k_p, and its permeability to water, k, as developed by Green et al.,[11] is used, then equation 9 can be transformed to yield the probability density function of k_p.

DESIGN APPLICATION

Conventional design procedures for clay-lined hazardous waste landfills generally require that their coefficients of permeability must not exceed some fixed value. This is usually a difficult condition to fulfill and a random sampling of the permeability of the constructed clay liner often produces values both above and below this value. By modelling the randomness associated with the coefficient of permeability, an alternative to the conventional design approach in which clay-liner permeability is treated as a single-valued quantity can be provided.

As an example, consider the case where a number of samples taken from the completed clay liner produced a mean value for the coefficient of permeability of $\bar{k} = 9.838 \times 10^{-8}$ cm/sec and a standard deviation of 11.641×10^{-8} cm/sec. If the design permeability is specified as 2.4×10^{-7} cm/sec, then the probability of reaching this requirement can be obtained by integrating equation 9; and is found to be 0.9. Similarly, the permeability value, k_{95}, at which one is 95% certain that any randomly selected point in the liner will have a coefficient of permeability less than it can be found from the following equation:

$$\int_{0}^{k_{95}} f_k(k_s) dk_s = 0.95 \tag{14}$$

where $f_k(k_s)$ is given by equation 9. The solution to equation 14 yields $k_{95} = 3.3 \times 10^{-7}$ cm/sec.

As is done for strength tests of steel and concrete, this approach would allow for the development of guidelines as to the percentage of permeability values randomly sampled from the completed clay liner that must fall within a certain permeability range to provide a given confidence in the required design permeability. The confidence level, of course, would depend on the nature of the chemical the landfill is required to retain.

SUMMARY

The primary criterion used in evaluating the suitability of hazardous waste landfills for containing hazardous waste is permeability, and many regulatory agencies have adopted regulations requiring clay-lined hazardous waste landfills to have a coefficient of permeability no greater than a fixed value. However, permeability is one of the most variable of engineering properties and its behavior is best described using probability theory. In this paper, the probability density function of the coefficient of permeability is derived from a microscopic view of flow through soils. In developing this model, the permeant was assumed to be a neutral agent (e.g., water). However, as is usually done in clay-liner design, this model can be modified to include the effect of chemicals on the clay.

REFERENCES

1. Morrison, A. "Can clay liner prevent migration of toxic leachate?" Civil Engineering Magazine, July 1981, pp. 60-63.
2. Dellaire, G. "Hazardous waste management in California: Lessions for the U.S.", Civil Engineering Magazine, April 1981, pp. 53-56.
3. Byer, H.G., Blankenship, W. and Allen, R. "Groundwater contamination by chlorinated hydrocarbons: causes and prevention," Civil Engineering Magazine, March 1981, pp. 54-55.
4. Pertusa, M. (1980) "Materials to line or to cap disposal pits for low-level radioactive wastes," Geotechnical Engineering Report GR80-7, Dept. of Civil Engineering, University of Texas, Austin.
5. Harr, M.E. (1977) "Mechanics of particulate media - A probabilistic approach," McGraw Hill, New York.
6. Cedergren, H.R. (1965) "Seepage, Drainage, and Flow Nets," J. Wiley & Sons, New York, pp. 23.
7. Olson, R.E. and Daniel, D.E. (1979) "Field and laboratory measurements of the permeability of saturated and partially saturated fine-grained soils" Geotechnical Engineering Report GE79-1, Dept. of Civil Engineering, University of Texas, Austin.
8. Brown, K.W. and Anderson, D. (1980) "Effects of organic chemicals on clay liner permeability - a review of the literature," Disposal of Hazardous Waste, edited by David Schultz, Proceedings of the 6th Annual Research Symposium, U.S. Environmental Protection Agency, EPA-600/9-80-101, Cincinnati, Ohio, pp. 123-124.
9. Mesri, G. and Olson, R.E. (1971) "Mechanisms controlling the permeability of clays," Clays and Clay Minerals, Vol. 19, pp. 151-158.
10. Anderson, D. and Brown, K.W. (1981) "Organic leachate effects on the permeability of clay liners," Land Disposal: Hazardous Waste, edited by David Schultz, Proceedings of the 7th Annual Research Symposium, EPA-600/9-81-002b, Cincinnati, Ohio, pp. 119-130.
11. Green, W.T., Lee, G.F. and Jones, R.A. (1979) "Impact of organic solvents on the integrity of clay liners for industrial waste disposal pits: Implications for groundwater contamination," Final Report to the U.S. EPA, Robert S. Kerr Environmental Research Laboratory, Ada, Okalhoma, June, 1979.
12. Ang, A. H-S. and Tang, W.A. (1975) "Probability concepts in engineering planning and design, Vol. 1, Basic Principles," S. Wiley & Sons, New York, pp. 82.
13. Parzen, E. (1962) "Stochastic Processes," Holden-Day, Inc. San Francisco, pp. 134.

314

CHAPTER 21

CONTAINMENT AND ENCAPSULATION OF
HAZARDOUS WASTE DISPOSAL SITES

Amir A. Metry
  Roy F. Weston, Inc.

INTRODUCTION

To securely contain low-level radioactive waste and contaminated soils, it was proposed that they be placed in an
encapsulation cell. This cell consists of an interconnected
cover and liner which totally encapsulate the waste. The
encapsulation and containment design was formulated to meet
the EPA criteria for remedial action at inactive uranium-
mill-tailings sites. Criteria of primary importance in the
design of the cell included those regulating radon gas
emission and groundwater contamination. The cover and liner
configuration that was recommended for use is shown in
Figure 1.

This chapter discusses state-of-the-art techniques for
containment and encapsulation of hazardous waste disposal
sites. Such techniques have been applied for remedial action
of inactive hazardous waste disposal sites containing chemical, toxic and low-level radioactive materials. The following discussion is based on a remedial action case history for
a large site containing low-level radioactive waste. The
technical approach, however, is applicable to other types of
hazardous and/or toxic waste.

COVER SYSTEM

Background

The cover, as an element of the encapsulation cell,
plays a very important role in protecting the environment
and public health. A properly selected, designed and constructed cover system will control potential releases of
radioactivity through air diffusion, surface and subsurface
migration, and other physical transport pathways.

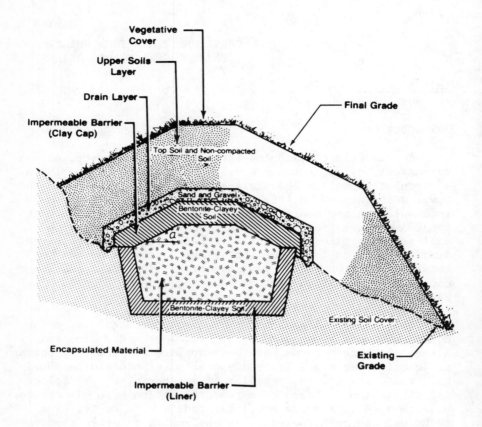

a – Slope Angle

FIGURE 1   PROFILE OF RECOMMENDED ENCAPSULATION
AND COVER CONFIGURATION

The evaluation and selection of cover systems for low-level radioactive waste is a function of various performance criteria and cover materials. A successful cover system will provide effective control of surface-water infiltration and radon gas emission, and will remain effective with minimum maintenance for 1000 years. The control of surface-water infiltration will minimize radionuclide leaching and subsequent transport.

Cover Material Evaluation

For the case history, 18 cover types were systematically evaluated based on 20 performance criteria. The covers were then ranked based on these criteria, and the best performer was identified. Six major classes of covers were evaluated: multilayer, asphalts, concrete, synthetics, natural soils and soil admixtures.

Table 1 illustrates the evaluation process for the various cover materials. If a cover material was given a positive performance rating, a plus sign appears in that criterion column. If a negative performance rating was given, a minus sign appears in the criterion column. It is clearly seen that the multilayered cover system shows the best performance. The multilayer cover system for the low-level radioactive residues is shown in Figure 2. The system consists of the following:

- top layer of noncompacted soil which will support vegetation

- a middle layer of coarse gravel or crushed rock

- a bottom layer of clay

Radon Attenuation in the Cover System

A primary purpose of the cover system was to reduce radon fluxes at the surface of the covered remedial action case history disposal site to 2 pCi/m^2/sec or less. It was necessary to design the cover to accommodate the highest radon flux anticipated from the waste encapsulation area.

Table I

Cover Material General Performance Criteria Evaluation

Performance criteria	Spray asphalt emulsion	Hydraulic asphalt	Synthetic -- CSPE	Synthetic -- PVC	Synthetic -- neoprene	Synthetic -- CPE	Concrete	low permeability native soils
Historical applications as a cover material	-	-	-	-	-	-	+	+
Trafficability	-	+	-	-	-	-	+	-
Impede water percolation	+	+	+	+	+	+	+	+
Radon gas control	+	+	-	-	-	-	+	+
Erosion control	+	+	+	+	+	+	+	-
Aid surface runoff	+	+	+	+	+	+	+	+
Desiccation	+	+	+	+	+	+	+	-
Freeze/thaw stability	+	+	+	+	+	+	+	-
Seismic stability	-	-	-	-	-	-	+	-
Crack resistance	-	+	-	-	-	+	+	-
Side-slope stability	-	+	-	-	-	-	-	-
Potential for side-slope seepage	+	+	+	+	+	+	+	+
Discourages rodent burrowing	-	+	-	-	-	-	+	-
Supports vegetation	-	-	-	-	-	-	-	+
Ease of construction	-	-	-	-	-	-	-	+
Probable 1000-year life	-	-	-	-	-	-	-	+
Cost of placement	+	-	-	-	-	-	-	+
Biological deterioration	-	-	-	-	-	-	-	+
Root penetration	-	-	-	-	-	-	+	-
Wave radiation gamma penetration	-	+	-	-	-	-	+	+

318

Table I
(continued)

Cover material	Historical applications as a cover material	Trafficability	Impede water percolation	Radon gas control	Erosion control	Aid surface runoff	Desiccation	Freeze/thaw stability	Seismic stability	Crack resistance	Side-slope stability	Potential for side-slope seepage	Discourages rodent burrowing	Supports vegetation	Ease of construction	Probable 1000-year life	Cost of placement	Biological deterioration	Root penetration	Wave radiation gamma penetration
Onsite soils	+	+	-	+	-	-	-	-	-	-	+	+	-	+	+	+	+	+	-	+
Soil admixtures (bentonite)	+	-	+	+	-	+	-	-	-	-	+	+	-	+	+	+	+	+	-	+
Bentonite	+	-	+	+	-	+	-	-	-	-	+	+	-	-	+	+	-	+	-	+
Well-graded gravel	+	+	-	-	+	-	+	+	+	+	+	-	+	-	+	+	+	+	-	+
Riprap	+	+	-	-	+	-	+	+	+	+	+	-	+	-	+	+	+	+	-	+
Silty-sand (soil)	+	+	-	-	-	-	-	-	-	-	+	+	-	+	+	+	+	+	-	+
Clayey-sand (soil)	+	-	+	+	+	+	-	+	-	-	+	+	-	+	+	+	+	+	-	+
Soil cement	-	-	+	+	+	+	+	+	-	+	+	+	+	-	-	-	+	+	-	+
Soil asphalt	-	-	+	+	+	+.	+	+	-	+	+	+	+	-	-	-	-	-	-	+
Multilayered -- grass/ topsoil/gravel/clay/soil	+	+	+	+	+	+	+	+	+	+	+	+	+	+	-	+	+	+	-	+

319

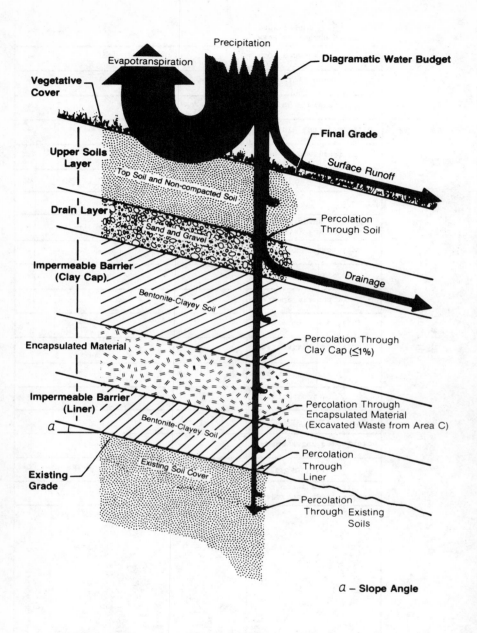

FIGURE 2   PROFILE OF RECOMMENDED ENCAPSULATION
AND COVER CONFIGURATION

320

The design cover thickness to reduce the design base flux to the 2-pCi/m^2/sec specification was computed using the following equation:

$$x_1 = -\ln\left(\frac{J_1}{J_o}\right)\frac{1}{\sqrt{\dfrac{\lambda\,p}{Dh}}} \tag{1}$$

where: x_1 = required cover thickness
J_1 = radon flux rate from covered materials
J_o = radon flux rate from uncovered materials
λ = radon decay constant (2.1×10^{-6}sec^{-1})
p = porosity of the cover material
D = effective radon diffusion coefficient for the cover material
h = dimensionless coefficient (~ 1 when $J \ll J_o$)

Analyses of the effects of various cover configurations on radon flux rates were conducted using a computer model developed by Rogers Associates Engineering Corporation [1]. Pertinent results are displayed in Table 2. For instance, from this table it can be seen that a 1500-pCi/m^2/sec flux rate from the encapsulation area can be controlled to the specified regulatory level of 2 pCi/m^2/sec with the use of a 10-foot multilayer cover system.

Table II

Radon Attenuation by Various Covers

Cover	Base radon flux (pCi/m^2/s)			
	100	500	1000	1500
	Radon emanation through cover system (pCi/m^2/s)			
3 ft soil	14.25	70.55	140.9	211.3
6 ft soil	1.173	5.175	10.18	15.18
3 ft clay/1 ft gravel/2 ft soil	2.354	11.10	22.22	33.25
3 ft clay/1 ft gravel/6 ft soil	0.2363	0.4940	0.8162	1.138

Functional Components of the Cover System

Vegetation and Upper Soils

Vegetation controls erosion and encourages soil water loss by evapotranspiration. Otherwise, erosion will ultimately degrade the cover and seriously reduce its effectiveness. The effect of vegetation quality on resultant percolation through the topsoil and underlying noncompacted soil was examined. The results for good grasses as opposed to poor grasses are shown in Figure 3. They were computed using the Hydrologic Simulation and Solid Waste Disposal Sites (HSSWDS) model developed by the EPA with the U.S. Army Corps of Engineers (2). Greater attenuation of percolation through the upper soil layers is achieved with greater total thickness as shown in Figure 4 (an example calculation for a site in the Pittsburgh, Pennsylvania area). Note that there is a significant reduction in percolation as a function of increased total thickness. Water budget results for the drain and clay layers follow.

Drain Layer

The drain layer consists of crushed rock or coarse gravel having a relatively large permeability, K_s, of 1 x 10^{-1} cm^3/sec. A drain layer thickness of one foot was used. The thickness requirement is a function of annual percolation rate, drain length, permeability and drain slope.

An example of calculating flow through the drain layer is given in Figure 5 (3). This figure shows a drain layer of thickness d(cm) overlying a low permeability material. The drain layer extends over distance, L. The saturated permeability of the drain layer is given by K_s. The annual percolation rate, e, is the amount of water, annually, that impinges on the drain layer. It is assumed that the percolation rate is constant with time. This is a valid assumption since seepage fluxes do not change rapidly with respect to time.

The height of the saturated water surface for the limiting case when $a = 0$ is given by (3) as follows:

$$h = \left(\frac{e}{K_s} \ (L-x) \ x \right)^{1/2} \tag{2}$$

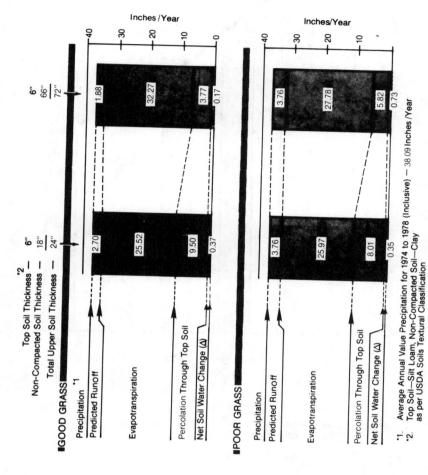

Inches/Year

40 — 30 — 20 — 10 — 0

Top Soil Thickness *2 — 6"
Non-Compacted Soil Thickness — 18"
Total Upper Soil Thickness — 24"

GOOD GRASS

6"
66"
72"

Precipitation *1
Predicted Runoff — 2.70 | 1.88

Evapotranspiration — 25.52 | 32.27

Percolation Through Top Soil — 9.50 | 3.77
Net Soil Water Change (Δ) — 0.37 | 0.17

Inches/Year

40 — 30 — 20 — 10 —

POOR GRASS

Precipitation
Predicted Runoff — 3.76 | 3.76

Evapotranspiration — 25.97 | 27.78

Percolation Through Top Soil — 8.01 | 5.82
Net Soil Water Change (Δ) — 0.35 | 0.73

*1. Average Annual Value Precipitation for 1974 to 1978 (Inclusive) — 38.09 Inches/Year
*2. Top Soil—Silt Loam, Non-Compacted Soil—Clay
as per USDA Soils Textural Classification

**FIGURE 3   COMPARATIVE WATER BUDGET RESULTS FOR VEGETATIVE
COVERING USING 'GOOD' GRASS AND 'POOR' GRASS**

323

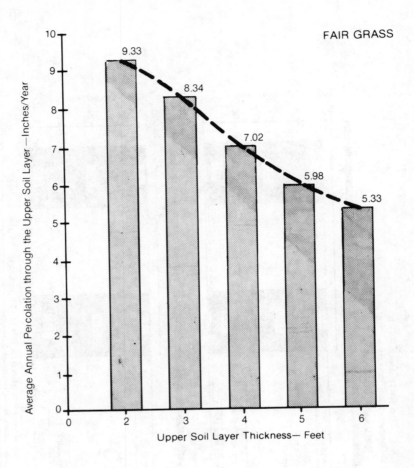

FAIR GRASS

1. Upper Soils Layer Consists of top soil (a constant thickness of 6 inches) and non-compacted soils

2. As per USDA Soils Textural Classification
   Top Soil—Silt Loam
   Non-Compacted Soil—Clay

3. Precipitation value (38.09 inches/ year) is the Average Annual Value Precipitation for 1974 to 1978 (inclusive).

FIGURE 4   GRAPH OF AMOUNT OF ANNUAL PRECIPITATION PERCOLATING THROUGH INCREASING THICKNESS OF UPPER SOILS LAYER

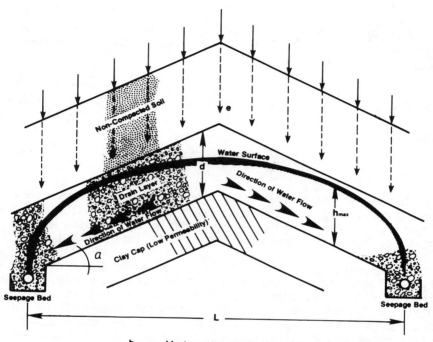

h_{max} — Maximum height of water standing in the Drain Layer

d — Drain Layer thickness

L — Distance between opposing laterals or seepage beds

e — Rate of water flow impinging on drain layer,
equal to percolation rate

a — Slope angle

SOURCE MOORE 1980

FIGURE 5   **DIAGRAM OF ASSUMED WATER SURFACE
PROFILE IN DRAIN LAYER.**

The maximum height of water in the drain layer, h_{max}, is given as:

$$h_{max} = \left(\frac{eL}{4K_s}^2 \right)^{1/2} \tag{3}$$

Setting the slope at some value greater than 0 ($\alpha > 0$) will accelerate the flow toward the collector system. h_{max} for $\alpha > 0$ is given by:

$$h_{max} = \frac{L\sqrt{C}}{2} \left[\frac{\tan^2 \alpha}{C} + 1 - \frac{\tan \alpha}{C} \sqrt{\tan^2 \alpha + C} \right] \tag{4a}$$

where:

$$C \equiv \frac{e}{K_s} \tag{4b}$$

Having a slope α, greater than zero is critical since in this case, if water were to cease impinging on the drain layer, the water would completely drain in a finite amount of time. If $\alpha = 0$, the drainage time is infinitely long.

The results of a sensitivity analysis for the previously mentioned Pittsburgh, Pennsylvania area site are given in Table 3 to examine the effects of percolation rate, drainage length and slope, and saturated permeability on the maximum height of water standing in the drain layer. Drain thickness requirements will increase as a function of an increase in annual percolation rate and decrease in permeability. Other parameters being equal, drain thickness requirements will decrease as a function of increasing slope.

The approach for estimating drain layer efficiency is based on saturated Darcy flow in both the drain layer and clay cap. The assumed geometry is given in Figure 6, at some time, t (3).

This approach postulates that at some initial time a rectangular slug of liquid is placed on the saturated liner to a depth, h_o. The liquid flows both horizontally along the slope of the system, and vertically into the clay liner.

Table III

Illustration of a Sensitivity Analysis
for the Pittsburgh, Pennsylvania Area Site

Drain layer length: 200 ft
$L = 2 \times$ drainage length = 400 ft = 121.92 m

Vegetation cover: "Fair" grass

Upper soil thickness	Maximum annual percolation rate	Slope 1%	5%	10%	15%
		Drain layer thickness (in.)			
$(K_s = 1 \times 10^{-1} \text{cm/sec})$					
24 inches	11.64 in./yr 9.4×10^{-9} m/sec	3.76	3.68	3.68	3.68
72 inches	9.47 in./yr 7.6×10^{-9} m/sec	3.37	3.31	3.31	3.31
$(K_s = 1 \times 10^{-2} \text{cm/sec})$					
24 inches	11.64 in./yr 9.4×10^{-9} m/sec	13.54	11.74	11.66	11.65
72 inches	9.47 in./yr 7.6×10^{-9} m/sec	11.93	10.54	10.48	10.47

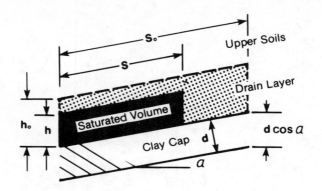

a GEOMETRY FOR CALCULATING
EFFICIENCY OF DRAIN LAYER

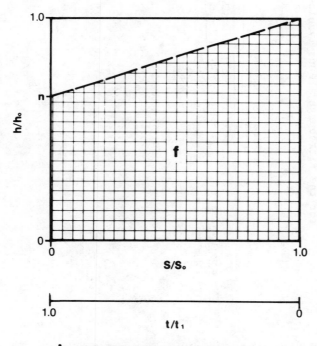

b DIAGRAM FOR COMPUTING
EFFICIENCY OF DRAIN LAYER

FIGURE 6   ASSUMED GEOMETRY FOR COMPUTING
DRAIN LAYER EFFICIENCY

The fraction of liquid moving into the collector drain system at time, t, is given (3), as follows:

$$\frac{S}{S_o} = 1 - \frac{t}{t_1} \qquad (5)$$

and the fraction of liquid seeping into the clay liner is given by:

$$\frac{h}{h_o} = \left(1 + \frac{d}{h_o \cos \alpha}\right)e^{-Ct/t_1} - \frac{d}{h_o \cos \alpha} \quad 0 \le t \le t_1 \qquad (6)$$

where:

$$t_1 = \frac{S_o}{K_{sl} \sin \alpha} \qquad (7)$$

$$C = \left(\frac{S_o}{d}\right)\left(\frac{K_{s2}}{K_{sl}}\right) \cot \alpha \qquad (8)$$

and

S = length of saturated volume at time, t (cm)

h = thickness of saturated volume at time, t (cm)

S_o = initial length of saturated volume = L/2 sec (cm)

h_o = initial thickness of saturated volume (cm)

K_{sl} = saturated permeability of the material above clay liner (cm/sec)

K_{s2} = saturated permeability of the clay liner (cm/sec)

α = slope angle of the system (O)

d = thickness of the clay liner (cm)

The efficiency of the liner is determined with reference to Figure 6(b) which plots h/h_o versus S/S_o and t/t_1. Equations (5) and (6) can be solved parametrically in t/t_1, to yield the line shown on the figure. (The line is actually a curve; however, for practical liner and drain layer configurations, it can be approximated as a straight line.) In this case, the efficiency of the system is given

by the area labeled "f." This area is most easily determined by calculating the value of h/h_o when $t/t_1 = 1.0$
(or $S/S_o = 0$). The term h/h_o is set equal to n and can be
obtained by solving Equation (9) with $t/t_1 = 1.0$:

$$ n = \left(1 + \frac{d}{h_o \cos \alpha}\right) e^{-C} - \frac{d}{h_o \cos \alpha} \qquad (9) $$

The value of n can be either positive or negative; however, most efficient designs will have $n > 0$. The efficiency
is given by either:

$$ f = \frac{1+n}{2} \qquad \text{for } n > 0 \qquad (10a) $$

or

$$ f = \frac{1}{2(1-n)} \qquad \text{for } n < 0 \qquad (10b) $$

Thus, the efficiency varies from 0 to 1.0.

Impermeable Barrier (Clay Cap)

The clay cap may be constructed either of one layer of
compacted soil; or two layers, compacted soil overlaid by a
compacted soil and bentonite mixture. The criterion for
barrier selection is permeability. Permeabilities of 10^{-6}
to 10^{-8} cm/sec are required for attenuation of radon as
well as reduction of seepage through contaminated materials.

In the early stages, the wetting process is described
by Equation (11) where the first term on the right side
dominates, shown as follows:

$$ \frac{\partial \tilde{\theta}}{\partial t} = D^* \frac{\partial^2 \tilde{\theta}}{\partial z^2} - K^* \frac{\partial \theta}{\partial z} \qquad (11) $$

Thus,

$$ \frac{\partial \tilde{\theta}}{\partial t} \cong D^* \frac{\partial^2 \tilde{\theta}}{\partial z^2} \qquad (12) $$

The D^* term represents capillary attraction. During
this stage of the wetting process, gravitational forces are
negligible as compared to capillary forces.

330

Imposing the following initial and boundary conditions:

Initial Condition --

$$\tilde{\theta} = \theta; \text{ for } Z > 0 \text{ and } t = 0$$

(Z is positive, downward)

At initial time (t = 0), assume that the moisture content is equal to θ, throughout the depth of the liner.

Boundary Condition --

$$\tilde{\theta} = \theta_s \text{ for } Z = 0 \text{ and } t \geq 0$$

At all times at the boundary (Z = 0), the moisture content is held at the saturation moisture content, θ_s.

The solution of Equation (12), having the initial and boundary conditions just given, is as follows:

$$\tilde{\theta} = \theta_i + (\theta_s - \theta_i) \text{ erfc } \frac{Z}{\sqrt{2} \, D*t} \tag{13}$$

The relationship for the cumulative amount of water entering the barrier soil at time, t, is as follows:

$$M_t = 2 (\theta_s - \theta_i) \sqrt{\frac{D*t}{\pi}} \tag{14}$$

and the quantity of liquid required to saturate the barrier to a depth, d, is given by:

$$M_t = (\theta_s - \theta_i) \, d \tag{15}$$

Equating Equations (14) and (15) yields:

$$t = \frac{\pi d^2}{4D*} \tag{16}$$

LINER SYSTEM

Background

The use of natural and synthetic materials of low permeability to line waste storage and disposal impoundments has been demonstrated to be a useful means of preventing leachate and waste liquid components from leaking and subsequently polluting ground and surface waters. These liner materials can also serve to prevent the migration of dangerous concentrations of radon and other gases from a waste containment site. Many liner materials are available from which the containment system for specific wastes may be chosen.

Two types of liner systems exist, active and passive. Active liner systems employ the use of leachate collection, and generally require considerable post-closure maintenance. An active liner system must also be constructed of highly impervious materials, and include a backup liner for quality assurance. Active liner systems have restricted life expectancies, and typically cannot be expected to provide a low maintenance 1000-year life.

A significant amount of information exists regarding the water resistance of lining materials, regardless of whether they are soils, asphalts or polymeric membranes. The contaminated materials may also contain other ingredients which could affect lining materials. It is, therefore, necessary to consider the totality of all constituents in a waste in assessing a liner material for a given application; the chemical composition of both the waste and the lining material must be considered.

This section, Liner System, will consider only passive liner systems because, with their use, a low maintenance service life can reasonably be expected. A variety of liner systems was considered for the encapsulation area, including asphalts, concrete, synthetics, natural soils and soil admixtures. Table 4 illustrates the systematic performance evaluation of these materials. The natural soils with possible soil admixtures (bentonite clay) were again chosen based on past experience and long service life. They are also desirable for their ability to provide controlled hydraulic flux and radiological attenuation. The cost of placement and ease of construction are favorable characteristics.

Table IV

Liner Material General Performance Criteria Evaluation

Liner material	Permits hydraulic controlled flux	Historical application as liner material	Seismic stability (2)	Crack resistance (3)	Radionuclide attenuation (5)	Vegetation penetration	Potential for damage to liner during placement	Ease of construction	Probable 1000-year life	Biochemical deterioration	Cost of placement
Spray asphalt emulsion	–	+	–	–	–	–	+	–	–	–	+
Hydraulic asphalt	–	+	–	+	–	+	+	+	–	–	–
Synthetic -- Hypalon	–	+	–	+	–	–	–	+	–	–	–
Synthetic -- PVC	–	+	–	+	–	–	–	+	–	–	–
Synthetic -- Neoprene	–	+	–	+	–	–	–	+	–	–	–
Synthetic -- CPE	–	+	–	+	–	–	–	+	–	–	–

Table IV
(continued)

Liner material	Cost of placement	Biochemical deterioration	Probable 1000-year life	Ease of construction	Potential for damage to liner during placement	Vegetation penetration	Radionuclide attenuation (5)	Crack resistance (3)	Seismic stability (2)	Historical application as liner material	Permits hydraulic controlled flux
Concrete	-	+	+	-	+	+	-	+	+	+	-
Low permeability/native soils (1)	+	+	+	+	+	-	-	-	+.	+	+
Soil admixtures (4) (bentonite)	+	+	+	+	+	-	+	-	-	+	+
Bentonite (4)	-	+	+	+	+	-	+	-	+	+	+
Soil cement	-	+	-	-	+	+	-	-	-	-	+
Soil asphalt	-	-	-	-	+	+	-	+	-	-	+

334

Functions of the Liner

The primary purpose and function of a liner system is to retard the physical movement of water into the natural environment. An optimal liner design would address the dual function of minimizing water (leachate) movement while passively treating any leachate that does migrate through the liner.

Water that permeates the clay cap will, in time, permeate the waste material and liner. The rate of water movement through the liner will, at saturation, equal that of the clay cap. Thus, water will not accumulate between the liner and the cap.

An ion-exchange barrier may be considered a means of controlling the migration of radionuclides in or into groundwater. This type of system could be constructed as follows:

- a curtain or barrier designed to intercept the flow of groundwater from a contaminated area

- a liner to be placed under a waste area designed to intercept any leachate that may be generated

Ion-exchange material may be comprised of natural soils (clays) or synthetic resins (zeolites, macroreticular polymers, gels, etc.).

The use of ion-exchange materials for control of radioactive wastes has been proposed in the literature (5, 6). The performance of various natural materials, e.g., expandable clays and zeolites, for adsorbing specific radioactive species has been reported. A recent literature search (4) for ion-exchange data associated with clays, zeolites and basalt identified 92 references to ion-exchange data on clays, 22 references for zeolites and 6 references on basalt.

Nowak (7) has proposed a model for radionuclide migration through an ion-exchange backfill barrier system. This type of modeling effort may also be applied to a liner. Nowak presented his model, beginning with its differential form, as follows:

$$\epsilon \frac{\partial C}{\partial t} + \frac{\partial S}{\partial t} + \epsilon v_g \frac{\partial C}{\partial x} - \epsilon D_L \frac{\partial^2 C}{\partial x^2} = 0 \quad (17)$$

where: C = liquid phase concentration, quantity of sorbing species per unit volume of liquid

S = concentration of species sorbed on the solid phase (quantity of sorbed species per unit volume of bed liquid plus solid volumes)

ϵ = effective porosity of bed (fraction of bed volume containing flowing liquid)

v_g = average interstitial velocity of flowing liquid

x = distance in bed along direction of flow and longitudinal diffusion

D_L = coefficient of longitudinal dispersion and diffusion combined

t = time

For the boundary condition,

$$C = C_o, \ x = 0, \ T > 0$$

and for the initial condition,

$$C = 0, \ x > 0, \ T = 0,$$

Crank (8) gives the solution for Equation (17) as follows:

$$\frac{C}{C_o} = 1 - \text{erf} \left[\frac{x}{2 \left(\dfrac{D_f t}{2 R_f} \right)^{1/2}} \right] \quad (18)$$

where: D_f = liquid phase molecular diffusivity

R_f = $1 + \dfrac{\rho BKd}{\epsilon}$

ρB = bulk packing density of solid sorbent, mass of solid per unit bed volume

K_d = distribution coefficient for a linear-sorption isotherm, the ratio of quantity of sorbed species per unit mass of solids to quantity of mobile species in the liquid phase per unit volume of liquid

Typical values for the parameters used in Equation (18) are presented in Table 5. The time to "breakthrough" for barrier walls with various characteristics is given in Figure 7. In developing these estimates, "breakthrough" is defined as $C/C^0 = 0.01$. As Figure 7 indicates, for those parameter values used, a barrier thickness ranging from less than 1.0 foot to approximately 6.5 feet would be necessary to attain a 1000-year design (i.e., at 1000 years of barrier life, the breakthrough concentration ratio, C/C_o, would be less than or equal to 0.01.

Table V

Typical Values of Physical and Chemical Properties
for the Ion-Exchange Barrier

K_d = 100 to 5000 ml/g

B = 2 g/cm^3

D_f = 10^{-4} to 10^{-6} cm^2/sec

ϵ = 0.25 to 0.40

X = 1 to 10 ft

The results for a clay barrier wall can be roughly applied to a clay liner system as well. A clay of the type to be used for the encapsulation-cell liner at the case-history site should have a K_D of about 500 ml/g and a D_f of about 10^{-5} cm^2/sec.

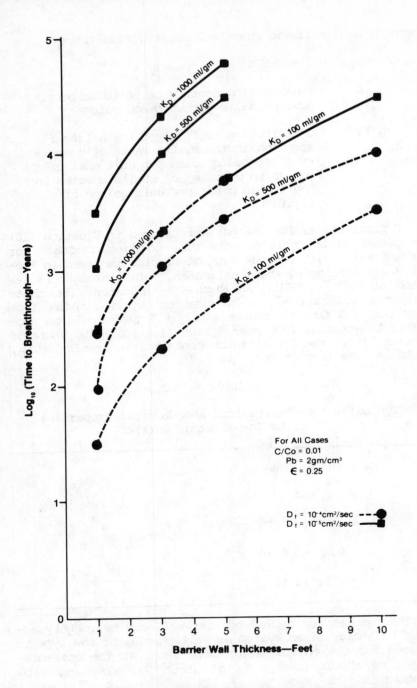

For All Cases
C/Co = 0.01
Pb = 2gm/cm³
ϵ = 0.25

D_f = 10⁻⁴cm²/sec - - - ●
D_f = 10⁻⁵cm²/sec ——— ■

FIGURE 7   GRAPH OF BREAKTHROUGH TIME AS A FUNCTION OF BARRIER WALL CHARACTERISTICS—HIGH AND LOW DIFFUSION RATES (D_f)

Liner-System Description

Two to three feet is the recommended thickness for the clay liner. This choice was made for several reasons: constructability, long-term ion-exchange capacity, and compactability with waste and contaminated material.

Bentonite combined with natural soils to produce a mixture of low permeability, or a native clayey soil may be used. The specific liner material can only be selected once the native soils are tested for permeability and cationic exchange capacity.

It should be noted that standard bentonite is susceptible to deterioration in an excessively low pH environment (9-11). The pH effects can only be assessed once the low-level waste of concern is tested.

A liner system used with a cover has the additional benefit of providing waste encapsulation. By tying the cover and liner systems together, the buried wastes can be completely sealed. Encapsulation allows more complete isolation of the disposed wastes, and therefore lessens any environmental impacts.

The attenuating capabilities and inherent long-term structural and physiochemical stability of soils are their outstanding characteristics. Relatively simple construction techniques, along with ready availability and accessibility, make soil an obvious choice as a liner material.

WASTE CONDITIONING

Waste conditioning is generally performed to meet one of the following three objectives:

- to improve the handling and physical characteristics of the waste

- to limit the solubility of various contaminants within the waste

- to decrease the surface area across which transfer and loss of contained contaminants can occur

A number of fixation and conditioning methodologies were considered for application, including the following:

- cement-based techniques
- lime-based techniques
- thermoplastic techniques
- thermosetting resins
- encapsulation techniques
- glass and ceramic-fixation techniques
- thermal stabilization
- extraction of contaminants

These techniques are all chemical (as opposed to physical). They may be used in the event waste material is found to have a low pH, which could damage a liner or cap made of bentonite clay and soil. Of the conditioning techniques considered, the lime-based techniques are the most applicable to such material. Fixation techniques using lime-type products usually depend on the reaction of lime with a pozzolanic (silicate-type) material, water, and the waste to produce a concrete-type material. The most common pozzolanic materials used in waste fixation are cement-kiln dust, fly ash and pulverized slag. The effectiveness of chemical fixation using this technique must also be demonstrated through bench-scale tests that simulate the actual process.

Waste conditioning may also imply physical conditioning to improve the physical properties (such as bearing strength, etc.) of the contaminated materials. This should result in a compactable material of optimum moisture content and density which is strong enough to support both the multilayered cover system and the temporary load of construction vehicles. Adequate support of the cover system from below is essential to promote long-term stability and integrity of the cover.

SUMMARY AND CONCLUSIONS

Containment and encapsulation of hazardous waste disposal sites could be achieved by utilizing a combination of the following:

- cover system for waste isolation, gas control and reduction of infiltration through waste material

- when applicable, use of a liner system as a back-up for the cover system and as a means for attenuation of leachate from waste material

- ion-exchange media (as part of the liner system or barrier) for attenuation of groundwater contaminants

- use of passive hydrogeologic isolation techniques including low permeability covers, liners, slurry walls and drainage channels

- in situ grouping, conditioning, solidification or fixation of waste material

- total management, surveillance, and monitoring of the site during and after remedial action activities

In situ containment and encapsulation offer a technically feasible, cost-effective means for control of contaminant release from inactive hazardous and low-level radioactive sites. When applicable, encapsulation is as effective for removal and off-site disposal of waste and contaminated material. However, the cost of in situ management is often a fraction of the latter approach.

The use of computer simulation for environmental simulation and optimization of design features an encapsulation system and has proven to be an invaluable tool. Techniques for evaluation of cover, liner and barriers are presented and illustrated in this chapter.

It is preferred to utilize passive rather than active techniques for encapsulating waste-disposal sites. Multi-layer cover systems, soil/clay liners, bentonite slurry walls and ion-exchange barriers represent the basic elements of remedial action that does not require perpetual post-closure operations.

REFERENCES

1. Rogers, V.C., G.M. Sandquist, and K.K. Nielson, March 1981. "Radon Attenuation Effectiveness and Cost Optimization of Composite Covers for Uranium Mill Tailings," Rogers and Associates Engineering Corporation (RAECO).

2. Perrier, E.R. and A.C. Gibson, September 1980. Hydrologic Simulation on Solid Waste Disposal Sites, SW-868, U.S. Environmental Protection Agency, Solid and Hazardous Waste Research Division, Cincinnati, Ohio.

3. Moore, C.A., September 1980. Landfill and Surface Impoundment Performance Evaluation Manual, U.S. Environmental Protection Agency, Solid and Hazardous Waste Research Division, Cincinnati, Ohio.

4. Benson, L.V., 1980. "A Tabulation and Evaluation of Ion Exchange Data on Smectites, Certain Zeolites, and Basalt," U.S. Department of Energy Contract No. W-7405-ENG-48, Lawrence Berkeley Laboratory, University of California.

5. Northrup, C.J.M., Jr., editor, 1980. Scientific Basis for Nuclear Waste Management, Volume 2, Plenum Press, New York.

6. Nowak, E.J., 1979. "The Backfill Barrier as a Component in a Multiple Barrier Nuclear Waste Isolation System," Sandia Laboratories Report, SAND 79-1109.

7. Crank, J., 1956. The Mathematics of Dispersion, Oxford Press, London, pp. 18-19 and 121-122.

8. Crim, R.G., 1979. "Stability of Natural Clay Liners in a Low pH Environment," Symposium on Uranium Mill Tailings Management, Fort Collins, Colorado, 19-20 November.

9. Morrison, A., July 1981. "Can Clay Liners Prevent Migration of Toxic Leachate?" Civil Engineering, American Society of Civil Engineers, New York.

10. van Zyl, D., and J.A. Caldwell, 1978. "Efficiency of a Natural Clay Liner for High Acidity Tailings Impoundment," Symposium on Uranium Mill Tailings Management, Fort Collins, Colorado, 20-21 November.

CHAPTER 22

ESTIMATES OF SEEPAGE THROUGH
CLAY COVERS ON HAZARDOUS WASTE
LANDFILLS

Steven J. Wright
Carol P. Miller
    Department of Civil Engineering
    The University of Michigan

INTRODUCTION

The problem of estimating the amount of leachate
generated in a hazardous waste landfill is of considerable
significance in the design of a leachate collection system
or in the assessment of potential environmental impacts.
The fundamental problem involved with such an estimate is
in the description of the complex interactions between such
factors as local meteorological conditions, flow in porous
media, evapotranspiration, etc. Although there are a variety
of regulations (e.g.[1]) controlling various aspects of landfill design, these typically specify design details such as
allowable liner thickness and permeabilities and are not
directly concerned with predictions of rates of leachate
generation. There are several methods currently available
for the estimate of leachate rates. A brief review of some
of these methods is presented along with an indication of the
shortcomings of each method. An alternative method based
upon a more mechanistic description of those processes which
are most important to the determination of leachate generation rate is presented. An example based upon a hypothetical landfill but with meteorological information from a
specific location (Cincinnati, Ohio) is used for this purpose.
The results are considered in terms of their implications
for design calculations which involve estimates of leachate
generation rates.

PROBLEM DESCRIPTION

The specific problem considered is not restricted to a
hazardous waste landfill, but, as indicated schematically in
Figure 1, is associated with a landfill overlain by a clay
cover. It is assumed that the clay layer is in turn covered
by a final cover soil upon which vegetation has been
established to prevent soil erosion. It is further assumed
that this cover layer is adequately graded so that water
does not pond above the clay layer, although it would be
possible to investigate the effects of surface ponding in
the model presented. Finally, it is assumed that the local
groundwater table is not in hydraulic contact with the landfill so that the only source of potential leachate to the
landfill is from precipitation. Local precipitation thus
provides the basic source input into any model description
of this problem. Possible flow pathways for the water that
falls as precipitation on the landfill site as indicated in
Figure 1 are:

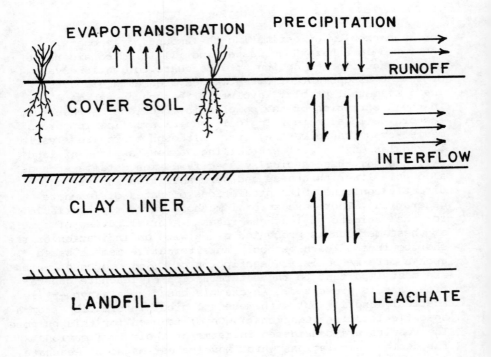

Fig.1. Schematic of Flow System

-Direct surface runoff
-Infiltration into the cover soil
-Uptake by evaporation and plants (evapotranspiration)
-Infiltration into the clay layer
-Horizontal interflow in the cover soil above the
 clay layer

Any complete mechanistic model of the moisture movement must
consider each of these processes. A review of previous
methods which consider the general problem reveals that
some of these processes have often been neglected, for
example, Lentz[2] considers only the problem of horizontal
interflow along with vertical infiltration into the clay
layer. Furthermore, no temporal variations are assumed in
any of the processes and the porous media flows are assumed
to be completely saturated. In order to solve the problem
a unit hydraulic gradient in the vertical is assumed in the
clay layer; results as presented below indicate this not to
be the case. Fenn, et al[3] present a water balance method
which considers surface runoff and evapotranspiration. This
method was not originally proposed for use with clay cover
layers, but has been used in some instances for this problem
in practice. The portion of the precipitation in this
method which does not compose the surface runoff or evapotranspiration components of the water balance is assumed to
be instantaneously available as leachate to the landfill.
Computations are made from mean monthly precipitation data,
but the rational method has been used to estimate the surface runoff. The use of this formula which is intended to
predict maximum runoff rates from individual storms is
questionable, and may significantly underestimate the amount
of surface runoff. A somewhat related analysis by Perrier
and Gibson[4] is intended for use with clay cover layers. The
Soil Conservation Service (curve number) method is used to
estimate the amount of surface runoff and precipitation data
is used on a daily basis. Although this application is
basically consistent with the SCS method formulation, there
are other significant shortcomings. The horizontal interflow component is ignored in this method and flow in the
vertical through the cover soil and clay layers is
accomplished by a phenomenological procedure which replaces
the mechanistic description of saturated-unsaturated flow
in these layers. So many unverified assumptions are made
with respect to the description of water movement in the
soil layers that the model must be regarded as uncalibrated
and subject to question.

An important consideration regarding the formulation of
any methodology for the analysis of a problem should be the

identification of the physical processes that are most
important to the specific situation. These can be determined most easily by comparing time scales associated
with moisture movement in each phase of the problem. For
example, there are individual precipitation events with time
scales of days and evapotranspiration variations with time
scales of months or seasons. A time scale for horizontal
interflow or vertical leakage through the clay layer can be
approximated from the following formula:

$$t_f = L/Ki \tag{1}$$

here L is the length of a flow path, K is the hydraulic
conductivity of the particular medium, and i is the hydraulic gradient driving the flow. If a unit hydraulic
gradient is assumed for simplicity in each case, an order of
magnitude estimate may be made of time scales associated with
the two flow processes. Typical clays used for landfill
covers are required to have hydraulic conductivities on the
order of 10^{-3}-10^{-4} m/day. A one meter thick layer would thus
have a time scale for vertical movement on the order of
1000-10000 days. Horizontal flow in the cover layer with K
of approximately 10 m/day over a path length of 100 m would
give a time scale of 10 days. This time scale should be
increased because actual hydraulic gradients will be considerably less than unity. In spite of this, the important
point is that the limiting process is the vertical movement
through the clay layer, and thus that any model for pre-
diction of leachate generation must include a realistic
description of the water movement in the clay layer. This
is in direct opposition to a similar problem without a clay
layer where a cover layer 1 m thick and hydraulic conductivity of 1 m/day, for example, would give evapotranspiration as the limiting process. Therefore, it is logical
to expect that the methods proposed by Fenn, et al and
Perrier and Gibson would be more suited for situations with
no clay layer and relatively rapid vertical flows. On the
other hand, the model formulation proposed in the present
study is specifically intended for applications when a low
permeability liner is present and should not be considered
for other unrelated applications.

MODEL FORMULATION

A numerical model was formulated consistent with the
above considerations. This model is still in the development stage and further refinements and modifications may be
expected in the future, although the present formulation is
regarded as generally valid. Since the flow in the clay

layer is not necessarily saturated, an unsaturated flow model was developed. A one-dimensional (vertical) approach is used in the present formulation; horizontal variations in flow behavior are assumed to be negligible. The one-dimensional flow equation may be obtained, for example from Bear[5]

$$C(\psi)\frac{\partial \psi}{\partial t} = \frac{\partial}{\partial z} \; K(\psi)(\frac{\partial \psi}{\partial z} +1) \tag{2}$$

Here, $\psi = P/\gamma$ is the pressure head, $K(\psi)$ is the hydraulic conductivity, and $C(\psi)=\partial\theta/\partial\psi$ relates the pressure to moisture content. An alternative to Equation 2 in terms of soil moisture content is not employed because the soil moisture diffusivity required in that formulation approaches infinity as saturation conditions occur and this situation is encountered frequently in most simulations. Material properties relating K and C must be specified along with appropriate boundary and initial conditions are required to complete the solution to Equation 2 and make estimates of leachate generation rates. Since the boundary condition at the top of the clay layer is assumed to be the most critical, the other formulations are discussed briefly and then more detail is outlined for possible upper boundary formulations.

Material Properties

Functional relations are specified for each of the material properties for ease in the numerical solutions although it would also be possible to supply tabular information from experiments for the same purpose. The relation between θ and ψ is subject to hysteresis effects and in principle, it is necessary to supply information on primary drying and wetting along with principal scanning curves. At present, it is assumed that the effect of these hysteresis effects on the solutions will be within the uncertainties of the estimation of the various parameters and can be neglected. Use of the drying curve should always result in an estimate of the unsaturated hydraulic conductivity that is equal to or greater than the actual value and thus provide an overestimate of the amount of vertical movement. This approach should be on the conservative side and is employed in the present formulation. Brooks and Corey[6] and White, et al[7] present a functional relation for $\theta(\psi)$ as

$$Se=(\psi\alpha/\psi)^A \tag{3}$$

The shape of this curve is indicated schematically in Figure 2 with Se defined as $(\theta-\theta_0)/(n-\theta_0)$ (n is the porosity

347

of the medium and θ_0 is the irreducible moisture content, also called field capacity). $\psi\alpha$ is the bubbling or air

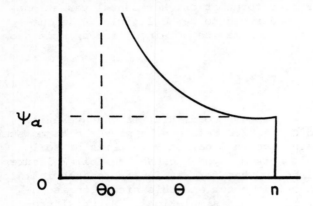

Fig.2. Suction Pressure Versus Soil Moisture Relation

entry pressure, while A is an empirical coefficient which varies from 7.0 for glass beads to 0.46 for a clay soil according to Brooks and Corey. White et al show a much narrower range in values with an average value of approximately 1.5 for several agricultural soils. Stakma[8] presents a theoretical argument which shows $\psi\alpha$ to be inversely proportional to grain diameter and on the order of several meters for clay media. Mitchell, et al[9] present a fairly commonly accepted relationship between K and Se:

$$K = K_0 Se^3 \tag{4}$$

Here, K_0 is the saturated hydraulic conductivity. Brooks and Corey present an alternative expression, but the above is taken as reasonable for the applications intended. With these relations, the following are obtained as functional relations for material property descriptions:

$$K(\psi) = K_0 (\psi\alpha/\psi)^{3A}$$

$$C(\psi) = -(\eta-\theta_0) (\frac{A}{\psi}) (\frac{\psi\alpha}{\psi})^A \tag{5}$$

Lower Boundary Condition

This boundary condition is formulated in a manner similar to that proposed by Eagleson[10]. It is assumed that at some distance below the bottom of the clay layer (assumed to be a few tens of centimeters at most) an equilibrium condition exists with a unit hydraulic gradient. Darcy's law for unsaturated flow thus states

$$q(z) = -K(\psi) \tag{6}$$

This formulation basically assumes that the clay layer
mediates the surface conditions to the extent that a quasi-
equilibrium exists in the relatively permeable landfill
material. This approximation should be fairly realistic
for clay layers with small hydraulic conductivities of the
type discussed in the problem description. This approach
requires the modeling of a few layers of the landfill
material, which, because it is assumed to be much more
permeable than the clay layer, should not have a significant
influence upon the solution.

Upper Boundary Condition

This boundary condition is taken at the interface
between the clay and the soil cover layer and must reflect
a consideration of all surface influences. Vertical flow
is assumed to occur instantaneously in the soil cover layer;
this is consistent with the discussion of relevant time
scales presented earlier. A general statement of mass
balance at this boundary must be

Precipitation - Surface runoff - Evapotranspiration =
Horizontal Interflow + Vertical Infiltrations into
the Clay Layer

The left hand side of this expression is assumed to be given
by some sort of consideration related to the surface hydrology and meteorology and variations of this formulation are
explored further below. The right hand side represents an
unknown partitioning of the moisture reaching the interface
and is presently based upon limitations associated with flow
in the clay layer. If the precipitation exceeds evapotranspiration demand, water is assumed to be available to
the clay layer at the computed rate. If this potential flux
is sufficiently large so as to force saturation at the interface, then horizontal flow at that level must begin to
develop. In the model, the specified flux is replaced by
a specified pressure (namely zero) condition under those
conditions. This ignores the pressure head associated with
the thickness of the horizontally flowing water at the interface, but this magnitude is so small relative to other
pressure variations in typical computations that it has
negligible effects upon the computed results. To include its
effect would require a three dimensional calculation and is
not considered to be justified at this point in time. The
alternate situation to the above is when evapotranspiration
demands exceed precipitation. Under this circumstance,

349

water is assumed to be transported upwards at the required
rate unless pressures drop as low as a predefined wilting
point. Israelson and Hansen[11] give magnitudes for wilting
point as on the order of 15 to 20 atmospheres. Possible
variations to this procedure might be to make the evapotranspiration to be a decreasing function of suction
pressure so that it approaches zero as the wilting point
is approached; this is probably more realistic. Another
possible modification is to account for water storage in the
soil cover layer; this is not presently done since a three
dimensional formulation would be required to describe this
exactly. Possible ways to handle this in a simplified and
realistic manner are being subjected to further investigation.

Numerical Solution

Complete details of the numerical solution are not
presented herein since it is a fairly standard procedure. A
finite difference scheme with the capability of computing
between fully explicit and fully implicit conditions was
formulated. In general, the fully explicit procedure will
be unstable for a typical choice of parameters, and thus is
not used; a centered difference Crank-Nicolson scheme is
typically employed. A double sweep iterative solution
procedure similar to that described by Rubin and Steinhardt[12]
is used. Time steps on the order of a few days are used in
the simulations and the clay layer is typically segmented
into 20 to 40 layers. This provides a fairly efficient
numerical procedure with simulations carried out for ten
years or so. The effect of the initial condition which is
some specified pressure profile is lost after the first year
or two of simulation in a typical computation.

RESULTS

The major uncertainties with respect to the specification of conditions for a particular simulation are in the
material properties of the clay media and in the exact description of the upper boundary condition. The question of
the material properties can only be resolved by experimental
studies on typical landfill cover clays. It should be
mentioned that the properties for the landfill material must
also be specified for the solution. The present formulation
of the upper boundary condition implies that the computation
of potential fluxes at the soil cover layer-clay layer interface can be completely uncoupled from the remainder of the
problem and that these are based purely upon hydrological
considerations along with meteorological data. One important
question is whether or not individual rainfall events need

to be modeled as proposed by Perrier and Gibson[4] or whether
longer term averages can be used such as the monthly average
precipitation data used by Fenn, et al[3]. Two essential
questions are whether the short term data is necessary to
estimate surface runoff and whether the short term data is
necessary to simulate the movement of moisture in the landfill. The key distinction is that if only the first case is
true, then it should be possible to examine hydrologic
records and determine a relationship between precipitation
and runoff (e.g. by a statistical approach such as is outlined by Eagleson[13]) and then these results can be used in
longer term descriptions in the model. Consistent with the
discussion of relevant time scales presented earlier, it is
logical to state that events with time scales of only a few
days are not important to vertical leakage through a clay
layer with a time scale of approximately 10 years as long as
a correct mass balance of inputs is maintained. Furthermore,
since a significant portion of the infiltration to the top
of the clay layer may exit as horizontal interflow, the
sensitivity to actual mass inputs may also not be too
significant. This is of considerable significance since it
implies that the computation of the potential fluxes which
are probably the least poorly understood of the overall flow
processes may not be that critical to the overall simulation
results. Also, significant simplications in data requirements are implied if only mean monthly precipitation-evapo-
transpiration data are required for realistic results. A
series of simulations were performed in order to attempt to
resolve some of these questions. A hypothetical landfill
located near Cincinnati, Ohio with soil properties as given
in Table 1 was selected as a test case. This site was
chosen because both Fenn, et al and Perrier and Gibson
include examples for this location. Since their methods
represent significantly different methods of estimating the
potential flux at the interface, reasonable conclusions can
be obtained regarding the significance of this part of the
model formulation. Two series of runs were performed:

A. A run corresponding to the Fenn, et al method for
the estimation of potential flux. This consists of inputs
of monthly flux data. The surface runoff is determined by
the average monthly precipitation and the rational method.
Evapotranspiration is estimated by the Penman method. Runs
as summarized in Table 1 include three different hydraulic
conductivity conditions and an additional run with the same
total annual flux, but distributed evenly over the simulation period at the annual average value.

B. A run corresponding approximately to the Perrier

and Gibson methodology. A more detailed evapotranspiration model is employed along with daily rainfall records. The runoff is estimated by the SCS method. The simulation used in this comparison used individual rain events (possibly lasting more than a day) from the precipitation data given for one year in the original report. The reader is referred to the original report for more details. These results are summarized in Table 2 with runs for three different hydraulic conductivities including the value used by Perrier and Gibson in their example. Also included is the same data except reduced to monthly format.

The results of the simulations in terms of annual average leakage rates for the different cases are presented in the respective tables. The results for the different hydraulic conductivities are presented to form a basis for comparison of the other results. Since Mitchell, et al[9] show the conductivity of a particular clay to vary by several orders of magnitude depending upon the water content at compaction, these presumably indicate the sort of uncertainty that may be inherent in the estimate of the material properties for a particular clay medium. Also included are some graphical representations of the leakage rates over the course of a year in Figure 3 and 4 for the two cases. A number of additional related test runs have been made that are not included in this presentation.

CONCLUSIONS

A number of interesting conclusions are implied by the above results and other similar numerical simulations. Although this work is still in progress, and conclusions are tentative, the following results have been observed:

1. The leakage rates through the clay layer are greater than the saturated hydraulic conductivity which is often stated to be the case for fully saturated flow. Leakage rates of four or five times this value can be computed. This is due to the fact that a greater than unit hydraulic gradient is predicted for circumstances with a saturated clay surface due to the large difference in clay and landfill properties.

2. If the total potential leakage into the clay is of the same order of magnitude as the saturated hydraulic conductivity or less, the liner will accept all downward flux. Since there is no horizontal interflow, the computation of surface runoffs and evapotranspiration are of extreme importance to the computations. However, it is not necessary to have these data on a per storm basis. In fact, the difference between using monthly data and average annual data is probably within the uncertainty of

Table 1

SIMULATION OF CONDITIONS PRESENTED BY FENN, ET. AL.

Material Properties

	Landfill	Clay Liner
ψ_a (m)	-.1	-10.0
A	1.0	1.0
Thickness (m)	-	1.5
$n - \theta_0$	.30	.22
K_0 (m/sec)	1×10^{-2}	5×10^{-10}

Leakage Rates (m/year)

	Fenn, et.al.	Present Simulation
1. Base Run*	.213	.07
2. Annual Average Precipitation Used	-	.10
3. Same as 2, $K_0 = 1 \times 10^{-9}$ m/sec	-	.16
4. Same as 2, $K_0 = 5 \times 10^{-9}$ m/sec	-	.20

*Base run used monthly average precipitation as provided in the referenced publication.

Table 2

SIMULATION OF CONDITIONS PRESENTED BY PERRIER AND GIBSON

Material Properties

	Landfill	Clay Liner
ψ_a (in)	-5.0	-200.0
A	1.0	1.0
Thickness (in)	-	18
$n-\theta_0$	.33	.19
K_0 (in/sec)	1×10^0	3.06×10^{-7}

Leakage Rates (in/year)

	Perrier and Gibson	Present Simulation
1. Base Run*	.6	.491
2. Same as 1, but monthly data		.490
3. Same as 2 $K_0 = 3.06 \times 10^{-8}$ in/sec		.374

*Base run used 5 day precipitation values and a 2.5 day time step. 1975 conditions in Cincinnati, Ohio were simulated using data presented in the referenced article. Annual precipitation for 1975 was 46 inches.

estimation of hydraulic properties of the clay. On the other hand, if the hydraulic conductivity of the clay is much less than the potential leakage rates, significant horizontal interflow occurs, and leachate generation rates are reduced.

3. The thickness of the clay liner only affects the time that it takes for a significant amount of leakage to begin to enter the landfill and has little effect upon the seepage rates.

Although some of these conclusions are somewhat obvious from a careful consideration of the problem, the analysis allows for quantification of the results. The conditions under which certain influences may be neglected may be more precisely delineated. Although more experimental work needs to be done to provide straightforward methods for determining the clay properties, the analysis makes it possible to determine those that most influence the results, so that efforts may be concentrated in those areas.

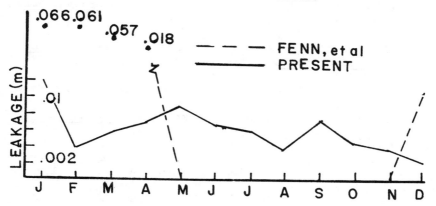

Fig.3. Simulations of Leakage Compared to Fenn et.al Result

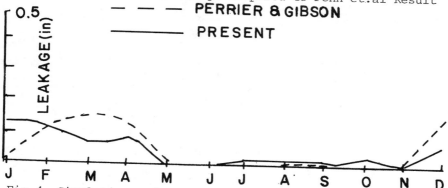

Fig.4. Simulations of Leakage Compared to Perrier and Gibson Result

355

REFERENCES

1. U.S.E.P.A. "Landfill Disposal of Soil Wastes, Proposed Guidelines", Federal Register, March 26, 1979, pp 18138-18148.
2. Lentz, J.J. "Apportionment of Net Recharge in Landfill Covering Layer Into Separate Components of Vertical Leakage and Horizontal Seepage", Water Resources Research, Vol. 7, August 1981, pp. 1231-1234.
3. Fenn,D.G., K.J.Hanley, and T.V. DeGeare "Use of the Water Balance Method for Predicting Leachate Generation from Solid Waste Disposal Sites", U.S. Environmental Protection Agency Report SW-168, 1975.
4. Perrier, E.R. and A.C.Gibson "Hydrologic Simulation on Solid waste Disposal Sites", U.S. EPA Report No. SW-868, Sept. 1980.
5. Bear, J. "Hydraulics of Goundwater", 1979 McGraw Hill, Inc.
6. Brooks, R.H. and A.T. Corey "Properties of Porous Media Affecting Fluid Flow", J. Irrigation and Drainage Div., ASCE, Vol. 92, IR 2, pp 61-88, 1966.
7. White, N.F., H.R. Duke, D.K. Sunada and A.T. Corey "Physics of Desaturation in Porous Materials", J. of the Irrigation and Drainage Division, ASCE, June, 1970, pp. 165-191.
8. Stakma, W.P. "The Relation Between Particle Pore Size and Hydraulic Conductivity of Sand Separates", Proceedings of the Wageningen Symposium, International Assoc. of Scientific Hydrology, 1968, pp. 373-382.
9. Mitchell, J.K., D.R. Hooper and R.G.Campanella "Permeability of Compacted Clay", Journal of the Soil Mechanics and Foundations Division, ASCE, July 1965, pp. 41-65.
10. Eagleson, P.S. 1978 "Climate Soil and Vegetation 3. A Simplified Model of Soil Moisture Movement in the Liquid Phase" Water Resource Research 14, pp.722-730.
11. Israelson, O.W. and V.E. Hansen "Irrigation Principles and Practices", 1950 John Wiley & Sons.
12. Rubin J. and R. Steinhardt "Soil Water Relations During Rain Infiltration: 1. Theory", Proceedings Soil Science Society of America, pp. 246-251, 1963.
13. Eagleson, P.S. "Climate Soil and Vegetation, 5, A Derived Distribution of Storm Surface Runoff", Water Resources Research, Vol. 14, 5, pp. 741-748, 1978.

CHAPTER 23

MONITORING, RECOVERY AND
TREATMENT OF POLLUTED GROUNDWATER

J. R. Quince
    O.H. Materials Co.

G. L. Gardner
    O.H. Materials Co.

INTRODUCTION

This paper presents information regarding underground
recovery and treatment systems that have been applied to
reclaim contaminated aquifers. Aquifer rehabilitation
systems are currently being considered, as an option for
remedial action, at many locations across the United States.
Published literature on this subject is very limited,
especially with regard to site specific restorations.
These systems have, however, been used in numerous cases
for recovery and treatment of contaminated groundwater
following accidental spills of hazardous materials and
contamination resulting from poor housekeeping at various
industrial facilities.

The concepts for recovery and treatment of contaminated
groundwater are well established. It is the application of
specialized techniques and equipment to real situations that
has been limited. Recovery can be accomplished using a
variety of techniques depending on site characteristics.
Treatment will depend primarily on the contaminant(s)
associated with the site. Effluent from the treatment
process can be effectively used for recharge to the aquifer
actively flushing contaminants to the recovery system. This
method creates a closed loop of recovery, treatment and
injection that can continually treat the groundwater until
the desired concentrations have been reached. Groundwater
and system monitoring is required to follow the progress
of the project. These systems are rapidly evolving to where
they will be applicable for many situations.

Four case histories are discussed briefly regarding recovery method, treatment methods used, and system efficiency at selected monitoring points. Some aspects of these projects have been disguised to maintain client confidentiality. All information presented is representative.

PRELIMINARY INVESTIGATION

The successful application of an aquifer rehabilitation system requires an understanding of existing site conditions. The site must be defined with respect to its underlying geology, groundwater description and contaminant characteristics. Many situations, for which aquifer rehabilitation may be appropriate, will have a variety of information available from previous investigations. This data should be collected, reviewed, and supplemented where necessary to provide the design criteria for the recovery and treatment system.

Information regarding site geology and groundwater occurence will be necessary to locate the recovery system. Geological information should detail the stratigraphy beneath the site including classical descriptions of soils and bedrock. The stratigraphy requires definition regarding horizontal extent and homogeneity. Groundwater occurence and aquifer characteristics are direct concerns for the recovery system design and application. Depth to static water levels will eliminate certain recovery methods while aquifer permeability may cause preferred selection of another. Hydrogeological data collected should include formation porosity, permeability, hydraulic gradient, groundwater velocity and direction, recharge/discharge information, and aquifer characteristics.

The quality and character of the contaminant(s), and their occurence in the aquifer system, will be necessary to determine the appropriate recovery method; select treatment system components; and provide an estimate of time for rehabilitation. The source of contamination commonly varies from a continuous release of material, such as leachate from a landfill, to a slug type release associated with accidental spills. The source should be removed or abated where feasible to prevent the continued release of pollutants to the aquifer. It makes little sense to renovate an aquifer which receives a continuous supply of pollutants.

Migration of contaminants within the groundwater flow system will depend in part on the character of the chemicals

358

involved. Many pollutants experience retardation in the aquifer due to attenuation and geochemical reaction during migration. Contaminants such as oil and gasoline, with a density ranging from 0.7 gm/cm^3 to about 1 gm/cm^3, will be immiscible in water producing a plume of material which floats on the top of the aquifer. Conversely, materials heavier than water (i.e., methylene chloride which is 1.3 times heavier than water) will sink to the bottom of the aquifer system creating special problems for detection and recovery. The extent and concentration of the contaminant should be defined both vertically and horizontally to enable optimum location of the chosen recovery system.

RECOVERY

The previous section has outlined the variety of information that will be used to select a method of recovery. Recovery techniques commonly applied use one or more of three principles; namely: gravity collection, suction lift or positive displacement. The majority of our current groundwater contamination problems relate to migration of contaminants to the water table aquifer.

Gravity collection is used through the application of interceptor trenches or french drains downgradient of the source. These methods provide for some degree of drawdown in the water table. This method of recovery is applied to contaminants in shallow flow systems associated with unconfined aquifers. Such systems have been employed for recovery of leachate migrating from an industrial landfill and collection of immiscible contaminants migrating on the groundwater surface. When dissolved constituents are involved, it may be necessary to monitor the groundwater downgradient of the recovery line to insure complete interception of contaminants. Gravity collection systems can be considered active or passive in nature, depending on their mode of operation. The system can be designed to recover groundwater with minimal changes in migration rates or to actively recover contaminants with accelerated flow rates. These systems are generally easily installed and require minimal maintenance.

Suction lift techniques (pneumatic recovery) have been used to effectively recover contaminated groundwater from shallow aquifers (0 to 25 feet deep). This method uses the direct application of a vacuum unit to a recovery well or recovery point system. The vacuum unit selected can be a common vacuum truck or specialized pumping equipment. Contaminated groundwater is drawn into the receiver using suction lift and transferred for processing through

appropriate treatment equipment. Care must be taken during
the installation of recovery equipment. Wells must be
located in the appropriate horizon and the bore hole annulus
sealed to insure efficient recovery.

Recovery point systems have proven very effective in a
variety of geological materials. Recovery point systems can
be installed in bore holes, driven, or jetted to the desired
depth. These systems employ numerous well points installed
perpendicular to the migration of contaminants. Individual
recovery points are spaced according to aquifer characteristics and installed using screen packing and bentonite
seals. Recovery points are connected to a main header
system using removable swing joints equipped with individual
valves. Swing joints are easily removed for individual
point samples and water level readings. Valves provide for
adjustment of recovery at each point according to the
desired drawdown. An important consideration is the ability
to monitor individual points and selectively recover from
areas indicating elevated concentrations which may result
from changes in lithology. Components of recovery point
systems are constructed of stainless steel, brass and
aluminum to resist corrosion from contaminants being
recovered. These systems are very adaptable and have proved
to be effective in intercepting contaminated groundwater
for treatment.

Positive displacement techniques use one or more wells
equipped with well pumps. Wells should be screened and
developed for optimum capacity. Pumping wells can be used
for shallow groundwater systems or aquifers located at
depth. Aquifer characteristics will be used to determine
appropriate well field configuration, spacing, and pump
specifications. In formations of low permeability, where
recharge to the well is slow, liquid level switches are
used to keep the well pumped down. The deep wells are
connected to a main header line which transports recovered
water to the treatment facility.

TREATMENT SYSTEMS

The type of treatment will depend primarily on the
contaminant(s) being recovered. Treatment systems may be
relatively simple, such as the use of carbon adsorption for
removal of a single contaminant, or extremely complex for
cases involving numerous contaminants such as leachate from
an industrial landfill or contaminated groundwater associated
with uncontrolled hazardous waste sites. In complex situations, a treatability study should be undertaken with
representative samples to determine appropriate treatment

components. Current technology provides for mobile treatment labs which can perform on-site treatability studies.

The background water quality should be examined for peculiarities which may affect treatment equipment. An example is the calcification which occurs in carbon filtration units due to hard water. This can create a loss of permeability through the filtration units and cause premature exhausting of carbon.

Actual treatment steps can be divided into three primary classifications: physical, chemical, and biological. An alternative "treatment" might be direct disposal of recovered material but transportation and disposal of large quantities of contaminated groundwater tends to be very costly. Actual treatment components selected can be applied in unlimited configurations depending on the degree of treatment desired.

Physical treatment techniques employ phase or component separation. Phase separation techniques are used for situations where there are distinct phase differences in the influent stream--be they gas, liquid, or solid. Phase separation techniques are relatively simple such as the use of oil separators for removal of an immiscible contaminant or the application of common filtering or settling techniques for removal of suspended solids.

Component separation techniques provide a more advanced (and more costly) form of treatment. Actual contaminants are separated from the waste stream through physical processes such as adsorption, stripping, ion exchange or ultra-filtration. These methods remove ionic or molecular contaminants from the waste stream for concentrated disposal.

Chemical treatment techniques provide for a change in the chemistry of the waste stream allowing removal of specific types of compounds. Reactive material is added to the waste stream to produce the desired change in chemistry. These methods include neutralization through pH adjustment, precipitation through the addition of flocculents, and oxidation or reduction through the addition of strong oxidizers or reducers. These methods create chemical transformations to destroy or detoxify hazardous components. Their most common applications have been for removal of metals through precipitation and for neutralization of corrosive wastes.

Biological treatment can be applied for biological detoxification of the waste stream. Such treatment uses enhanced microbiological activity to convert contaminants to nontoxic by-products. This process enhances the natural activity of microorganisms through adjustments of temperature, dissolved oxygen, nutrients; and, where appropriate, bacterial population. Biological techniques include activated sludge, aeration lagoon, trickle filters, anaerobic digestion, composting, and waste stabilization. Biological treatment is an emerging technology in which new strains of bacteria are being isolated for selective removal of contaminants.

The treatment system components can be selected from a variety of current and emerging technologies. The actual treatment configuration will be different for each case, depending on the contaminants to be removed and the degree of removal required. Current wastewater technologies need to be adapted for use in aquifer rehabilitation programs where the removal limits are generally lower than for wastewater treatment.

RECHARGE

Treatment system effluent can be used as an integral part of an aquifer rehabilitation program. Effluent is used for recharge to the aquifer to flush contaminants and/or for hydrodynamic controls. Effluent can be discharged to the aquifer downgradient of recovery if contaminant concentrations have been reduced to acceptable limits. Most commonly, it has been used to recharge the aquifer upgradient thereby creating a closed loop of recovery, treatment, and recharge. This recovery-recharge method creates a defined area which is hydraulically separate from the aquifer. This is particularly appropriate in special cases where treatment does not reduce contaminant concentrations to given discharge criteria during initial volume exchanges. Generally, 10 to 15 volume changes are considered necessary to remove contaminants from the affected system.

Aquifer recharge can be accomplished using a variety of techniques. If the aquifer represents the sole location of contamination, recharge may be through wells or well point systems located in the aquifer. Commonly though, contaminants have migrated from a source on the surface via specific paths. If these migration routes can be defined, it becomes appropriate to use the recharge to flush contaminants to the recovery system. This can prove particularly effective in removing residual contaminants from the vadose zone. This flushing action is accomplished through

362

the use of surface spraying and jetting; infiltration
trenches or ponds; buried injection lines; or injection
point systems. The quantity of water needed for injection
will vary such that some effluent may be discharged off
site or additional water sources may be required to suppli-
ment injection. Recharge techniques, used in conjunction
with an active recovery system, provide for accelerated
groundwater flow rates, many times the native velocity,
thereby rapidly flushing contaminants from the aquifer.

Recharge can be effectively used to establish groundwater divides. These can be strategically located to abate
migration of contaminants during the rehabilitation. The
divide can be located downgradient of the plume to reverse
local flow directions and allow for recovery of contaminants
which might otherwise be considered lost.

A further extention of this technique has been the
addition of biodegradation agents in recharge waters.
Many compounds will readily biodegrade under natural conditions over a given period of time. Hydrocarbons, and a
number of organic chemicals, tend to be retarded in the
geological environment thereby requiring extensive flushing
to reduce contaminant concentration. In-place biodegradation provides an effective means for removal of biodegrad-
able contaminants. Microorganisms may be available in the
native environment or they can be introduced through the
application of commercial biodegraders. The chemistry of
the recharge water is adjusted to produce optimum conditions
for microbial activity. Recharge water is adjusted with
respect to its pH, temperature, dissolved oxygen content,
nutrient concentration, and when appropriate microbial
populations. Biodegrading agents tend to move through
the aquifer as a front breaking down contaminants into
nontoxic by-products through the natural metabolism of the
organisms. Certain species of microorganisms (i.e., hydrocarbon biodegraders) are facultative in nature thereby
having the ability to biodegrade contaminants under aerobic
or anaerobic conditions. This is a very desirable phenomenon for aquifer rehabilitation where oxygen supplies are
rapidly depleted. Biodegradation of subsurface contaminants is an emerging technology which will likely evolve
to be an effective means for aquifer rehabilitation.

MONITORING

Monitoring aquifer rehabilitation programs can be
divided into two main concerns: groundwater monitoring
and systems monitoring. Indicator compounds are identified
during preliminary investigations or gross analyses

363

performed, such as TOC, to reduce analytical costs.
Complete pollutant scans can be performed periodically to
insure that the indicator compound is providing meaningful
data and to gain data for other contaminants recovered on
complex projects. An on-site laboratory is appropriate for
rapid turnaround of samples. Numerous checks and balances
are generally required on large projects and rapid analyses
prove to be beneficial.

Groundwater monitoring is performed on a regular basis
to determine changes in water quality and monitor physical
changes due to the injection-recovery system. Data
collected will indicate the effectiveness of the recovery
system, hydrodynamic barriers, and the rehabilitation
program in general. The monitor well network will be
different for each situation but should monitor changes
within the zone of influence as well as water quality
outside the perimeter of the affected area. Monitor wells
penetrating underlying aquitards must be properly installed
to insure representative samples and prevent downward
migration to clean areas.

APPLICATIONS

These recovery and treatment systems outlined have
been applied to a number of situations during the past
five years. This section discusses information acquired
from four different projects.

The most commonly used system has been the underground recovery and treatment system employing pneumatic
recovery and the well point recovery system. This system
was used to decontaminate groundwater associated with two
uncontrolled hazardous waste sites in New Jersey. Groundwater removed from these sites had an extremely complex
chemistry due to extensive leaching of different contaminants from storage containers and lagoons. This system has
also been used at several different transportation accidents
involving spills of chemicals such as acrylonitrile,
orthochlorophenol, chloroform, and other organics. In all
cases chemicals migrated to the groundwater and created a
defined plume of contamination. Contaminated groundwater
was thus recovered, treated and discharged to abate migration.

Shallow well systems have been used in different cases
for the recovery of hydrocarbons following accidental
gasoline leaks at commercial businesses. In one case,
a biodegradation program was implemented to remove residual
gasoline adsorbed to silty material after the free gasoline

was recovered. Another treatment system is currently being used to remove trichloroethylene from groundwater using pumping wells located at depths of over 100 feet. The trichloroethylene is removed using activated carbon adsorption and processed water is discharged to local storm sewers. These systems have been used for recovery of contaminated groundwater from geological materials ranging from gravel to clay, fractured bedrock or combinations of the above.

The time required to renovate a contaminated aquifer and the associated costs will be very different for each case. Such projects can run from weeks to years in length and costs may range from tens of thousands to millions of dollars. Numerous factors will be considered in estimating the time frame that will be required to efficiently clean a polluted groundwater system. The hydro-geological environment, the character of the contaminant and the time since release of the pollutants will be primary considerations. Each situation will have its complexities that require solution, but these systems do work and are applicable to many subsurface contamination problems. Emerging treatment technologies indicate promise for providing alternative cost effective means for cleaning our groundwater contamination problems. As these systems are applied more often, we can expect to receive valuable information regarding the effects of contaminants on the subsurface environment.

Case History #1

Following a train derailment, involving spillage in excess of 30,000 gallons of a volatile carcinogenic liquid, several methods of recovery were needed to control the spread of the pollutant through a complex geologic formation. In this situation, the accident occurred in a location which allowed rapid movement of the contaminant through the subsurface soil and percolation out of an embankment of fill material below the railroad track. Below this fill area, a fractured coal bed promoted further subsurface migration. A layer of siltstone below the coal bed acted as an impermeable barrier preventing further downward migration.

Pure product, leaching from the embankment of fill material, entered city storm drains which emptied into a surface stream providing further migration of the pollutant. This storm drain had apparent fractures and open joints providing a conduit for the pollutant to enter the groundwater system 1,000 feet from the original spill area. Due

to these circumstances over 1.8 million cubic feet of soil
and groundwater became polluted in a short period of time.

Due to the complexity of this project, several means
of recovery and containment for various locations were
necessary. Table I outlines methods which were utilized,
their respective circumstances, and concentrations
encountered. Locations of recovery sources and the wide
area of contamination are illustrated in Figure 1.

The combination of these recovery methods contained
the pollutant preventing further migration. Prior to installation of each system a carefully planned monitoring program
began to profile existing conditions of soil and groundwater.

Soil samples were obtained from borings at the original
spill area to provide data regarding placement of several
series of recovery points. Initially this system provided
removal of highly concentrated material from the unsaturated
area of the site. In later stages, this system was used
in conjunction with shallow injected water to flush contaminants from the soil. The compound demonstrated a
high affinity for the soil and fill material causing this
phase to be very time consuming.

Pure material which flowed through the sewer and into
the stream was removed immediately. After the majority of
the pollutant was removed from the stream, equipment was
installed in the creek bed to remove water containing small
quantities of the compound due to sediment contamination.
This process involved approximately 90 days to clean 1,000
linear feet of streambed at concentrations of greater than
300 mg/liter, to less than 0.5 mg/liter.

Due to the topography of the site in relation to the
depth of contamination, a gravity collection trench and an
additional series of production wells were installed downgradient from the original spill area. The main function of
this combination system was to provide a cutoff at the
fringe of the contaminant plume and recover this water for
treatment. Since vacuum-operated wells at the original
spill zone were limited to depths of 20 feet; this system
proved to be highly effective in recovery of groundwater
below the 20 foot depth. Groundwater movement in the
formation was slow and the water yield ranged from 10,000
to 30,000 gallons per day depending on weather conditions
and the flushing activities up gradient.

In addition to these three recovery operations, localized conditions of high contamination were recovered by

Table I. Recovery Methods Applied At Different Locations to Contain and Recover Contaminated Groundwater

Recovery Method	Media	Concentration Range
Vacuum Recovery System I	Unsaturated Zone Original Spill Site	>2,000 - <10 ppm
Gravity Collection Trench	Downgradient at Coal Clay Interface	500 - <10 ppm
Submersible Well Pumps	Localized Wells in Residential Areas	600 - <10 ppm
Vacuum Recovery System II	Recovery of Groundwater Downgradient From Site to Depth of Impermeable Barrier	400 - <10 ppm
High Capacity Pumps	Surface Stream	>2,000 - <.5 ppm
Vacuum Recovery System III	Creek Sediments	>2,000 - <.5 ppm

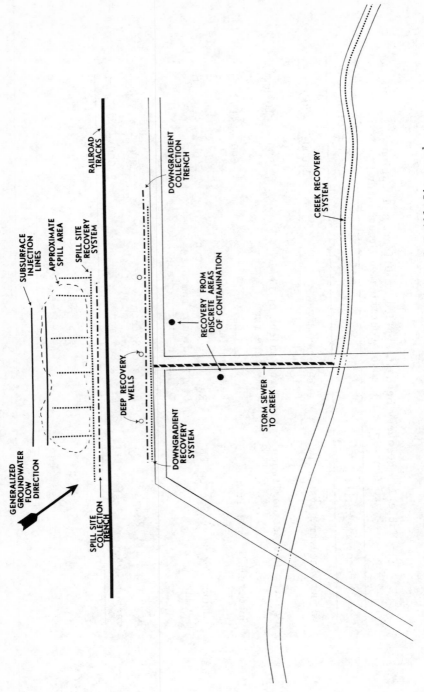

Figure 1. Site Map Delineating Original Spill Site and
Various Recovery Methods Following Initial Contaminant Migration

368

means of pumping wells for extended periods of time. This method was used especially in areas which were contaminated by means of the storm sewer leaking during the spill. Surrounding monitor wells indicated that contamination did not exist there indicating that these local "hot spots" were not related to the primary plume.

Two of the recovery sources are illustrated in Figures 3 and 4 designating reductions in concentrations during the course of the project.

All of the contaminated groundwater recovered was treated at a site near the spill area. This treatment system consisted of mobile equipment sized to meet the volume of water recovered daily. A schematic of the system is described in Figure 2. Overall recovery ranged from 20,000 to 120,000 gallons per day depending on weather conditions and site activities. Water recovered from any phase was transferred to a central bulking point and pumped into tankers which transported the water to the treatment facility.

Concentrations of the influent water ranged from 72,000 to 1 mg/liter. To meet regulatory requirements, this water had to be treated to less than .1 mg/liter prior to discharge. Since the material was highly volatile, aeration and heat stripping were used with high efficiency to reduce the amount of granular activated carbon normally needed. The system was completely sealed to prevent any pollution of the ambient air.

During the cleanup, over sixty monitor wells were sampled on a daily basis. Samples could be analyzed quickly using an on-site laboratory thereby making it possible to reduce or change the system based on current data. Over 60,000 samples were analyzed in the on-site laboratory during the project to monitor conditions at the site. The large number of samples taken did not affect analytical costs since the lab was placed on site at fixed cost.

Overall, this was a very complex and time intensive project. To meet regulatory requirements, soil and groundwater required extensive flushing, recovery, and treatment. This system proved to be successful in accomplishing the tasks described to concentrations of less than 10 mg/liter. At that level, recovery remains effective but time and cost considerations need to be evaluated to reduce concentrations to less than .1 mg/liter.

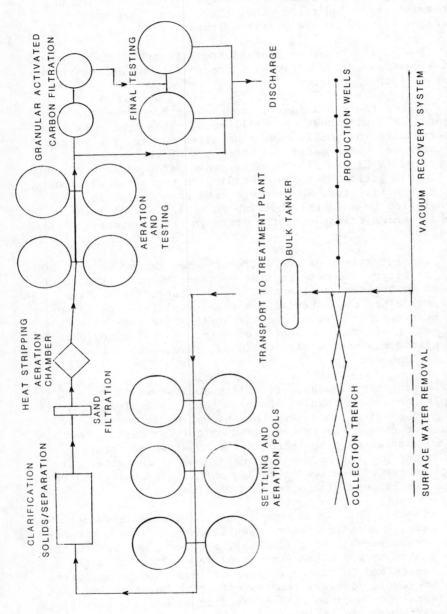

Figure 2. Treatment System Schematic – Case History #1

370

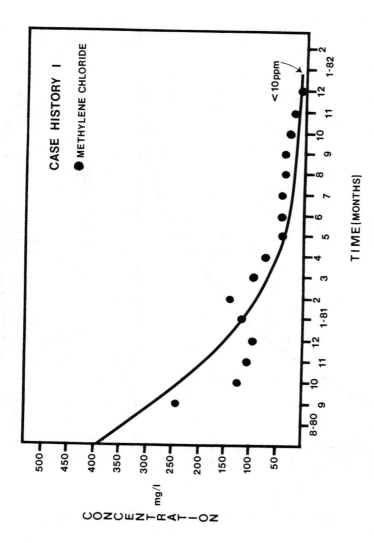

Figure 3. Reduction in Contaminant Concentration for
the Collection Trench Downgradient of the Spill Site

371

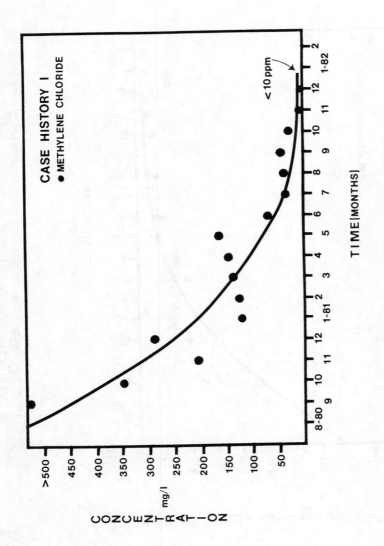

Figure 4. Reduction in Contaminant Concentration for the On-site Recovery System Influent

A transportation accident occurred in the Midwest involving many leaking storage tankers, resulting in unrestrained flow of 100,000 gallons of various chemical compounds. This spill affected an immediate surface area of 250,000 square feet. Due to the volatile nature and toxicity of several compounds, residents living in the immediate vicinity were evacuated from their homes. Emergency cleanup crews responded to the scene to contain and recover material preventing further spread of the contaminants.

Three of the five chemicals penetrated the soil at a rapid rate, limiting recovery during the surficial cleanup. Once the primary surface cleanup was completed, a hydrogeological survey began to evaluate the effects of the spill on the underlying formations.

The surface area and outer perimeter was gridded and soil borings were taken using a drill rig equipped with hollow stem augers and split spoon samplers. Samples were taken at increments of 50' at depths of 1', 3', 5', 7', and deeper if needed. The subsurface soil consisted of a thick silty clay extending to a depth of greater than 50'. This formation prevented migration into the main aquifer and limited containment and recovery to the perched water table within the upper clay layer.

Monitor wells surrounding the outer perimeter of the spill area were used initially to detect any migration of contaminants and later to monitor the ability of the recovery system to contain the material. In addition, residential wells and surface waterways within a one mile radius of the site were monitored. All three chemicals involved were tested for at a mobile laboratory on site. This provided rapid turnover of samples making it possible to evaluate new data daily.

The initial study indicated that migration of the chemicals in any lateral direction was quite limited due to the low permeability of the soil. Soil borings indicated that the material had migrated downward to depths of 6 to 8 feet. Residential wells and waterways which were monitored, demonstrated that no detectable contamination occurred from the spill event.

The recovery system used in this case, involved physical removal of the groundwater, and formation of a cone of depression to prevent migration. Based on the samples taken, this system was placed within the contaminant plume and along the outer perimeter of the affected area. Over 200 recovery wells were installed to depths of 6 to 10 feet and connected to a common header system to transfer groundwater to a central location. Initial recovery produced 75,000 gallons per day. Concentrations varied from 2,000 to 10,000 parts per million (as a total of all three organic chemicals) during the startup of the recovery system.

Water recovered by the system required treatment to remove these contaminants so the effluent water could be reinjected into the ground. This injection flushed contaminants from the soil and replaced water depleted during recovery activities.

The treatment system consisted of several modules, each designed to treat the three chemicals in an efficient manner. A schematic of the system is presented in Figure 5. The system was completely closed, due to the volatility of the compounds, to prevent ambient air contamination. Contaminated air generated during the process was treated and reused for air stripping. Pretreatment proved to be highly effective in conserving granular activated carbon which was used as a final phase of treatment.

Monitoring of site activity and overall progress during the recovery activities consisted of the following procedures:

1. Water table elevations of monitor wells and piezometers were obtained twice daily to check flow gradients.
2. Contour maps were developed to examine elevations and record the evidence of a cone of depression within the spill site.
3. Water samples were obtained from site monitor wells daily to delineate the plume size and observe overall reductions.
4. Various water samples from the treatment system were obtained on a daily basis to review the efficiency of individual phases within the system.
5. Background samples were obtained for a bio-feasibility study regarding the application as a treatment alternative.
6. Water samples obtained from individual recovery wells within the complete system showed a more detailed description of the plume and areas of reduced or slow cleanup (hot spots).
7. Replicate soil analyses at different periods to monitor cleanup from injection and removal.

374

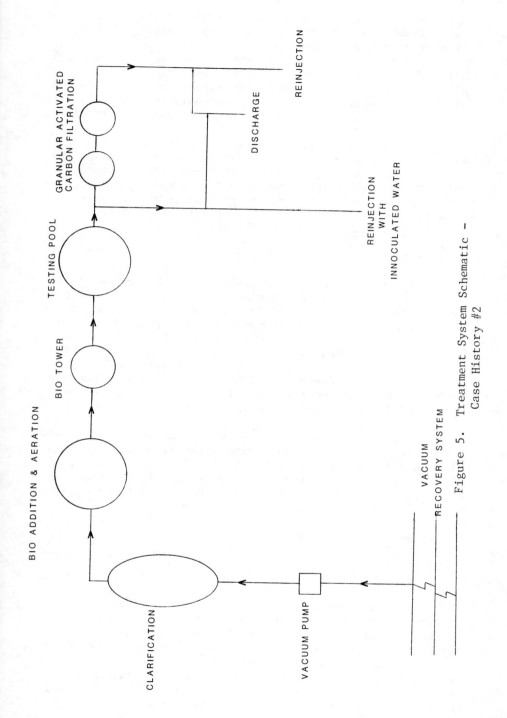

Figure 5. Treatment System Schematic –
Case History #2

Once levels of the three compounds reached concentrations less than 200 parts per million, a biological degradation program began based on the feasibility study which indicated that the materials could be biodegraded in situ. Water was innoculated with special bacteria, nutrients, and air prior to re-injected into the subsurface. This method of treatment decreased the time necessary to reach levels of concentration approved by the regulatory agencies involved.

Figures 6 and 7 show progress and reductions at key monitor wells and the recovery system influent, during the actual periods of aggressive recovery-injection. When extremely low concentrations were reached which met regulatory approval, cleanup ceased. At this point, monitoring of wells continued until it was verified that the contamination remained below detectable levels.

Case History #3

This project involved a groundwater cleanup necessary due to the discharge of a myriad of components from leaking drums, tanks, and other contaminated sources. Prior to any monitoring or groundwater recovery, the source of contamination was removed. This removal included buildings, tanks, drums, and a concrete slab covering 80% of the site. This particular site had a wide range of compounds including polychlorinated biphenyls, poisons, carcinogens, flammable organic materials and toxic pesticides. The exact nature of all products stored at the facility remains unknown.

Subsequent to the surface cleanup, a large scale monitoring program began to determine existing subsurface conditions and the necessity of subsurface reclamation. The site was adjacent to a river and was situated in the center of an industrial zone. Primary monitoring included examination at the following areas:

1. adjacent river
2. soils, on-site and off-site
3. monitor wells on-site
4. control wells

This investigation involved analysis for priority pollutants by gas chromatography, GC/MS, TOC, and atomic adsorption methods outlined by the Environmental Protection Agency. Results from the study revealed high levels of volatile organic compounds, heavy metals, and elevated concentrations of total organic carbon when compared with control samples obtained off site. Table II outlines data ranges from the sources tested for TOC values.

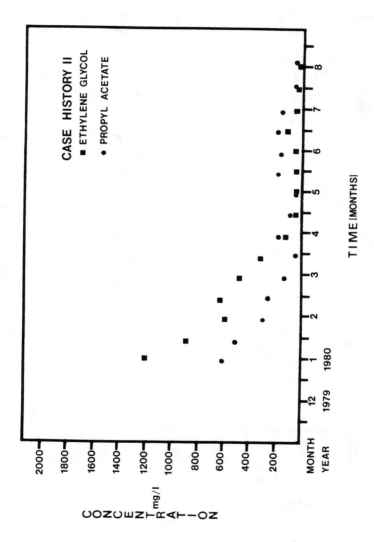

Figure 6. Reduction in Contaminant Concentration for the
On-site Recovery System Influent

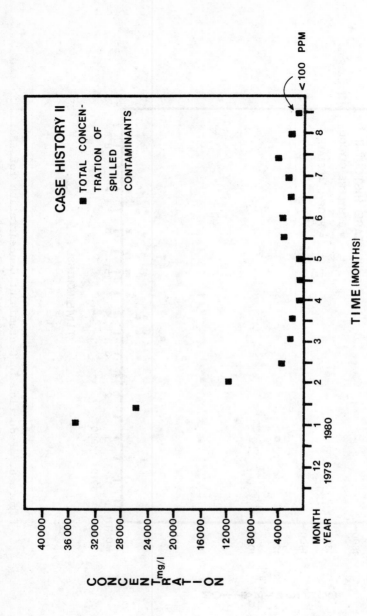

Figure 7. Reduction in Contaminant Concentration for
a Monitor Well Within the Spill Plume

Table II. Location of Monitoring Points and Respective Concentrations

Monitor Point Location	Concentration Range TOC Levels (ppm)
Adjacent River Upstream	<10 - 20
Adjacent River At Site	20 - >400
On-Site Soil	5,000 - >50,000
Control Soils	1,000 - 10,000
Site Monitor Wells	1,000 - >5,000
Control Wells	<50 - 125

During the course of the project, data consisted of TOC analysis which provided quick turnaround time for samples and reduced analytical costs considerably. Predominant compounds (organic and inorganic) evident in the initial screening were checked periodically to monitor reductions of these specific indicators.

Data from the soil borings and monitor wells dictated placement of the recovery system. The soil consisted of permeable fill material underlaid by a thick clay. Fill varied in depth across the site from 10 to 15 feet below the surface. Tidal effects from the adjacent river caused a large amount of water to be exchanged daily. The main purpose of this recovery system was to remove the majority of the pollutants and prevent migration into the adjacent river.

A vacuum operated recovery system was installed consisting of five lines of recovery wells extending from each end of the site. Over 180 recovery points were placed along these lines. Recovery rates within this system ranged from 50,000 to 120,000 gallons per day. In winter weather, the complete system was insulated to prevent freezing. To maximize recovery, this system operated 24 hours a day. During periods of low tide, water recovery was reduced proportionately.

Since the polluting components varied in composition and required specialized treatment to reduce each class of components, an elaborate treatment system was constructed. This process involved several descrete functions, each designed to perform in series or separately, depending on analytical data regarding the waste stream. The steps involved and a schematic of this system are presented in Figure 8.

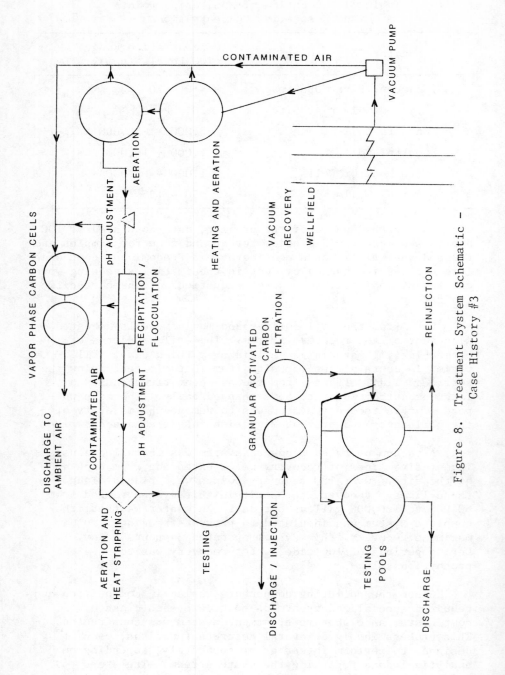

Figure 8. Treatment System Schematic –
Case History #3

Monitoring each part of the system enabled operators to make adjustments to the system reducing the treatment necessary for the waste stream. A reduction of the groundwater TOC averaged over 70% throughout the system. Many volatile compounds were reduced to non-detectable levels in the heated air stripping chamber prior to carbon filtration.

Figures 9 and 10 illustrate the overall reduction in TOC values for the recovery system and two wells located within the site. These reductions varied and these wells were selected since concentrations remained consistently high during initial testing.

After five months of recovery, over 5.5 million gallons of groundwater were treated. A 90% reduction in total organic carbon was noted in groundwater recovered from the beginning of the project until completion. This system provided a cost effective cleanup of contaminated water and treatment at a lower cost than for transporting to a treatment facility. Levels of contamination were significantly reduced during the time allotted for cleanup activities.

Case History #4

This project involved an industrial facility which stored many organic chemicals in surface and subsurface tanks. Due to leaks, overfills, and spillage groundwater contamination problems developed over a prolonged period of time. The site geology consisted of unconsolidated glacial deposits ranging from clay to fine sands. This sand and clay layer was underlaid by fractured limestone bedrock. During subsurface tank installation the overburden was removed to the fractured bedrock. Sandy backfill replaced the soil removed during excavation for the underground storage area.

Monitor wells, located downgradient from the storage areas, indicated the presence of several chemicals at low but regulated concentrations. To prevent contamination from entering the aquifer below the limestone, containment and recovery measures were taken.

A vacuum operated recovery system consisting of forty recovery wells and ten deeper wells equipped with pumps were installed to remove the contaminated soil water. Recovery rates were low due to the volume of water available. Soil which was not excavated required flushing by injection and surface spraying to remove the pollutants.

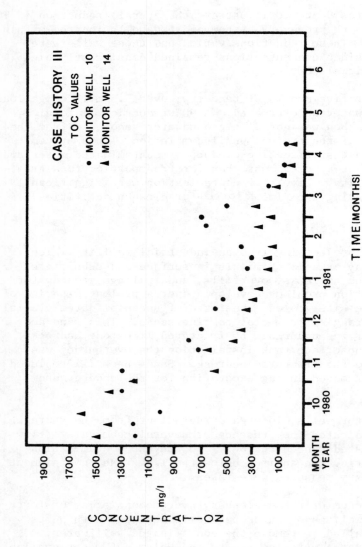

Figure 9. Reduction in Contaminant Concentration for Two Monitor Wells Within the Plume

382

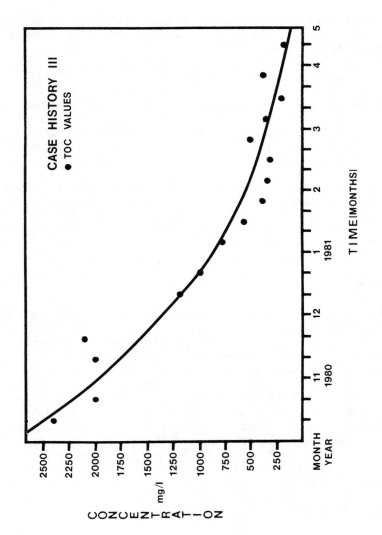

Figure 10. Reduction in Contaminant Concentration for the On-site Recovery System Influent

383

In this case, the pollutants were easily volatilized so air stripping could be used as an effective treatment method. All water recovered was treated by air stripping and inoculated with hydrocarbon degrading bacteria. This treatment proved to be highly efficient so granular activated carbon treatment was unnecessary. The treatment system is presented in Figure 11.

During the project monitor wells within the study area were checked on a weekly basis to derive the effectiveness of the recovery system. Daily testing of bacteria and nutrient levels indicated steady increases until optimum conditions existed for treatment. Following the addition of commercial biodegrader, testing of the effluent showed a significant decrease in the levels of organic contaminants after a 36 hour exposure time. Once the acceptable level was reached, the inoculated water was re-injected to flush the soil or discharged. Recharge water provided for effective in situ biodegradation of residual contaminants.

Figures 12 and 13 show the decreases in concentrations of organic constituents at a monitor well and through overall recovery. When a 95% reduction in contaminants was reached, operations were terminated. The established bacterial population was expected to continue to biodegrade residual contaminants in the soil.

CONCLUSION

This paper has presented an overview of techniques currently applied to recover and treat polluted groundwater. The given site must be well defined, through preliminary studies, to determine appropriate recovery and treatment methodology. Monitoring points established during preliminary investigations should be supplimented as necessary to insure effective monitoring of the recovery, treatment and injection system. Treatment system components should be monitored regularly to maximize their efficiency. Treatment system effluent is effectively used for aquifer recharge creating a closed loop system of recovery, treatment, and recharge. A limited number of applications indicate 10 to 15 volume exchanges are required to effectively reduce contaminants to acceptable standards. This will be an astronomical quantity of water in some cases.

Pollutant removal from groundwater systems indicates an asymptotic relationship such that the majority of the contaminant is removed in early phases of such projects. In situ recovery and treatment techniques including biodegradation have proven more effective (and environmentally aesthetic)

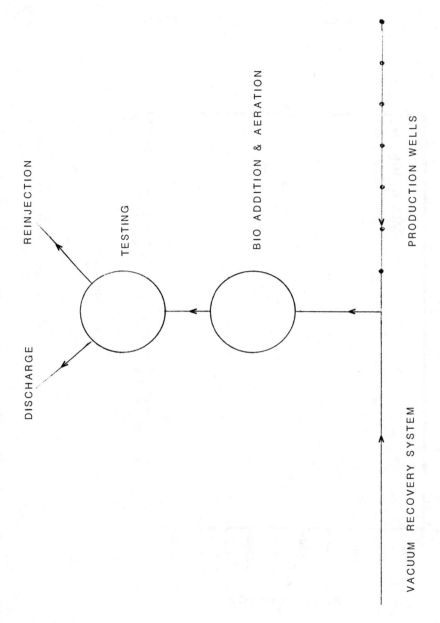

Figure 11. Treatment System Schematic –
Case History #4

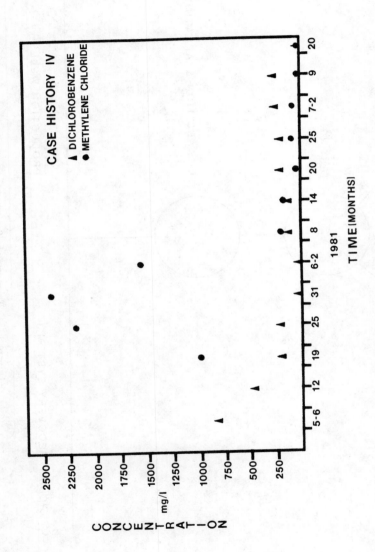

Figure 12. Reduction in Contaminant Concentration for a Monitor Well Within the Contaminated Zone

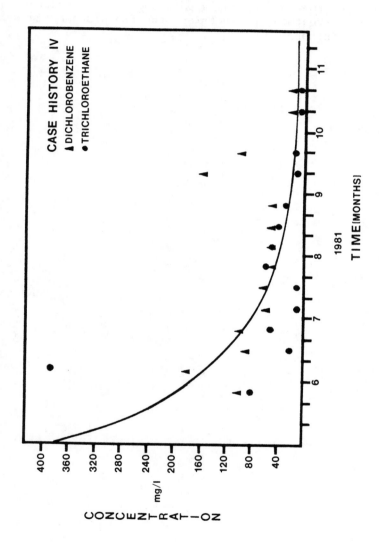

Figure 13. Reduction in Contaminant Concentration for
the On-site Recovery System Influent

than excavation and disposal of contaminated material.

These systems will undoubtedly see continued use and evolution in the near future.

ACKNOWLEDGMENTS

We are indebted to our colleagues at O.H. Materials Co. for their information regarding past and present underground recovery and treatment projects and their assistance in preparing this paper.

CHAPTER 24

GROUND-WATER MONITORING AT AN INDUSTRIAL WASTE DISPOSAL SITE IN PENNSYLVANIA

Nancy L. Cichowicz
Richard M. Cadwgan
John R. Ryan
          Environmental Research & Technology, Inc.

INTRODUCTION

A hydrogeologic investigation is being conducted at a former industrial waste disposal site in Pennsylvania. Coal-coking sludge, agricultural chemical residue, and desulfurization waste are included in the variety of industrial wastes known to have been disposed of in trenches. The 35-ac site is located on the tip of an island in a river. A preliminary site investigation had already determined that ground water was contaminated with several organic chemicals and that the direction of ground-water flow was toward the end of the island into the river.

The hydrogeologic investigation is being conducted in two phases. The purpose of Phase I is to determine a preliminary estimate for the discharge rate of organic contaminants to the river via ground water. Because the discharge rate is dependent upon the hydraulic gradient of the water table, the purpose of Phase II is to better define the hydraulic gradient by continuously monitoring ground and river water levels.

HYDROGEOLOGY

The island on which the site is located is a dissected river terrace that was deposited by an ancestral river. The terrace is partly submerged by existing impoundments on the river, but remnants of it flank both sides of the river at approximately the same elevation as the island. The terrace consists of approximately 60 ft of fluvial deposits which overlie sandstone and shale bedrock. Shown in cross-section (Figures 1 to 4), the upper 30 ft of the island consists of silty clay and silty fine sand, while the lower 30 ft is primarily sand and gravel.

Approximately the lower 40 ft of the fluvial sediments is saturated. Perched ground water exists locally in the upper fine-grained sediments, though the aquifer is essentially unconfined. Geologic data show that the river has truncated the fine-grained sediments and eroded 10 to 20 ft into the sand and gravel deposits present beneath the site. The aquifer beneath the site is in direct hydraulic connection with the river.

PHASE I - PRELIMINARY HYDROGEOLOGIC ASSESSMENT

A multi-level sampling system was installed during November 1980 to permit a three-dimensional assessment of ground-water quality conditions. Twenty-one multi-level monitoring locations were equipped with gas-driven samplers to permit collection of ground water at 46 locations beneath and adjacent to the site. In addition, five existing monitoring wells installed during the previous site investigation were equipped with gas-driven samplers. Collection of ground water from these samplers is accomplished using inert nitrogen to pressurize polyethylene tubing and force the water to the surface.

Ground-water sampling was conducted on three occasions between January and July 1981. The first sampling round was most extensive in terms of samples collected and chemical analyses performed. Samples, field blanks, and duplicates were analyzed primarily for the organic priority pollutants (as identified by the US Environmental Protection Agency) and five pesticides.

Information about ground-water flow beneath the site was derived from six sets of water-level measurements taken between January and August 1981. Some additional water-level measurements were available for previously existing monitoring wells. River elevations were obtained from lockmaster records at two nearby dams and from a gage installed near the site. Generally, the contour maps of ground-water elevation showed a gentle, elongated mound in the water table, with elevation usually highest near the center of the site. The maximum observed relief of the mound above the river was about one foot. Ground water flows predominately toward the north and south shores of the island at an average gradient of 0.0025.

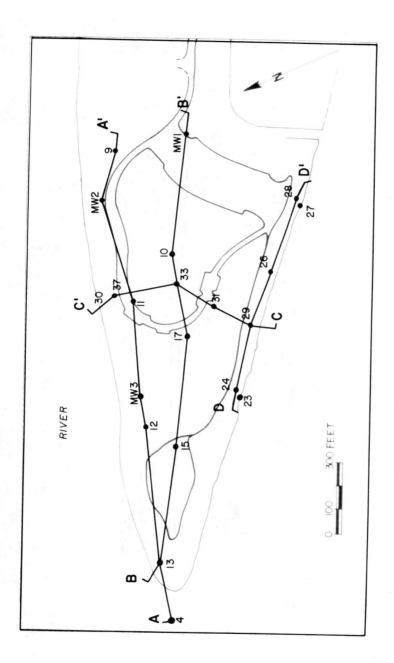

Figure 1. Location of cross sections A-A', B-B', C-C', and D-D' shown in Figures 2, 3, 4, and 5.

RIVER

0 100   300 FEET

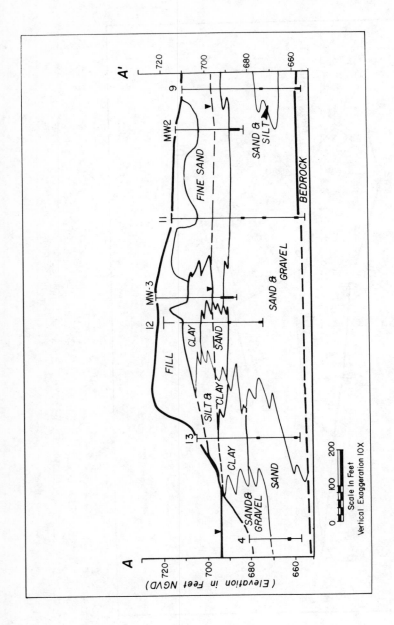

Figure 2. Cross section A-A' (MW series borings installed during previous site investigation.)

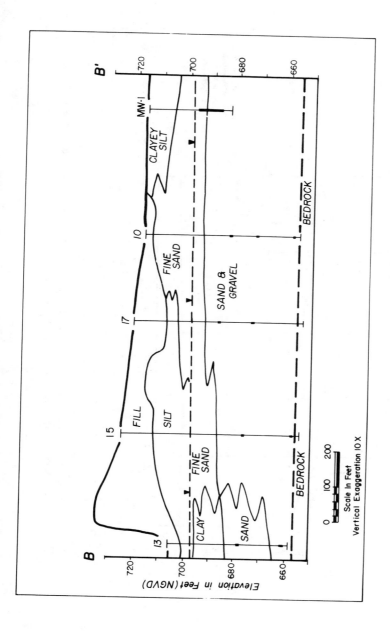

Figure 3. Cross section B-B' (MW series borings installed during previous site investigation.)

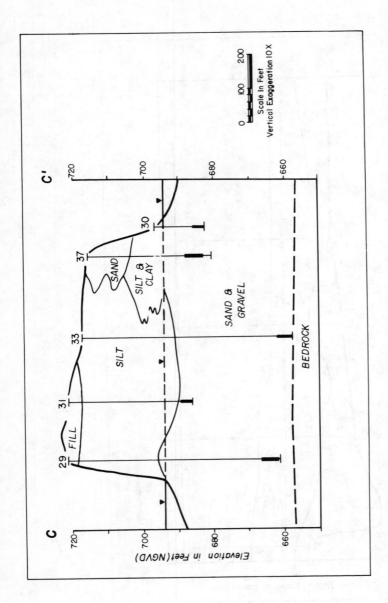

Figure 4.   Cross section C-C' (MW series borings installed during previous site investigation.)

394

Using the data on ground-water quality and flow, an estimate of the mass of contaminants being discharged to the river could be determined. A simplified model of ground-water flow for the site was developed. Cross sections with contaminant concentration isopleths were prepared for the northern and southern discharge area of the site and for a section through the width of the island. As an example, Figure 5 shows the distribution of benzene along the southern discharge area of the site. In addition to these cross sections, several assumptions were made concerning the local hydrogeology:

o   average annual recharge of 9 in/year,
o   average hydraulic conductivity for sand and gravel of 2×10^{-3} cm/sec,
o   average aquifer porosity of 30 percent, and
o   area of discharge of 98,230 ft^2.

Each cross section was further divided into cells to account for the heterogeneous distribution of contaminants. Concentrations were assigned to cells within the shoreline cross sections based on analytical results for benzene, toluene, xylene, phenol, 2,4-dichlorophenol, 2,4,6-trichlorophenol, and three pesticides. Figure 6 shows the dimensions of the flow cells and benzene concentrations assigned to the southern shoreline discharge area.

A comparison of the total mass of each of the contaminants listed above shows that benzene is the dominant compound being discharged to the river via ground water. The discharge rate is estimated to be less than 2 lb/day. The total mass of the other contaminants being discharged is estimated at less than 1 lb/day.

PHASE II - CONTINUOUS WATER-LEVEL MONITORING

The simplified model of ground-water flow for the site was developed using an average hydraulic gradient for the water table with the direction of flow toward the river. However, the determination of the discharge rate of contaminants is primarily dependent upon the hydraulic gradient. It was speculated that rapidly changing river elevation could affect the gradient and direction of ground-water flow beneath the site. Therefore, in November 1981, Phase II of the hydrogeologic investigation was initiated. The purpose of Phase II was to better define

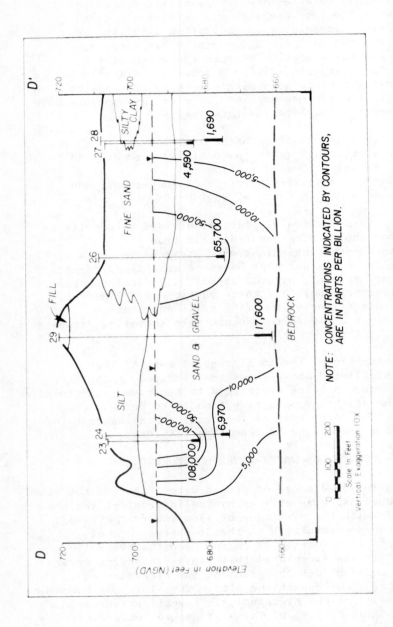

Figure 5. Cross section D-D' with benzene concentrations detected during July 1981 sampling round.

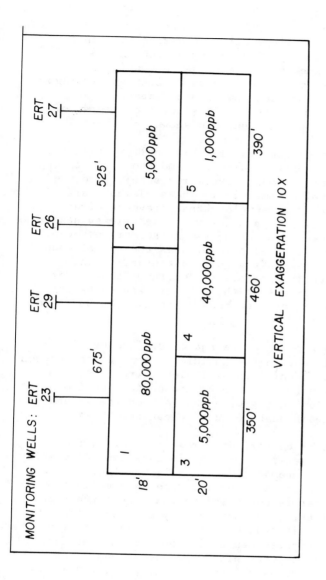

Figure 6. Dimensions of flow cells and benzene concentrations assigned to model of southern shoreline discharge area.

the hydraulic gradient. This was accomplished by continuously monitoring ground and river water levels at the site.

Equipment was installed in two open standpipe wells on the site and one stilling well in the river to monitor real-time water levels. The equipment consists of electrical resistance water-level sensors which are connected by wire to automatic data collection, storage, and satellite transmission components located in an existing building on the site. A schematic diagram of real-time monitoring components is shown in Figure 7.

The water-level sensors are thin, helical resistors one and one-half inches in diameter and about 17 ft long. Internal resistance elements are strip steel, gold, and spiral-wound nickel chrome wire. These elements are enclosed within a Mylar envelope that prevents direct contact with water. The protective jacket flexes easily as the exterior liquid level changes. As submergence depth increases, the Mylar envelope depresses the spiral-wound nickel-chrome wire against a gold contact strip and shortens the effective resistance length of the sensor. Because adjacent spirals of the wire are spaced exactly 1.0 cm apart, water level changes as small as 0.03 ft can be detected.

The sensors are linked to and controlled by centralized equipment in the building. Connection between sensors and control units is by 12-lead wire, with a maximum separation between sensor and control equipment of about 1500 ft. Connecting wire was laid on the ground surface and buried beneath 6 in of clean soil.

At the control center, data may be obtained from manual read-out boxes that provide instant digital display of water levels. Automatic data acquisiton functions are controlled by a microprocessor inside a data collection platform (DCP) that is connected to the manual read-out units. The DCP queries sensors hourly and stores data. Every three hours, it transmits data via satellite from the site to the ERT computer center in Concord, Massachusetts. Each transmission includes nine water level and three battery voltage readings. A small solar panel provides power to keep a 12-volt battery fully charged most of the time. Extended periods of cloudy weather and thick morning fog at the site occasionally require that a battery charger augment the solar panel.

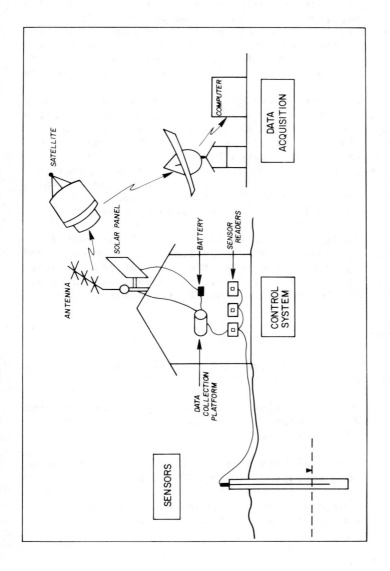

Figure 7. Schematic representation of continuous water-level monitoring components.

RESULTS

For a preliminary review of the data, hourly measurements of water levels for each monitoring location were averaged over a day and plotted over a two-month period. Figure 8 shows daily-averaged water elevations for the three monitoring locations between November 20, 1981 and January 26, 1982. On most days, the elevation of the river is below the elevations recorded at the ground-water monitoring wells. An average hydraulic gradient of 0.0015 toward the river has been calculated using those elevations.

The three peaks on the graph occur as a result of a sudden increase in the river elevation. On these occasions the direction of ground-water flow was toward the site. The steepest hydraulic gradient as calculated from water-level measurements taken December 24, 1981 was 0.0058. These reversals in the direction of ground-water flow have been observed during low-flow conditions of the river. The impact of high-flow conditions upon ground-water elevations has not yet been observed.

Water levels at the three locations will be monitored for one year. Following an evaluation of the frequency and magnitude of gradient reversals, the model of ground-water flow for the site will be refined, as necessary. In turn, the discharge rate of contaminants to the river will be modified.

CONCLUSIONS

Phase I of the hydrogeologic investigation of a former industrial waste disposal site has shown that:

o   ground-water flows predominately toward the north and south shores of the island at an average gradient of 0.0025.

o   benzene has been detected in ground-water samples in concentrations up to 108,000 ppb.

o   the discharge rate of benzene to the river via ground water is approximately 2 lb/day.

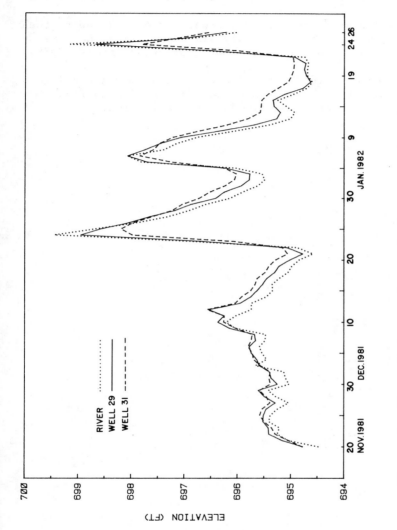

Figure 8. Plot of river and ground-water elevations between November 20, 1981 and January 26, 1982. (Locations of wells 29 and 31 are shown in Figure 1.)

Preliminary results of the Phase II investigation have shown that:

o   the hydraulic gradient of the water table is typically toward the river and over a two-month period during low-flow conditions averaged 0.0015.

o   the direction of ground-water flow was toward the site for approximately six days during the same two-month period at a maximum gradient of 0.0058.

o   water levels should be continuously monitored for one year to determine the impact of high-flow conditions upon ground-water elevation.

Janet A. Barker-Stonerook
Burgess & Niple, Limited

INTRODUCTION

A groundwater assessment is an evaluation of the chemical and physical qualities of a particular groundwater. The groundwater quality assessment program should determine whether hazardous waste or hazardous waste constituents have entered the groundwater, the rate and extent of migration of hazardous waste or hazardous waste constituents in the groundwater, and the concentrations of hazardous waste constituents in the groundwater. The assessment should also determine if groundwater may be affected by a past, existing, or future practice.

A groundwater monitoring program is an evaluation of an existing facility's impact on the quality of the groundwater.

The Resource Conservation and Recovery Act (RCRA) requires a groundwater monitoring program (40 CFR Part 265.90) be initiated by owners or operators of a surface impoundment, land treatment, or disposal facilities for hazardous wastes as part of the groundwater monitoring regulations. Owners and operators of certain facilities are required to prepare an outline of a groundwater quality assessment program. On February 23, 1982, the US Environmental Protection Agency (US EPA) delayed the compliance date until August 1, 1982 for quarterly reports and the assessment outline. The delayed date was promulgated to allow US EPA to eliminate and/or streamline these requirements.

The groundwater assessment for a proposed disposal site is performed to establish existing quality, quantity, and potential for development. The approval of a proposed hazardous waste site is partially contingent upon the findings of the groundwater assessment. Groundwater quality data generated during the assessment serve as a baseline to determine if significant changes occur relative to disposal activities.

Groundwater assessments should involve at least several sets of samples taken over a period of time. Groundwater assessments for proposed facilities generally require a minimum of three months to complete from the time the wells are installed to the report phase. While it is more desirable to obtain data over different seasons for a year, developers and/or owners of proposed sites generally are interested in a rapid turnaround time. A similar amount of time is typical for an existing facility, if no significant groundwater contamination is discovered. If an existing facility is causing groundwater degradation, then the assessment phase may involve several years of sampling and additional wells.

Groundwater assessments typically include the following:

- Review existing water quality data to determine local groundwater quality. Review existing test boring data. Sources include data from state agencies, published literature, owner records, and records from past engineering studies.

- Perform test borings and laboratory testing on earth materials obtained during drilling including standard penetration test, gradation analyses, permeabilities, moisture content, liquid/plastic limits, and soil quality analyses.

- Install monitoring wells at those locations where geologic and groundwater conditions indicate the best information may be obtained. Initially, one upgradient and two or three downgradient wells are installed.

- Estimate rate of groundwater movement based on transmissivities, both from aquifer pumping tests and laboratory permeabilities of aquifer and aquitard materials.

- Collect water samples from completed monitor wells and analyze in laboratory for specified para- meters. Analyze the data to determine the extent and rate of migration of hazardous waste constit- uents in groundwater based on water quality results and rate of movement.

- Prepare a final report to evaluate all results to determine if additional test borings, soils analysis, monitor wells, or water samples are required.

This paper will present case histories of groundwater assessment performed for a variety of clients. Groundwater contamination is a sensitive issue. Therefore, the clients and exact locations are not identified.

GROUNDWATER ASSESSMENT FOR LANDFILL DEVELOPMENT

General

A major industry initiated a feasibility study for the potential development of a solid or hazardous waste disposal facility on 34 ac. of land. A groundwater assessment was performed to determine if hydrogeologic conditions were favorable for this development, including soil permeability and groundwater quality and quantity.

The proposed site was located in a poor groundwater yield area of the Mahoning River basin where maximum groundwater yields are reported in the range of 0 to 25 gpm. The poor groundwater yields are a result of the relatively impervious clay and silt strata in the unconsolidated material and impervious shales in the bedrock. Three soil borings taken at the site encountered interbedded silts, clays, silty sand, and some minor isolated lenses of sand and gravel.

Due to the poor groundwater availability in the immediate vicinity of the proposed site, both in the unconsolidated material and in the bedrock, and due to the relatively impermeable nature of the silt and clay layers, the site had potential for development as a waste disposal facility.

Monitoring Wells

An 8 in. diameter observation well (Monitoring Well I) was installed at the site using a cable tool drilling rig. The well was cased to bedrock and completed as an open hole in the bedrock as shown on Figure 1. No groundwater was encountered in the overburden. The yield was estimated to be about 2 gpm and the well could be bailed dry.

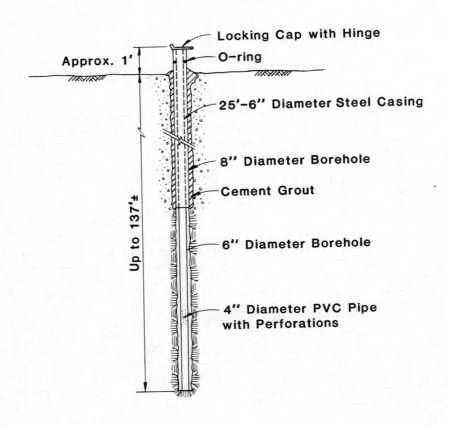

Figure 1. Typical Deep Monitoring Well

Two of the three soil borings were converted to monitoring wells by the installation of 2 in. diameter polyvinyl chloride casings with a 3 ft. section of slotted screen as shown on Figure 2. The holes were backfilled with Volclay pellets to the first sand or water bearing seam. The screen was set and sand was placed around the screen. The pipe was then grouted in using Volclay from just above the screen to the ground surface.

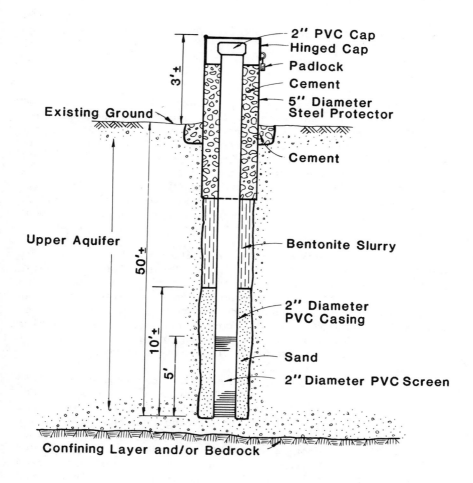

Figure 2. Typical Shallow Monitoring Well

Gradient

Static groundwater levels and pumping water levels were taken in the monitoring wells during September 1980 to April 1981. The general direction of the groundwater gradient in the vicinity of the proposed site is towards the Mahoning River.

The depth at which groundwater was encountered during drilling ranged from 25 to 42 ft. below the surface. Upon completion, static groundwater levels ranged from about 2 ft. to 8 ft. below ground level. These water levels were much higher than those encountered in the wells when originally drilled due to the artesian condition present on the site.

Quality

Well 1 was drilled into bedrock, and the water came from a layer of sandstone within the shale bedrock. The bottom of the well casing was set at 37.3 ft. below the surface for Well 1. Wells 2 and 3 were installed in silty sand layers above the bedrock. The screen was set between 51 to 54 ft. below the surface in Monitoring Well 2 and at a depth between 31 to 34 ft. in Monitoring Well 3. The silty sand layer is probably not continuous between Wells 2 and 3. Therefore, each of the three wells are set in different waterbearing layers and yielded water of different quality.

The Safe Drinking Water Act established primary drinking water standards for the protection of public health. These standards have been incorporated in Ohio EPA and US EPA solid and hazardous waste regulations in an effort to protect groundwater resources in the vicinity of disposal sites. The average concentrations of all samples from the monitoring wells were within primary drinking water standards tested and are shown in Tables I through III.

After bailing, water samples recovered from each of the wells periodically contained high suspended solids concentrations, particularly at Monitoring Well 3, a 2 in. diameter well. Federal protocol requires that the heavy metal analyses be performed on unfiltered samples which have been acidified. Upon acidification, substantial suspended solids concentrations remained in the samples which affected the laboratory analyses. These solids probably interfered with the analyses, causing higher than actual readings.

A set of samples from each of the monitoring wells was collected at a later date to assess the effects of suspended solids and preservation with nitric acid. Each sample was split into four subsamples as follows:

- Unacidified/Unfiltered
- Acidified/Unfiltered
- Unacidified/Filtered
- Acidified/Filtered

Table I. SUMMARY OF SAMPLES FROM MONITORING WELL 1

Parameter (mg/l)	Maximum	Minimum	Mean
COD	90	6	29
Nitrate, NO_3-N	0.6	0.06	0.29
Residue, Total	1,700	210	595
Residue, Suspended	720	30	150
Residue, Dissolved	330	160	235
Arsenic Total, As	0.006	0.001	0.003
Barium Total, Ba	0.200	0.200	0.200
Cadmium Total, Cd	0.010	0.001	0.006
Chromium Total, Cr	0.020	0.001	0.010
Copper Total, Cu	0.010	0.005	0.008
Iron Total, Fe	6.400	0.180	3.440
Lead Total, Pb	0.060	0.002	0.026
Manganese Total, Mn	0.200	0.070	0.133
Mercury Total, Hg	0.0011	0.0002	0.0005
Nickel Total, Ni	0.027	0.001	0.013
Selenium Total, Se	0.010	0.001	0.007
Silver Total, Ag	0.010	0.001	0.006
Zinc Total, Zn	0.090	0.001	0.034

Table II. SUMMARY OF SAMPLES FROM MONITORING WELL 2

Parameter (mg/l	Maximum	Minimum	Mean
COD	210	16	102
Nitrate, NO_3-N	0.3	0.0	0.10
Residue, Total	2,000	420	620
Residue, Suspended	100	15	60
Residue, Dissolved	460	390	415
Arsenic Total, As	0.018	0.004	0.010
Barium Total, Ba	0.220	0.200	0.203
Cadmium Total, Cd	0.020	0.001	0.008
Chromium Total, Cr	0.020	0.001	0.009
Copper Total, Cu	0.020	0.003	0.010
Iron Total, Fe	1.300	0.050	0.500
Lead Total, Pb	0.030	0.003	0.018
Manganese Total, Mn	0.180	0.030	0.083
Mercury Total, Hg	0.0014	0.0002	0.0006
Nickel Total, Ni	0.066	0.001	0.031
Selenium Total, Se	0.010	0.001	0.006
Silver Total, Ag	0.010	0.001	0.004
Zinc Total, Zn	0.520	0.010	0.203

Table III. SUMMARY OF SAMPLES FROM MONITORING WELL 3

Parameter (mg/l)	Maximum	Minimum	Mean
COD	500	16	112
Nitrate, NO_3-N	6.3	0.04	1.33
Residue, Total	3,800	890	1,500
Residue, Suspended	3,100	20	580
Residue, Dissolved	1,030	870	880
Arsenic Total, As	0.010	0.001	0.004
Barium Total, Ba	0.200	0.200	0.200
Cadmium Total, Cd	0.020	0.001	0.010
Chromium Total, Cr	0.040	0.001	0.018
Copper Total, Cu	0.030	0.006	0.014
Iron Total, Fe	2.700	0.220	1.050
Lead Total, Pb	0.040	0.014	0.030
Manganese Total, Mn	0.410	0.050	0.201
Mercury Total, Hg	0.0018	0.0002	0.0006
Nickel Total, Ni	0.110	0.006	0.040
Selenium Total, Se	0.010	0.001	0.009
Silver Total, Ag	0.010	0.001	0.005
Zinc Total, Zn	0.300	0.081	0.080

Analyses were completed for arsenic, cadmium, and lead which were the parameters suspected to be most affected. The results are presented in Table IV. Samples that were unfiltered and preserved with acid typically yielded higher arsenic and lead concentrations. No clear pattern was discerned on the effects of preserving or filtering samples for cadmium analysis; generally, however, the highest concentrations of cadmium were observed in the unfiltered and unpreserved samples. In conclusion, lead and arsenic concentrations, as shown on Tables I through III, are higher than actual, even though the average concentrations were within drinking water standards. The two exceedances of the cadmium standard are also probably high due to suspended solids in the samples.

Table IV. EFFECTS OF FILTERING AND ACIDIFICATION

Parameter (mg/l)	Monitoring Well 1	Monitoring Well 2	Monitoring Well 3
Arsenic			
Filtered with acid	0.0005	0.010	0.001
Filtered no acid	0.0006	0.004	0.004
Unfiltered with acid	0.044	0.012	0.190
Unfiltered no acid	0.006	0.014	0.141
Cadmium			
Filtered with acid	0.011	0.012	0.013
Filtered no acid	0.008	0.011	0.030
Unfiltered with acid	0.011	0.004	0.007
Unfiltered no acid	0.020	0.006	0.030
Lead			
Filtered with acid	0.008	0.020	0.017
Filtered no acid	0.030	0.004	0.007
Unfiltered with acid	0.080	0.050	0.270
Unfiltered no acid	0.050	0.020	0.250

The results of the groundwater assessment program indicate that goundwater in the area of the proposed site is within primary drinking water standards tested and has low potential for development. Therefore, the hydrogeologic conditions are favorable for approval of the proposed site as a waste disposal facility.

GROUNDWATER ASSESSMENT FOR ABANDONED LANDFILL

General

An industry sought to determine if there was any imminent hazard from an abandoned landfill to local groundwater supplies. The 85 ac. abandoned landfill had been developed in a buried valley transversed by a major river with sand and gravel deposits serving as a productive aquifer. The landfill received industrial and residential wastes for ten years in the 1960's. Some wastes, now classified and listed as hazardous material, may have been deposited in the abandoned landfill.

Monitoring Wells

Three shallow monitoring wells (one upgradient and two downgradient) were installed in order to acquire information on soil classification, soil permeability, depth to groundwater and water quality. The monitoring wells consisted of 2 in. polyvinyl chloride (PVC) threaded casing and 2 in. PVC threaded well screen with No. 10 slot openings as shown previously on Figure 2. A strong chemical and septic odor was noted while drilling two of the borings downgradient of the landfill. The well screens were set in the downgradient wells in the first predominantly sand and gravel water bearing layer encountered after noting the odors. This was selected in order to monitor the groundwater which appeared to be the most affected by disposal activities.

Washed sand was installed in the annular space between the boring walls and the well screen. A bentonite clayseal was installed near the surface and a steel protector casing was grouted in place at the surface. The monitoring wells were completed with a locked cap.

Each well was developed by pumping water out of the well until the water was clear. No chemicals were added to the wells to aid in development. Additional development consisted of air surging.

The water levels in the monitoring wells were measured on three occasions and in local private wells on one occasion. The direction and gradient of groundwater flow was determined utilizing the three newly drilled monitoring wells and three private wells east of the landfill and agreed with what was known the local geology and hydrology in the region. The calculated direction of flow was to the southwest towards the major river at approximately 65 degrees west. The calculated gradient was 0.01. The direction and gradient of the groundwater flow was determined during a low flow stage in the nearby river.

The upgradient monitoring well showed no evidence of any groundwater degradation. There was no indication of any groundwater degradation due to leachate from the landfill in ten private or industrial wells sampled in the vicinity.

The two downgradient monitoring wells indicate high concentrations of volatile organics, conductivity, chemical oxygen demand, and pH. Groundwater from these wells exceeded the toxic limits of these parameters if ingested by humans. The two monitoring wells are located within a 50 and 20 ft. horizontal distance from the landfill and were completed at depths of less than 22 ft. during sampling and drilling activities, a strong chemical and/or septic odor was noticed originating from the groundwater. Recommendations included the installation of two deep monitoring wells and the analysis of groundwater to determine if contaminated groundwater is migrating into the lower part of the aquifer.

Additional shallow monitoring wells should be installed to determine if contaminated groundwater is seeping into the stream.

GROUNDWATER ASSESSMENTS FOR SURFACE IMPOUNDMENTS

Case 1

General

A manufacturing industry sought to assess the impact of its lagoon system on the quality of the groundwater and to delist from hazardous to nonhazardous the waste deposited in the lagoons. The US EPA granted the delisting request. The major objectives of this project included describing the on-site geologic and groundwater conditions as identified through the drilling and test borings and the installation of monitoring wells, and describing the existing groundwater quality in the monitoring wells.

Monitoring Wells

Five monitoring wells were installed in the test borings previously shown on Figure 2. They consisted of 2 in. dia. PVC casing and screen. Saturated conditions were noted in thick, lower sand and gravel deposits that were encountered in four of the five borings at depths varying from 60 to 83 ft. The upper 60 to 70 ft. of underlying deposits were found to be sandy, clayey silts of low permeability. Water level measurements were taken on July 6 and August 20, 1981 in the monitor wells. For Monitor Wells 2 through 5, the levels on a single occasion varied as little as 0.05 ft. indicating an extremely low gradient for the area. The measured water level in Monitor Well 1 is approximately 35 ft. higher than those in the other four monitoring wells. Originally, it was believed that the fine sand and silt deposit developed in Monitor Well 1 at a depth of 52 ft. was physically connected to the sand and gravel deposits developed in the other wells. Because this water level is extremely higher than the others, and the character of the deposits is different, this fine sand and silt is interpreted as representing an isolated lense of limited areal extent.

The water level measurements taken in the monitoring wells were used to determine the direction of groundwater movement. Based on calculations using the two sets of water level measurements, the direction of groundwater movement was found to vary from toward the northeast to toward the

413

southeast. As explained above, the water levels from Monitor Well 1 were not used in this determination; however, it should again be mentioned that the range in the water levels in the four wells varied from 0.05 to 0.15 ft. and this is an extremely low gradient. Any slight variation in reading a water level measurement could make quite a difference in the resulting determination on groundwater movement.

Due to potential intermittent pumpage in the area, five monitor wells were installed to ensure an upgradient well and three downgradient wells exist at all times. With the low gradient present in the area, very slight fluctuations in the water levels could change the status of the upgradient well. Five wells were also considered necessary because of the acreage involved at the facility.

Quality

Water samples were collected from the five monitor wells using a bailer. The results of the laboratory analyses are summarized in Table V.

The constituents analyzed on these water samples were only those considered relevant in the listing given in the Hazardous Waste Regulations, Part 265.95 (45 FR 33240, May 19, 1980). Additional constituents included are cyanide and nickel because of their importance in the original listing of the wastewater treatment sludge contained in the lagoons as a hazardous waste.

Review of the analytical results from the monitor wells shows that the sample from Well 1 and the first sample from Well 3 exhibit the best overall water quality. Based on the metal analyses only, Monitor Well 5 and the first sample from Monitor Well 3 show the lowest concentrations. The sample from Monitor Well 2 had the greatest number of high concentrations of metals. this well was determined to be upgradient from the surface impoundments and should represent ambient groundwater quality.

Based on the results of the constituents analyzed and the hydrogeologic investigation, there is no indication that the surface impoundments are causing or will cause contamination of the groundwater.

Table V. GROUNDWATER ANALYSES

Parameter (mg/l)*	Monitor Well 1 7-6-81	Monitor Well 2 7-6-81	Monitor Well 3 7-6-81	Monitor Well 3 8-20-81	Monitor Well 4 7-6-81	Monitor Well 5 7-6-81
Chloride	14	4	30	40	34	36
Conductivity, umhos	300	245	365	280	375	410
Cyanide	0.1	0.1	0.1	0.48	0.1	0.1
Fluoride	0.86	0.31	1.05	0.54	1.10	1.18
Nitrate	3.82	0.31	6.20	3.44	3.34	5.24
pH, S.U.	7.1	7.3	7.6	8.1	6.9	7.2
Phenols	0.009	0.041	0.034	0.06	0.015	0.015
Sodium	9.5	3.5	22.4	23	26.4	28.4
Sulfate	300	88	470	77	490	490
Arsenic	0.004	0.009	0.008	0.001	0.004	0.003
Barium	0.04	0.001	0.070	0.050	0.120	0.070
Cadmium	0.001	0.040	0.010	0.008	0.002	0.010
Chromium	0.001	0.05	0.017	0.015	0.050	0.040
Iron	1.00	2.70	0.72	0.140	1.60	1.20
Lead	0.008	0.025	0.015	0.07	0.018	0.004
Manganese	0.03	0.12	0.06	0.05	0.12	0.05
Mercury	0.0002	0.0002	0.0002	0.0003	0.0002	0.0002
Nickel	0.020	0.009	0.050	0.001	0.070	0.02
Selenium	0.002	0.001	0.002	0.01	0.002	0.004
Silver	0.008	0.002	0.005	0.001	0.005	0.003

*Unless otherwise noted

Case 2

General

An industry that utilized surface impoundments for treatment and storage of its industrial wastewaters sought to assess the groundwater quality and flow on their property.

Monitoring Wells

Six wells were drilled, three of which were 50 ft. deep and consisted of 2 in. diameter galvanized steel pipe and a stainless steel well screen. The other three wells consist of 6 in. and 8 in. steel casing and stainless steel screens. The industry maintained four deep pumping wells and three previous monitoring wells.

Drilling of the deep production wells encountered three sand and gravel aquifers separated by till or clay. Installation of the shallow monitoring wells encountered pockets or lenses of water bearing sand and gravel separated by till or clay. The deep wells were set in the aquifer above the bedrock.

Quality

Parameters selected for analysis were selected because they would adequately describe the process wastes and include: biochemical oxygen demand, chemical oxygen demand, conductivity, iron, pH, total phosphate, potassium, sodium, and solids. Additional parameters were analyzed infrequently to supplement the data.

The 5-day biochemical oxygen demand concentrations indicate that high organic loads are present in the groundwater in the shallow monitoring wells nearest the lagoons. Lower concentrations in the other wells indicate that the high organic loads have dissipated with greater distances and depths from the holding lagoons. The low concentrations in the production wells are normal for natural groundwater quality and do not indicate anything unusual.

The chemical oxygen demand concentrations indicate that substances creating a high chemical oxygen demand are present in the groundwater near the lagoon. The lower concentrations in the production wells are slightly higher than expected for natural groundwater, but this may be due to the low number of natural groundwater samples analyzed

416

for chemical oxygen demand, since this item is not routinely analyzed.

The conductivity concentrations indicate that higher than anticipated levels are present in the shallow well points nearest the lagoons. The concentrations in the production wells are what would be expected for natural groundwater quality.

The shallow wells are of distinctly poorer quality than the deep production wells as Table VI illustrates.

The wells listed in the table were selected on the basis of their location and depth (whether screened in a shallow or a deep aquifer) and in an attempt to represent the range of concentrations beneath the industry's property. The shallow wells show much higher concentrations of 5-day biochemical oxygen demand, chemical oxygen demand, total organic carbon, conductivity, and total solids. Slightly higher concentrations exist for sodium, color, and possibly phenols. Slightly lower values were recorded for pH. These trends agree with the trends recorded for samples of lagoon influent.

Monitoring Wells (MW) 5 and 3 were the poorest in quality with MW 5 being slightly worse. MW 5 was located just east of the lagoons, and MW 3 was located just north of the lagoons. MW 11 and 13 are similar with the exception of biochemical oxygen demand and chemical oxygen demand concentrations. MW 7 generally displays the best quality of the shallow wells. This is probably due in large part to its greater distance from the lagoons than the other shallow wells in the same groundwater flow path.

To obtain an indication of any changes in groundwater quality which may have occurred over the last several years, the average concentrations of common parameters in samples analyzed by the Ohio Department of Health from 1971 through 1974 were compared with the average concentrations determined from the present study. Well 4 was sampled by the Ohio Department of Health, but the well was destroyed prior to 1979. MW 13 was installed in a similar location in 1979.

When comparing with data from 1971 through 1974 of the Ohio Department of Health, MW 5 has the poorest quality, but interestingly enough has improved in the parameters tested in the seven parameters compared within the past ten years. MW 3 and MW 4/13 have apparently slightly deteriorated in quality over that same period, with the exception of chemical oxygen demand in MW 3 which has improved slightly.

417

Table VI. AVERAGE CONCENTRATIONS IN SELECTED WELLS
FEBRUARY–NOVEMBER 1979

Well	BOD_5 (mg/l)	COD (mg/l)	Conductivity at 25° C. (umhos)	pH	Total Phosphate PO_4-P (mg/l)	Potassium (mg/l)	Sodium (mg/l)	Total Solids (mg/l)
PW 1*	3	29	693	7.1	0.02	3	11	449
PW 4*	3	52	614	7.0	0.03	2	11	372
MW 5	65	386	1,106	6.8	0.94	8	68	1,411
MW 7	8	198	364	7.9	0.69	7	11	728
MW 12*	4	78	431	8.3	1.01	6	10	337
MW 13	26	202	715	7.7	0.25	6	49	624

PW = Production Well
MW = Monitoring Well
*Deep Well

MW 7 now shows significantly higher chemical oxygen demand, total phosphate, and total solids concentrations. Taken as a whole, these results indicate no substantial change in groundwater quality in the vicinity of the holding lagoons over the past ten years.

The results of this study indicate that the general direction of groundwater flow was east to west. Water is seeping from the holding lagoons and moving toward three of the production wells. The shallow groundwater in the vicinity of the holding lagoons has concentrations of certain parameters above what is considered normal for natural groundwater quality in the area. In general, the shallow groundwater quality exhibits no substantial change over the past ten years. The deep groundwater quality, as represented by the plant production wells, appears to be unaffected at this time by any contamination. The possible explanations for this are a combination of dilution by groundwater underflow and adsorption by the clay material separating the aquifers.

If the industry were to increase pumpage from its existing production wells, groundwater would move toward the wells at faster rates leading to more rapid deterioration of at least the shallow groundwater quality. The significance of this is that any groundwater contamination which may be present in the plant area may be contained in the area as long as all groundwater users in the area continue to withdraw water at or near their present rates. If the industry's withdrawal rate were to be substantially reduced or were to cease, then any groundwater contamination would not be contained and would follow natural groundwater flow paths which would be somewhat south and west. The natural groundwater flow paths would still be subject to influence by other pumping wells in the region which might divert the water toward them.

If the lagoons were to be eliminated and all surface ponding and spillage were contained, this would virtually eliminate the sources of contamination. However, the saturated soil would still be contaminated.

Recommendations included continuing to sample and measure water levels in the shallow wells in the vicinity of the lagoon, plus sample all of the pumping wells, at least four times a year. Routine analyses which should be performed on these samples include chemical oxygen demand, conductivity, and pH, as these parameters were found to be the most significant indicators of groundwater quality

419

changes in this study. There is no known contamination based on the analyses of the deep aquifer.

CONCLUSIONS

Each waste disposal site is a unique case. Background information on area geology, hydrogeology, soils, and topography should be consulted and utilized before the installation of any wells. Proper installation of wells is essential to eliminate potential migration of surface water into the well or seepage along the casing from other zones than that being tested.

Sampling techniques may need to be adapted to specific wells to obtain a representative sampling. Care must be taken to ensure that any standing water in the monitoring well is emptied prior to sample collection and that a fresh sample is collected.

It is important to minimize agitation and aeration of the sample in transferring the sample from the pump, bailer, or other procurement means to the proper containers. Excessive aeration und agitation in open containers can alter the stability of some constituents in the samples. It is recommended that the water samples be transferred directly to clean sample containers. If it is necessary to remove suspended silt or clay from the sample by properly developing the monitoring wells or in the case of older wells, it is advisable to use either field filter apparatus or carefully pour the water into a clean, large capacity container. After the suspended matter has settled, the clear water can be decanted into the sample container and sent to the analytical laboratory.

CHAPTER 26

PRACTICAL ASPECTS OF LAND
TREATING HAZARDOUS WASTES

John Charles Nemeth
CH2M HILL

INTRODUCTION

In the past ten to fifteen years, an awareness of, and reaction to increasing environmental degradation and accelerating costs associated with the treatment and/or disposal of wastes have combined to cause a concomitant rise in level of efforts to discover acceptable solutions. The solution modes most often fall into one, or perhaps, a combination of approach techniques for dealing with society's sanitary and industrial wastes. These techniques are either the simple, but temporary, "fix it" methods such as landfilling, intermediate technology methods such as land treatment, or the ever increasingly important high-technology recycling and reuse methods. Land treatment, the intermediate technology, has certain elements in common with each of the other extremes but, in general, the only significant commonality shared with landfilling is the use of the land. As the term intimates, land treatment is a method whereby various kinds of wastes are actually altered. The physical and chemical processes available in soils to accomplish this are numerous but fall into one of three general categories: (1) adsorption and chemical modification of metallic species, (2) controlled migration of certain anions, and (3) chemical/biodegradation of organic compounds to innocuous by-products.

Land treatment of wastes is really a rediscovered technique. The primary difference between current practice and that used in the past is that the treatment capability of soils is managed for optimal reduction of waste volumes and treatment of waste constituents. Thus, this chronologically-old but technically-emerging, cost-effective method is beginning to assume an increasingly significant role in, not only domestic, but industrial and hazardous waste management and treatment. Because treatment is accomplished, wastes can be used to improve soil quality in some cases, water recycling can be accomplished, and instances of crop production are possible; thus, land treatment is naturally aligned closely with recycle/reuse technologies. Regardless of how efficient the processes might be, waste creation will continue and land treatment is likely to be a part of our waste-treatment technology for the foreseeable future.

The area of industrial/hazardous waste management has evolved in the past 30 to 40 years from a primary focus on disposal with treatment having been somewhat secondary, to very sophisticated treatment-oriented systems. A very broad spectrum of industries, including petroleum, chemical, tanning, electroplating, food processing, pulp and paper, photographics, pharmaceuticals, and many others, has, over the years, successfully employed land treatment. In fact, with the exception of radioactive wastes, and despite certain regulatory prohibitions, largely unfounded and speculative in nature, an acceptable land-treatment system can be designed for any industrial waste constituent.

The scepticism that part of the waste-management community has for the land-treatment concept stems from a lack of understanding of a cardinal principal. Land-treatment success does not derive from nomographed, standardly approached, and iteratively used plans. Each potential land-treatment system is an individual situation for which design basis is a complex interaction of site conditions (soils properties, seasonal climate, topography, hydrology, etc.) and waste-stream characteristics (quantity, quality, and variability in each). Design criteria, therefore, depend on the variability of nature as well as the waste, perhaps, even more so. Thus, in this technology, the lead roles belong to the soil scientists, hydrologists, and agronomists, who, as a team, must reconcile the quantity and quality of waste generated to the assimilative capacity of the site with regard to land area needed, application rates, operation and maintenance, and monitoring.

This then is a very important practical aspect of land treating hazardous wastes--use expertise in the areas of specialization that will enhance treatment as its possibilities exist on the site(s) available, given the waste to be handled. Briefly, the following matched list represents the basic needs.

o Soil Scientist   Soil chemistry & physics
                             (water relations)

o Hydrologist   Surface water control &

o	Agronomist/Micro-biologist	Cover crop and microbial ecology management
o	Soils (Geotechnical) Engineer	Determine depth to bedrock and water table, subsurface tests and characterization for the hydrologists and soil scientists, and competent observation-well installation.
o	Chemist/Toxicologist	Chemical compatibilities between the waste(s) streams and between these and the environment.
o	Chemical Engineer	In-plant control and modification to enhance treatment success by identifying possible quantity or quality changes in certain feedstocks or processes.
o	Irrigation Expert	Hardware engineering and Design

If you are going to consider land treatment, then remember that the medium you intend to base your system on is the soil--that upper, surface few feet of the earth which is the soil scientist's province.

GENERAL METHOD APPROACH AND MODES

Methodology

Just as many method approaches to land-treatment design probably exist as there are practitioners. Despite the variations; however, long-term goals are relatively constant. I tend to think of land-treatment methodology in terms of two distinct but related goals: (1) determination of the design basis and (2) selection of the optimal management system. The first is based on treatment facility goals and constraints in relation to:

o	site assimilative	soils, physical and

capacity	chemical; geology, hydrology, and quality of surface & groundwater, vegetation & soil microbiology, climate/air quality.
o waste-stream evaluation	quantity/rate and quality/ limiting factors.

Given these data, the two-step method proceeds with the calculation of design loading rates for the hydraulic levels and chemical constituents, in other words, the design basis. One or some set of the constituents will be identified as land-area determining factors (LAD's). These factors define not only the land area needed but the type of management system to best treat the waste. Thus, cost to treat may be calculated and the second step of the method is completed with the selection of the optimal treatment system. This may involve comparison with other treatment ideas and/or the modification of in-process procedures to extend land-treatment facility life or reduce the treatment-site land area.

To provide the most cost-effective and reliable system, monitoring is a must. With it, the predicted performance of the design basis and of operation and maintenance effectiveness can be directly measured.

Other direct benefits of monitoring are: (1) learning information that identifies modifications of the system operation which will allow continued, efficient treatment on an uninterrupted basis, and (2) at closure, the record is available upon which logical decisions regarding final land-use disposition can be made, and the nature of site-closure requirements, if any, can be logically defined. A minimum monitoring and record-keeping program is detailed further in a subsequent section.

Modes

A brief comment on modes of land treatment is relevant here. Two basic modes are available to the industrial/ hazardous waste land treater--low-intensity and high-intensity loading. The former mode is based upon a design that, in the extreme, allows no environmental degradation and wastes that are, therefore, rendered non-hazardous by chemical alteration or by storage below critical thresholds. The high-intensity mode may center on volume reduction and only partial chemical-constituent treatment with surface soils being ultimately removed for other treatment or storage or, perhaps, a capping of the site might be done. On the other hand, high-intensity application may be accompanied by high-intensity management that also results, finally, in environmental conditions acceptable for a multitude of post-closure land uses.

FACILITY DESIGN, OPERATION AND MAINTENANCE, AND MONITORING CONSIDERATIONS

As inferred in the previous section, the interactive relationship of the site and wastes determines the specific configuration of the facility and the way in which it will be operated, maintained, and monitored. Since it is highly unlikely that any two facilities would be the same, these general considerations and options are offered to hopefully aid in the goal of environmentally sound management.

Such generalizations may tend to leave most scientists, engineers, and/or incipient facility operators wanting more details. Land-treatment systems, however, are so individualistic that attempts at collating information on various design and O&M aspects would likely, because of the nearly unlimited environmental variations possible, result in an oversized, unusable document. Such cookbook attempts usually bog down in detail, when it is philosophy of method and a knowledge of the principal kinds of situation-specific factors that are most useful. The remainder of this paper will, therefore, center on a brief discussion of general facility layout suggestions, followed by a minimum checklist of land-treatment facility considerations.

Some General Design Considerations and Suggestions

Facility systems may include such components as waste storage ponds, access roads, berms and ditches, runoff storage ponds, and the land-treatment plots. The actual shape, position, and size of various components is dependent largely upon topography and the nature of the waste. The latter will probably determine the application method; sprinklers, injectors, spreaders, etc. to control certain waste characteristics as well as enhance incorporation with the soil. Topography will dictate the specific equipment along with trafficability related to the actual waste loadings and normal precipitation levels. Since hazardous waste land-treatment plots are almost universally bermed for surface-flow diversion and containment of runoff in perimeter ditches, access roads for hauling and application equipment can be constructed as part of the berms between plots or along the peripheries. The hydrologist-chemist-soil scientist team should control facility configuration (layout) and drainage design, based on hydrologic design storm events and level of control necessary for the wastes. Beyond this, construction should be monitored to assure that top soil from the intended plots not be used in construction of berms, etc.

Perimeter ditches are placed to collect runoff for routing to a treatment plant, for reapplication after temporary storage, or, depending upon quality, direct discharge to a receiving body of water. Part of the routine operation and maintenance of a facility is ditch mainte-nance. It can be crucial when storm events require use of all the design capacity of waste storage lagoons, ditches, and runoff-storage impoundments, as well. Good drainage, thus the prevention of developing anaerobic conditions, even if only temporary, can be a principal factor in avoiding odor problems after extended or heavy short-term rains.

The number of treatment plots needed and their sizes will depend upon waste quantity and quality. If large, noncon-tinuous applications are planned, then one plot is probably operationally sufficient. Conversely, there are several conditions under which multiple plots are favored:

o when the need for continuous, year-round application where a set period of time and parcel of land of a certain extent are required to treat some fraction of the annual waste load.

o when variable wastes are used that come from differing sources or plant processes and may not be physiochemi-cally compatible with each other or with their joint behavior in the soil.

o when plot soil capacity for a given LAD is reached, forcing new plots to be opened. The first plot could be used for other wastes if the available streams are variable, but with a consistent waste stream, expended plots must be closed.

The location of storage impoundments, plots, etc. should be set to minimize potential for offsite migration of odor-causing substances. Thus, buffer zones should be adequate to control normal conditions. Regarding odors, however, I believe that the well designed, operated, main-tained, and monitored facility will avoid such airborne pollutant migrations and others, such as surface and groundwater, under all but the most catastrophic of condi-

426

tions. Buffer zones employing vegetation and/or fences can have positive dispersal effects on air movements, but I suspect their greatest value, given the just previously stated constraints, is aesthetic.

LAND-TREATMENT FACILITY DEVELOPMENT CHECKLIST

The preceding sections give some indication of the high level of complexity involved in land-treatment technology. Referred to as simple, this treatment method is simple only in the positive aspects of equipment simplicity and the fact that it is a low energy consumer. While we could all, given sufficient time and inspiration, derive a checklist of applicable needs and concerns regarding land-treatment facility design, it is sometimes useful to have one's thought processes piqued. The following broad-category checklist is offered for your use and, hopefully, your modification and expansion.

General Checklist

1. Identify Facility Nature and Conceptual Design
   - Desired goals and benefits (waste treatment, volume reduction, both)
   - Constraints (ultimate planned landuse at closure, single source-waste or multisource wastes, continuous or not, and waste compatibilities)
   - Mode selection (high-versus low-intensity application or an intermediate variation)

2. Site Evaluation (discussed elsewhere)

3. Waste-stream Evaluation (discussed elsewhere)

4. Quantification of Site Assimilative Capacity (What are the land-area determing factors(s)--LAD's--and what particular LAD properties are going to dictate level and method of control? See 1 above, Constraints also.)

5. Facility Design and Layout
   - Single or multiple plots, waste-storage ponds
   - Buffer-zone needs
   - Erosion and evaporation control (grading and sealing)
   - Surface and groundwater control (runoff and leachate control with berms, ditches, runoff collection basins, terraces, contours)
   - Contingency treatment or storage (surface runoff or leachate collection)

6. Waste Distribution and Site Maintenance

- Equipment (see 4 and 5 above) types
- Periodicity

7. Crop Management
   - Crop type (forest, pasture grass, microbes along with no cover)
   - Moisture relations (drainage, irrigation)
   - Nutrient balance (fertilizer, pH control)
   - Harvesting techniques
   - Cultivation (aeration and waste distribution)

8. Scheduling and Record Keeping
   - Waste received/applied
   - Site maintenance
   - Crop management
   - Site inspection
   - Security and health/safety
   - Monitoring (waste, physical and chemical; soil, incorporation zone and below; surface water; groundwater, soil-film water by field-moist extraction or lysimeter; cover crop; soil microbial populations; and air)

9. Selection and Training of Facility Operators

The continuum of effort described in this list is, of course, composed of many more details and potential complexities. The last item deserves some further mention; however. One of the greatest potential sources of failure is the low attention usually paid to the quality and training of the individuals who manage and operate systems. A land-treatment facility will not manage itself. Just as those having the greatest influence on design of the facility and its operation, maintenance, and monitoring planning should have soils, hydrologic, and agronomic expertise, so should the manager. I recommend that the training or experience be equivalent to, at least, the normal 2-year associate of arts degree given in many schools of agriculture or community college systems. Further, routine, perhaps annual, monitoring by outside experts, usually the "designers," should be accomplished after they provide an intensive training program.

By now, the major factors that I believe result in successful land-treatment systems, hazardous/industrial or otherwise, should be apparent. Unencumbered by the details presented previously, those key items are:

(1) Technically competent, environmentally sound design.

(2) Careful, routine operation and maintenance based on a comprehensive monitoring program.
(3) Proper training and quality assurance monitoring of manager/operator staff.

CHAPTER 27

DEVELOPMENT OF GENETICALLY ENGINEERED MICROORGANISMS TO DEGRADE HAZARDOUS ORGANIC COMPOUNDS

G. E. Pierce
Battelle's Columbus Laboratories

INTRODUCTION

The ability of microorganisms to degrade a diverse number of organic compounds is well documented in the literature. Both naturally occurring compounds, as well as many synthetic compounds, are degraded. Some of the compounds degraded by microorganisms are: methane,[1,2] fatty acids, benzene,[3,4] naphthalene[5,6] and various petroleum fractions.[7,8] Many synthetic compounds, such as 2,4-dichlorophenoxy acetic acid (2,4-D)[9,10]; 2,4,5-trichlorophenoxy acetic acid (2,4,5-t)[11]; DDT[12,13]; 2-chloro-4-(ethylamino)-6-(isopropylamino)-s-triazine, (Atrazine)[14]; and some isomers of polychlorinated biphenyls (PCB),[15-17] are also degraded.

All of these compounds are also degraded by environmental microorganisms (i.e. "wild-type" microorganisms that have been isolated from the environment). If environmental microorganisms exist that can degrade these compounds is there a need for employing genetic engineering?

Although many compounds can be degraded under laboratory conditions, some of them persist in the environment.

Several problems encountered when microorganisms are used to degrade hazardous organic compounds may be addressed by using genetic engineering. The four major areas that may benefit from genetic engineering include stabilization, enhanced activity, multiple degradative activities, and health, safety and environmental concerns.

It is not sufficient to base the development of a microbial system which will degrade hazardous organic compounds on genetic engineering alone. Knowledge of microbial physiology and biochemistry, chemical engineering, processing engineering, etc. must be integrated with developments achieved through genetic engineering if a successful system is to become a reality.

Genetic engineering, itself, cannot be limited to the implementation of recombinant-DNA procedures, but must include all relevant techniques for strain improvement, such as classical mutation of wild type strains and in vivo gene manipulation techniques (e.g., protoplast fusion, conjugation, transduction, etc.).

POTENTIAL OF GENETIC ENGINEERING

Stabilization

The ability of microorganisms to degrade some hydro-carbons has been shown to be encoded on extrachromosomal-DNA (plasmids).[18-21] If these plasmids are lost, the ability to degrade that hydrocarbon will also be lost. Cells of Pseudomonas putida mt-2, which contain the TOL-plasmid (encodes the genes for toluene degradation) will lose the TOL-plasmid and, thus, the ability to degrade toluene if they are repeatedly subcultured on benzoate. The stability of toluene degradation in these cells is therefore low when cultured on benzoate. This lack of stability of biochemical traits is not an isolated phenomenon. Many environmental strains with unique biochemical activities can lose these traits quite easily. We have observed that bacterial strains capable of 2,4-D degradation (Tfd-plasmid encoded)[20] will lose the ability to degrade 2,4-D if repeatedly subcultured in the absence of 2,4-D.[21]

Loss of plasmids that confer special biochemical activities does not result in a loss of viability. Thus, it can be inferred that their maintenance is not a requirement for life. Selective pressure must be present to maintain these plasmid-encoded traits.

Genetic engineering can be used to successfully stabilize plasmid-encoded degradative traits. Hybrid plasmids[22-24] can be constructed that will stabilize these traits. Insertion of desired genetic elements into a region where replication is under the control of a conserved element is one method to successfully stabilize a normally unstable degradative trait.

Enhanced Activity

The more copies of a gene which exist in a cell, the more enzyme can be produced and therefore, the higher the level of activity produced. The copy number (i.e., number of copies of a plasmid in a cell) of degradative plasmids is generally low (1-3 copies).[20] However, the copy number of other plasmids is higher[25,26] or can be caused to be increased (amplified).[26,27] Assuming the plasmid which encodes the degradative trait is not amplifiable, increased copy number could be achieved by hybridizing this trait with one which is a high copy number trait or one which is capable of amplification.

Genetic elements, in addition to containing that DNA responsible for protein, also contain regions referred to as promoters which allow the protein to be synthesized. The position of this promoter has a great influence on the amount of protein synthesized. Optimization of promoter location could greatly enhance the activity of a degradative trait.[28]

Multiple Degradative Activities

The ability of strains of <u>Pseudomonas</u> to degrade camphor, toluene or octane is encoded respectively by the CAM, TOL and OCT plasmids.[29] In naturally occurring strains, only one of these plasmids will ever be found in a pseudomonad. The reason for this is that these plasmids belong to the same incompatibility group.[30] Incompatible plasmids are mutually exclusive (in other words, no two plasmids belonging to the same incompatibility group can exist together in a cell). Chakrabarty, by fusing the CAM, TOL and OCT plasmids, was able to construct a novel microorganism with multiple degradative traits. Recombinant—DNA techniques could be similarly employed to insert elements representing multiple degradative traits into a single plasmid.

It may also be possible to construct microbial strains which exhibit novel or not previously expressed degradative traits. Vandenbergh et al.[31] were able, by transferring a degradative trait to another pseudomonad, to construct a recombinant strain which had degradative traits not expressed previously in either the donor or the host microorganisms. Recently, Chakrabrty has claimed the production of novel degradative traits in strains using plasmid assisted molecular breeding.

Health, Safety, and Environmental Concerns

Whenever a microorganism is considered for use in an industrial process, the potential for adverse effects should be examined and addressed. This potential for adverse effects applies equally to wild-type and engineered strains. It is unlikely though that recombinant—DNA techniques, as applied to microbial degradation, will result in additional biohazards.

Genetic engineering may be employed to construct strains with a minimized potential for adverse effects. The use of nontransmissible plasmids minimizes the

likelihood of a plasmid escaping its host and transferring
to another microorganism. Also, by reducing the size of a
plasmid to the smallest size possible without affecting
the desired trait, superfluous genetic information and
cryptic genes will be removed, reducing the likelihood of
transferring undesirable genetic information. In addition, a host microorganism could be constructed which is
enfeebled so that it can only survive under defined
conditions and is incapable of survival in the environment. Containment systems can also be constructed which
limits the potential for escape.

FACTORS THAT MAY LIMIT THE SUCCESS OF
GENETIC ENGINEERING AS APPLIED TO
MICROBIAL DEGRADATION

Although genetic engineering promises significant
improvements, there are several factors which if not taken
into account, could jeopardize a genetic engineering
program in the area of degradation. The three major
factors that must be addressed are compatability and
coordination, unfavorable environmental conditions and
adequate containment.

Compatibility and Coordination

As mentioned previously, the problem of plasmid
incompatibilty must be considered. The plasmid fusion
technique of Chakrabarty or the construction of hybrid-
plasmids by recombinant-DNA techniques are both possible
approaches to overcome this problem.
Once a degradative trait has been transferred to and
cloned in a new host, it must then be expressed (i.e., the
necessary biologically active proteins must be
synthesized). A host must therefore be employed which
will allow the trait to be expressed. A promoter must
also be present which will allow adequate gene transcrip-
tion and translation to occur.
Assuming the cloned gene can be properly expressed,
other problems may be encountered. It is imperative that
the pathway or reactions encoded by the inserted plasmid
be compatible with the biochemical pathways encoded by the
host chromosome. Pseudomonas putida mt-2, which harbors
the Tol plasmid, can degrade toluene.[32] The Tol plasmid
can be transferred to E. coli, but no toluene degradative
phenotype will be expressed. It has been shown in many
cases that degradative plasmids are responsible for only
the first several biochemical reactions of the degradative
pathway[20], while the remainder of the pathway is encoded
by chromosomal genes. Therefore, E. coli strains that do

not have the chromosomal genes normally found in Pseudomonas putida mt-2 (Tol+) are unable to degrade toluene even when the tol plasmid is present.

Coordinating biochemical pathways and their regulation mechanisms is further complicated in pseudomonads because there are three documented pathways for the degradation of aromatic compounds (ortho and meta cleavage and gentisate pathways).[33]

Therefore, successful strategies should be based not only upon selection of an appropriate host, but on sound genetic and physiological characterization of the host and the inserted element.

Environmental Conditions

A genetically engineered (or wild-type) microorganism, which demonstrates a high degradative ability under laboratory conditions, may not be successful when used in the field. Some potential reasons for failure to degrade in the field are as follows. A microorganism with a unique degradative trait, that might be considered to confer some selective advantage, would have to compete with the resident microflora to establish a permanent position in the microbial community. If the assumption is made that there are $10^8 - 10^9$ bacteria per gram of soil, it can then be seen that a very large number of bacteria would have to be added to this soil to achieve a reasonable probability of establishing a "new" bacterium in the community.

Assuming a "new" microorganism could be introduced and established in a given ecosystem, additional problems would also have to be addressed (e.g., temperature, pH, substrate concentration, oxygen tension, etc.). Most microorganisms have operating parameters which are relatively well defined. Because it is likely that at any given time one or more of these parameters are likely to be outside the optimal range for activity, the majority of the time the microorganism will have a reduced activity, (e.g., a bacterium which degrades n-alkanes optimally at 30°C in an aqueous medium would most likely perform poorly if exposed to an oil spill under North Sea conditions).

Because of generally unfavorable operating conditions and the difficulty which is likely to be encountered in establishing a "new" bacterium in an ecosystem, broadcast application of genetically engineered microorganisms to treat toxic-chlorinated hydrocarbons is not likely to be widely successful given our current state of knowledge in this area.

Containment

It is probable that engineering systems can be fabricated
which provide additional containment and also provide a
means whereby operating conditions could be controlled and
optimized. Such a system could be developed and applied
to aqueous treatment schemes (e.g., landfill leachate,
plant effluent, holding ponds, collected runoff, municipal
water, etc.). These treatment systems would most likely
incorporate additional features, such as pretreatment to
concentrate toxicants to levels which would support a
higher microbial activity.

SUMMARY

There are considerable benefits to be derived from
the successful application of genetic engineering to the
microbial degradation of hazardous compounds and
pollutants. The benefits to this area will most likely
include: stabilization of activities, increased rate of
activity, increased spectrum of activity, and development
of safe microbial systems. Currently, there are technical
restraints or problems to achieving these goals. These
restraints will be overcome as better cloning-vector
systems for pseudomonads and other degradative bacteria
become available. Also, with more information regarding
the physiology, biochemistry, and regulation of these
degradative bacteria the realization of these goals will
come closer to reality. Finally, before successful
"engineered" microorganisms are a reality, processes that
can utilize these microorganisms must also be developed.

REFERENCES

1. Sönhngen, N. L. Sur le role du methane dans la vie organique. Rec. Trav. Chim. 29:238-274 (1910).
2. Söhngen, N. L., Benzin, Petroleum, Paraffinol und paraffin als kautschukc durch mikroben. Zentr. Batteriol-Parasitenk., Abt II. 40:87-98 (1913).
3. Stewart, J. E., R. E. Kallio, D. P. Stevenson, A. C. Jones, and Schissler. Bacterial hydrocarbon oxidation. J. Bacteriol. 78:441-448 (1959).
4. Dagley, S., P. J. Chapman, D. T. Gibson, and J. M. Wood. Degradation of benzene nucleus by bacteria. Nature. 202 (1964).
5. Davies, J. I. and W. C. Evans. Oxidative metabolism of naphthalene by soil pseudomonads. Biochem. J. 91:251-261 (1964).
6. Cerniglia, C. E. and S. A. Crow. Metabolism of aromatic hydrocarbons by yeasts. Arch. Microbiol. 129:9-13 (1981).
7. McKenna, E. J. and R. E. Kallio. The biology of hydrocarbons. Ann. Rev. Microbio. 19:183-208 (1965).
8. Walker, J. D., H. F. Austin, and R. R. Colwell. Utilization of mixed hydrocarbon substrate by petroleum-degrading microorganisms. J. Gen. Appl. Microbiol. 21:27-39 (1975).
9. Bollag, J. M., C. S. Helling, and M. Alexander. 2,4-D Metabolism. J. Agric. Food Chem. 16:826-828 (1968).
10. Loos, M. A. Indicator media for microorganisms degrading chlorinated pesticides. Can J. Microbiol. 21:104-107 (1975).
11. Rosenberg, A. and M. Alexander. Microbial metabolism of 2,4,5-trichlorophenoxy acetic acid in soil, soil suspensions, and axenic culture. J. Agric. Food Chem. 28:297-302 (1980).
12. Goring, C.A.I., D. A. Laskowsk, J. W. Hamaker, and R. W. Meikle. Principles of pesticide degradation in soil in: Environmental Dynamics of Pesticides, R. Haque and V. H. Freed (Eds). Pleunum, New York. 135-172 (1974).
13. Chacko, C. I., J. L. Lockwood, and M. Zabik. Chlorinated hydrocarbon pesticides: Degradation by microbes. Science. 154:893-895 (1966).
14. Kaufman, D. D. and P. C. Kearney. Microbial degradation of s-triazine herbicides. Residue Rev. 32:235-265 (1970).
15. Ahmed, M. and D. D. Focht. Degradation of polychlorinated biphenyls by two species of Achromobacter. Can J. Microbiol. 19:47-52 (1972).

16. Ahmed, M. and D. D. Focht. Oxidation of polychlorinated biphenyls by Achromobacter PCB. Bull, Environ. Contamin. Toxicol. 10:70-72 (1973).

17. Pal, D., J. B. Weber, and M. R. Overcash. Fate of polychlorinated biphenyls (PCBC) in soil-plant stems. Residue Rev. 74:45-98 (1980).

18. Chakrabarty, A. M., G. Chou, and I. C. Gunsalus. Genetic Regulation of Otane Dissimulation Plasmid in Pseudomonas. Proc. Nat. Sci. 70:1137-1140 (1973).

19. Reinwald, J. G., A. M. Chakrabarty, and I. C. Gunsalus. A transmissible plasmid controlling camphor oxidation in Pseudomonas putida. Proc. Nat. Acad. Sci. 70:885-889 (1973).

20. Fisher, P. R., J. Appleton, and J. M. Pemberton. Isolation and characterization of the pesticide-degrading plasmid pJP1 from Alcaligenes paradoxus. J. Bacteriol 135:798-804 (1978).

21. Pierce, G. E., T. J. Facklam, and J. M. Rice. Isolation and characterization of plasmids from environmental strains of bacteria capable of degrading the herbicide 2,4-D, in Developments in Industrial Microbiology (LA. UnderKofler, Ed). Society for Industrial Microbiology, Arlington, VA. (In press) (1980).

22. Windlass, J. D., et al. Improved conversion of methanol to single-cell protein by Methylophilus methylotraphus. Nature. 287-396-401 (1980).

23. Bassford, P. et al. Genetic fusions of the lac operon: A new approach to the study of biological processes, in the Operon. J. H. Miller and W. S. Rezinikoff (eds). Cold Spring Harbor Laboratory. Cold Spring Harbor, New York. 245-262 (1980).

24. Jacoby, G. A. Classification of plasmids in Pesudomonas aeruginosa, in: Microbiology-1977, D. Schlessinger (Ed). American Society for Microbiology, Washington, D.C. 221-224 (1977).

25. Novick, R. P., R. C. Clones, S. N. Cohen, R. Curtiss III, N. Datta, and S. Falkow. Uniform nomenclature for bacteria plasmids: a proposal. Bacteriol. Rev. 40:168-189 (1976).

26. Clewell, D. B. Nature of Col E_1 plasmid replication in Escherichia coli in the presence of chloramphenicol. J. Bacteriol. 110:667-676 (1972).

27. Chang, A.C.Y. and S. N. Cohen. Construction and characterization of amplifiable multicopy DNA cloning vehicles derived from the P15A cyrptic miniplasmid. J. Bacteriol. 134:1141-1156 (1978).

28. Shine, J. and L. Dalgarno. Determinant of cistron specificity on bacterial ribosomes. NATURE 254:34-38 (1975).

29. Chkrabarty, A. M. Plasmids in _Pseudomonas_. Ann. Rev. Genet. 10:7-30 (1976).

30. Korfhagen, T. R., L. Sutton, and G. A. Jacoby. Classification and physical properties of _Pseudomonas_ plasmids, in: Microbiology-1978, D. Schlessinger (Ed). American Society for Microbiology, Washington, D.C. 221-224 (1978).

31. Vandenbergh, P. A., R. H. Olsen and J. F. Coloruotolo. 1981. Isolation and genetic characterization of bacteria that degrade chloroaromatic compounds. J. Appl. Environ. Microbiol. 42:737-739.

32. Worsey, M. J. and P. A. Williams. Metabolism of toluene and xylenes by _Pseudomonas putida_ (_arvilla_) mt-2: Evidence for a new function of the TOL plasmid. J. Bacteriol. 124:7-13 (1975).

33. Doelle, H. W. Bacterial Metabolism (2nd Ed). Academic Press, New York. 499-516 (1975).

CHAPTER 28

HAZARDOUS WASTE MANAGEMENT:
LEGAL IMPLICATIONS OF CLEANING
UP INACTIVE SITES

Richard D. Fox, Esq., P.E.
Patrick E. Gallagher, Esq.
Camp Dresser & McKee

GENESIS

Solving problems related to the historic disposal of hazardous waste materials has become a high priority, national environmental concern. Federal legislation enacted in December, 1980, entitled the Comprehensive Environmental Response, Compensation, and Liability Act (CERCLA), focuses directly on procedures to be undertaken or guided by the federal government in response to this urgent environmental problem. This legislation, popularly known as Superfund, also delineates the parties liable for costs incurred in cleaning up these problem disposal sites.

It is significant to note that CERCLA was enacted by Congress prior to the present Administration's newly formulated "federalism" policy. Thus, legislation drafted in anticipation of a strong federal role is being implemented under an Administration that is advocating a strong state role. CERCLA also is significant in that Congress deliberately excluded claims for personal injury and property damage incurred by individuals, leaving them to fend for themselves using common law and existing statutes.

Current policies for implementation of CERCLA and the early stage of its development have placed the states and responsible parties (individual generators, transporters, and disposal site operators and owners) at a critical decision point in the overall cleanup process. This paper will focus almost exclusively on legal issues of concern to state governments and responsible parties at this decision point.

It should be noted that the following presentation of legal issues is by no means complete. An effort has been made to provide information of sufficient breadth to be of general guidance to state agencies and responsible parties. Many of the issues presented need to be examined in much more detail before important decisions are made. Additional issues will certainly be raised as more in-depth analyses of a given decision are undertaken. The general issues and framework discussed herein will hopefully provide a good starting point in such analyses.

A synopsis of legal terminology and the overall legal structure that concerned parties must deal with are first presented. The legal maneuvers through which each high-priority hazardous waste site may be expected to pass is then described, and a check-list of issues is provided to assist individual deliberations. While it is hoped that the enclosed checklist and synopsis of legal structure and terminology will be helpful, the singular purpose of this paper is to cultivate a recognition that a very complex, far-reaching decision is awaiting resolution by hundreds of potentially responsible parties and by a majority of state governments.

LEGAL TERMINOLOGY

Briefly described below are key legal terms used frequently in this paper:

o Responsible Parties. A unique aspect of CERCLA is its broad definition of parties liable for cleanup costs and natural resource damages related to releases from inactive hazardous waste disposal sites. So-called "responsible parties" for a given inactive site include the following:

   - present owner and operator of the site;
   - owner or operator of the site at the time of disposal;
   - generator who arranged for disposal or treatment, via a transporter or otherwise; and
   - transporter who selected the disposal site.

This broad definition of responsible parties leaves little room for any party somehow connected with a given site to escape responsibility for its cleanup. Because generators are included as responsible parties, the number of such parties for a single site can be exceedingly large. Out of some 1,500 initial notice letters that will be forwarded by the EPA to allegedly responsible parties, for example, about 1,000 relate to only six sites.

o <u>Joint and Several Liability</u>. This is when one party is required to pay the full damages resulting from the combined actions of himself and other parties. Under the common law, when more than one party is the cause of harm, it is generally the plaintiff's burden to prove the specific harm caused by each party and to apportion damages accordingly. However, when the harm caused by several parties is indivisible, meaning that it is not theoretically possible to apportion damages to the different parties, then the burden of proving apportionment shifts to the multiple parties and the plaintiff can hold one party responsible for the total. That party is said to be subject to joint and several liability. Because of the potentially large number of responsible parties under CERCLA for a given site, the question of joint and several liability is extremely important. CERCLA itself is silent on this issue, leaving its resolution to the common law on a case-by-case basis.

o <u>Third-Party Liability</u>. Parties not directly involved in a given lawsuit are referred to as third parties, and any liability running from the parties to the lawsuit to such third parties is called third-party liability. In the context of CERCLA litigation involving both federal and state governments and responsible parties, third party liability refers to the potential for private parties to pursue claims of personal injury and property damage against such responsible parties. If only the federal government is involved in CERCLA litigation, however, the state could be considered a third party.

LEGAL STRUCTURE

The basic legal structure governing hazardous waste disposal, like most environmental programs, consists of comprehensive legislation, administrative regulations, and common law. The following paragraphs briefly summarize major federal hazardous waste and related legislation, selective state statutes, and in very general terms, the common law. Hazardous waste regulations are not addressed because they generally govern present and future disposal activities, not the cleanup of inactive sites.

Major Federal Statutes

Two key federal statutes regulate a wide variety of activities regarding hazardous waste: the Resource Conservation and Recovery Act of 1976[1], as amended in 1980 (RCRA), and CERCLA[2]. RCRA governs present and future generation, transportation, storage, treatment, and disposal activities. It is the source of the U.S. Environmental Protection Agency's (EPA) authority for the many hazardous waste regulations issued to date. CERCLA, on the other hand, principally looks to past disposal activities, being concerned in large part with present releases from inactive hazardous waste sites. Facilities falling under RCRA are for the most part outside the reach of CERCLA.

RCRA grants to EPA broad authority for developing and enforcing regulations covering hazardous wastes from its source of generation to its final disposal site (commonly referred to as "cradle-to-grave" regulation). The statute outlines broad objectives and various kinds of requirements that EPA regulations must address. Such regulations, which vary for generators, transporters, treaters, storers, and disposers, include such responsibilities as reporting; record keeping; labeling; packaging; financial guarantees; monitoring; contingency plans; closure; post-closure; and design, construction, and operating standards.

An owner or operator of treatment, storage, or disposal facilities must obtain a permit under RCRA. Facilities existing on November 19, 1980 may achieve interim status, meaning that the RCRA permit requirement is considered met until action is taken by EPA on permit applications. New facilities, however, cannot be constructed until a RCRA permit is issued. The EPA also has broad powers under RCRA to order monitoring, testing, and analysis when a facility may present a substantial hazard to human health or the environment, and may seek injunctive relief when a imminent and substantial endangerment to human health or the environment is caused by improper hazardous waste handling.

CERCLA differs from RCRA in many ways. By addressing inactive sites, CERCLA's principal aim is to correct problems caused by past hazardous waste disposal activities. It squarely places liability for the costs of cleaning up such sites and for damages to natural resources on past and present owners and operators of each site and upon generators and transporters that used them. A $1.6-billion fund is established (to be accumulated largely by taxes on crude oil and certain chemical feed stocks[3]) to finance cleanup action by governmental agencies. The fund is not intended as an insurance policy to assume liability for inactive sites, but rather as a ready source of funds for speedy governmental action. Cleanup costs incurred by government agencies are recoverable from responsible parties under strict liability provisions.

The guiding force for governmental cleanup of hazardous waste sites under CERCLA is the National Contingency Plan[4]. CERCLA sets forth specific activities to be addressed in the plan, including methods for: (a) discovering and investigating inactive sites; (b) evaluating and remedying releases; (c) assuring that governmental actions are cost effective; and (d) determining priorities among sites throughout the United States. All cleanup actions taken by governmental agencies under CERCLA must be consistent with the National Contingency Plan. Responsible parties are liable for cleanup costs incurred by governmental agencies only if such actions are consistent with the plan.

In addition to the liability, funding, and authority for governmental action to cleanup inactive hazardous waste sites, CERCLA imposes two reporting requirements. First, immediate notification of spills and other releases of hazardous substances into the environment must be made to the National Response Center. Federally permitted releases and other releases less than reportable quantities are exempt from such notification requirement. Secondly, notification must have been filed with EPA by June 9, 1981, identifying inactive disposal sites. Such notification must have identified the site; the type and amount of hazardous substances; and any known, suspected, or likely releases. Again, facilities with a federal permit or interim status are excluded.

For both RCRA and CERCLA, key threshold questions revolve around the definition of hazardous wastes and hazardous substances. Wastes or substances that are not hazardous have the luxury of ignoring the demanding requirements of CERCLA and RCRA. The RCRA hazardous waste definition incorporates four key characteristics: toxicity, reactivity, ignitability, and corrosivity. Hazardous substances under CERCLA span a broad universe, including substances within the RCRA definition of hazardous wastes. CERCLA hazardous substances expand the RCRA definition to include materials that are not wastes and a host of hazardous materials designated under other federal statutes. For purposes of authorizing governmental response, the universe of substances covered by CERCLA is expanded even further to include pollutants or contaminants that may present an imminent and substantial danger to public health or welfare.

Three other major federal statutes have applicability to the handling of hazardous wastes. The Safe Drinking Water Act[5] protects underground sources of drinking water and gives EPA emergency power to act to protect public health when an imminent and substantial endangerment exists. The federal Clean Water Act[6], primarily intended to protect navigable surface waters, provides for injunctive relief against a pollutant discharge that is presenting an imminent and substantial endangerment to health and welfare.

The last major federal statute, the Toxic Substances Control Act[7], addresses the need to regulate the manufacture, distribution, and use of toxic chemical products. It requires testing of new chemicals and of new uses of known chemicals that may present an unreasonable risk of injury to health or the environment. In contrast to RCRA and CERCLA, this statute generally regulates toxic materials at their source. Specific provisions, however, address the disposal of polychlorinated biphenyls (PCB's), a toxic substance freqently found at inactive disposal sites addressed by CERCLA.

Selected State Statutes

Essentially every state has enacted RCRA-type legislation. Most states have implemented their own permitting and licensing procedures, but many state closure and financial responsibility requirements are presently in limbo because of uncertainty with the federal requirements. State RCRA statutes generally parallel, both in form and in implementation progress, federal efforts. The extent to which state RCRA statutes have been used to address both historic and prospective hazardous waste problems reflects to a great extent the aggressiveness of the particular state agency involved.

Superfund legislation at the state level, in contrast to RCRA-type legislation, has been enacted in only a few states. Several states appear to be awaiting further implementation of the federal Superfund program. Imposition of a new tax at the state level is generally not well received, particularly when many states perceive increased financial burdens from the new "federalism". Concern regarding the scope of the federal government's exclusive right (preemptory right) to collect such taxes to support a

446

cleanup fund also may have put a damper on state legislative action. The following paragraphs briefly summarize state "superfund" statutes.

California. The California Hazardous Substances Account Act[8] became effective in September, 1981. The taxing structure of California's statute utilizes the quantity of wastes disposed of rather than feedstock as a tax basis. This tax structure provides a positive incentive to reduce the quantity of material generated for disposal. The waste tax approach was possible in California, because it, unlike the nation as a whole, possessed a manifest system capable of serving as a basis of tax charges. The 10-year fund has a maximum ceiling of $10 million.

The California Superfund, unlike CERCLA, provides for the payment of 100 percent of out-of-pocket medical expenses and 80 percent of property damage for victims who can demonstrate that they have no other source from which these damages may be recovered. Funds expended for such expenses cannot exceed $2 million per year. It should be noted that the California law does not appear to have created a new statutory cause of action for victims against responsible parties. Victims are entitled, however, to use the full spectrum of common law remedies.

The California law places a duty on the state to seek monetary recovery from responsible parties when Superfund monies are used. It permits apportionment of cleanup costs when a responsible party and the fund are used jointly and the cleanup costs can be reasonably allocated. (CERCLA is silent on apportionment and the federal government has indicated its intent to seek joint and several liability where appropriate.)

Kentucky. Kentucky's hazardous waste legislation[9] preceded the federal Superfund and thus was based almost entirely on RCRA. However, Kentucky's legislation establishes a fund to cleanup abandoned hazardous waste sites when other resources are not available. The tax supporting this fund, like California's, is based on the quantity of hazardous waste materials transported to long-term disposal. The tax will be assessed to hazardous waste generators until mid-1984, at which time it is to be levied against hazardous waste disposal facilities. The fund has a statutory floor of $3 million and a ceiling of $6 million. The Kentucky statute mandates recovery of cleanup costs from responsible parties where possible. It specifically notes that, although the Kentucky Department of Natural Resources is given a cause of action to recover

cleanup costs, it is intended to neither broaden nor limit
the basis of liability, burden of proof, or extent of
liability of any other person.

New Jersey. The New Jersey legislation[10] also preceded
CERCLA and was drafted using RCRA as a basis. An emergency
response cleanup fund is established by taxing hazardous
materials transferred. The fund's floor and ceiling are $40
million and $50 million respectively. The New Jersey
statute imposes strict liability, jointly and severally, for
cleanup costs on any responsible party. The statute does
not provide for personal injury compensation but does permit
recovery for losses to personal and real property. The New
Jersey legislation, like Kentucky, California, and the
federal Superfund, grants the state subrogation rights for
any monies paid out of the fund.

A 1981 amendment to the New Jersey statute broadens the
definition of cleanup costs to include the costs of
providing alternate water supply[11]. In late 1981, the New
Jersey electorate approved a $100-million bond issue to
further underwrite the costs of cleaning up the state's
inactive hazardous waste sites.

Common Law

The body of law developed by courts to resolve disputes
not covered by statutes, called the common law, is at an
early stage of development for hazardous waste management.
Federal statutes, as briefly reviewed above, provide a
comprehensive scheme for addressing the problem of inactive
hazardous waste sites, raising the possibility that federal
common law in this area is preempted. State statutes
dealing with inactive hazardous waste sites, in contrast,
are not abundant. The principal federal statute (CERCLA),
however, provides for, and to some extent requires, action
by the states. A major gap in statutory law occurs when it
comes to recovery by private parties of damages for injury
to persons and property. The body of law available to
resolve disputes in this regard, the state common law, is
generally considered inadequate for this purpose, because of
the high cost of litigation and the difficulty of proving
fault and causation for specific injuries.

Common law theories of trespass, nuisance, negligence,
and strict liability represent potential grounds for
recovery of compensation for personal injuries and property
damage. The trespass and nuisance theories are limited by
the requirement that the plaintiff have a possessory

448

interest. This largely eliminates recovery for injuries to the health of the community and to parties not owning or leasing land within a given site's direct zone of influence.

The theory of negligence is open to a much broader class of plaintiffs, but at its heart is the concept of fault, which must be proven by the plaintiff. A potentially more difficult hurdle is the need to prove causation. This can be especially troublesome when multiple defendants are involved and when an attenuated chain of events connect the defendant's actions with the plaintiff's injuries. In addition, a laundry list of defenses (depending on the state) are available to insulate the defendant, including the statute of limitations, independent contractor liability, contributory negligence, assumption of risk, and state-of-the-art technology.

The last common law theory, strict liability, is similar to negligence with one major exception: there is no need to prove fault. Strict liability holds that a person engaging in an abnormally dangerous or ultra-hazardous activity is liable for injuries caused by that activity regardless of fault. The actor assumes the risks inherent in engaging in such an activity. The applicability of this doctrine will vary from state to state.

Even if a plaintiff is successful in asserting a theory of liability and proving damages, an insolvent or financially weak defendant will offer little relief. All too frequently, when owners and operators of inactive hazardous waste sites are called upon to respond to damages, a thinly capitalized corporation whose principal asset is the site itself is the only responsible party. The common law theory of piercing the corporate veil to hold stockholders personally responsible may be helpful in such a situation. However, this is only rarely accomplished, and in order for it to be helpful, there needs to be major assets in the hands of a few stockholders. Generators tend to be more solvent, but asserting the common law theories of liability and proving damages are much more difficult.

THE SUPERFUND DECISION

Background

The EPA, largely through a series of consultant contracts and by use of one of CERCLA's notice requirements, has identified several thousand hazardous wastes disposal sites which may require cleanup action. EPA evaluations

449

continue on many of these sites to identify the extent and
imminency of the environmental and health risks presented.
The EPA has identified responsible parties for certain sites
and recently announced its intent to send an initial wave of
so-called "notice letters" to some 1,500 such parties.

Exhibit 1 illustrates that the alleged responsible
parties receiving notice letters, the EPA, and the state
governments in which these hazardous waste sites are
located, face a critical decision point in the federal
effort to cleanup inactive hazardous waste sites. Two basic
options are open: a voluntary cleanup route and an
involuntary route, with the latter possibly involving both
administrative and judicial action.

An alleged responsible party selecting the voluntary
route will probably be asked to accept the EPA's general
standards for cleanup and to execute an agreement committing
dollars and other resources to the cleanup effort. The
involuntary route provides two distinct courses. If the EPA
(or the state under a cooperative agreement) believes that
urgency at the site merits immediate attention, Superfund
monies may be expended and after-the-fact judicial action
initiated to recover cleanup costs. If the EPA (or the
state) determines, on the other hand, that site cleanup is
less urgent, it may elect to issue an administrative order
or seek judicial action before spending Superfund monies.

The voluntary and involuntary routes provide distinct
advantages and disadvantages for the parties involved. It
is prudent for federal and state regulatory agencies and the
alleged responsible parties faced with this decision to make
a reasoned and calculated judgment as to whether the
voluntary or involuntary course of action is in their best
interests. New questions of law, as well as scientific,
economic, political, public relations, environmental,
social, and other factors influence this judgment. The
emphasis here is on the voluntary versus involuntary
decision point within its legal context.

Issues to be Weighed

Implementation of CERCLA has proceeded to a point where
several hundred alleged responsible parties and nearly every
state are faced with an immediate, far-reaching decision
concerning the role that each wishes to play in the cleanup
of EPA-identified sites. In this regard, fiscal and legal
analysis should be prepared by responsible parties to assess
the boundaries of their exposure. State agencies should

evaluate the potential for initating site cleanup with and without federal and state Superfund dollars to better assess their negotiating posture with an alleged responsible party. Political, social, and environmental issues should be weighed carefully by both responsible parties and governmental agencies.

Exhibit 2 (found at the end of this chapter) presents a general listing of legal issues that should be weighed by all parties faced with selecting a course of action for responding to problems caused by an inactive hazardous waste site. The issues are grouped as follows:

I. Exposure: What is the extent of liability for cleanup under federal and state law?

II. Negotiation: How can exposure be minimized by settlement?

III. Third Party Claims: What exposure exists regarding private claims for personal injuries and property damage?

Each of the basic legal issues are discussed briefly with emphasis on the perspectives of a state agency and a responsible party. While this discussion is by no means complete, it does illustrate the "negotiating" mentality that may be required to address this decision.

Cleanup Exposure Under Existing Law

Strict Liability/Other Theories. Under CERCLA, strict liability for governmental cleanup costs and natural resource damages related to inactive hazardous waste sites extends to so-called responsible parties. A key threshold question, therefore, is whether a given generator, transporter, site owner, or site operator fits under CERCLA's definition of responsible parties, and whether sufficient evidence is available to sustain a court determination to that effect. A few state Superfund statutes similarly impose strict liability for the cost of cleanup, but others leave the basis of liability to existing state statutes and common law. The existence of a strict-liability theory places great emphasis on the threshold question of who can be shown to be a responsible party and will significantly ease the burden of proof (since fault is not at issue). Recent hazardous waste decisions in federal court imply (although not conclusively) that federal common law will not be available, a limitation likely to be of concern only in interstate situations.

Other theories of liability for cleanup are abundant, but are not generally attractive from an enforcement standpoint when compared to strict liability. EPA's ability under CERCLA to issue an administrative order backed by penalties ($5,000 per day of violation) and treble damages (up to three times actual cleanup costs) is a powerful enforcement tool, but it appears limited to situations involving an imminent and substantial endangerment. State agencies and responsible parties should consult the relevant state statutes regarding the power to issue administrative orders and the penalties for enforcement. If such orders can be issued in a broader context than CERCLA's imminent and substantial endangerment limitation, they may substantially supplement CERCLA's strict liability provisions.

An area of particular concern to responsible parties regarding other theories of liability is the broad range of remedies potentially available. Extraordinary remedies such as receivership and punitive damages, for example, may be available under state law. The availability and the applicability of such remedies to the specific circumstances of a given site represent major factors in evaluating the full exposure of a responsible party.

Joint and Several Liability. The legislative history of CERCLA indicates that Congress entertained joint and several liability and deliberately left out express statutory language to that effect. The present position of the federal government is that such refusal to insert an express joint and several liability provision in the statute is not a congressional mandate that joint and several liability is unavailable. Rather, the federal position is that Congress intended governmental agencies and responsible parties to look to existing statutes and common law in the host state. The federal government has indicated an intent to pursue enforcement activities seeking joint and several liability on this basis.

It is expected that joint and several liability issues will be litigated in the near future, although final resolution of such litigation may require years. State agencies and responsible parties should review appropriate state legislation and common law to assess the applicability of joint and several liability to the site under consideration. Some state superfunds, such as the New Jersey statute, expressly provide for joint and several liability for cleanup of abandoned hazardous waste sites. The availability of joint and several liability may be a significant aid to state agencies in their pursuit of voluntary cleanup agreement and a strong inducement for responsible parties to agree to pay their fair share in return for a release of such broad liability.

Personal Assets. Most of the alleged responsible parties identified to date have been corporate entities. The corporate structure generally insulates the personal assets of its stockholders, prohibiting state agencies from obtaining recovery from the personal assets of stockholders when the corporation itself is insolvent. Courts have, in extraordinary cases (e.g., where the corporate structure is viewed as a sham), pierced the corporate veil and tapped the personal resources of its stockholders. While such a remedy is extraordinary, its potential in a given situation may play a significant role in making a decision to seek voluntary or involuntary cleanup actions.

Standard of Performance. If governmental agencies are to encourage voluntary cleanups in lieu of judicial remedies, commitments required of responsible parties must have finite limits. The difficulty in drafting a standard of performance is that the basic goals of the governmental agencies and the responsible parties may be diametrically opposed. For example, the responsible party may prefer an agreement specifying expenditure of a given sum of money for the performance of specific activities such as monitoring, installation of wells, etc., to place a well-defined limit on its obligations. In contrast, governmental agencies may prefer a more general obligation such as "whatever activities are necessary to meet certain water quality criteria."

For governmental agencies and responsible parties to reach a voluntary agreement, the standards of performance to be included within the agreement will require compromise, considering the general hierarchy of standards of performance given on Exhibit 2. A factor to be weighed heavily by state agencies is the precedent-setting value of its initial voluntary agreements. A principal concern of responsible parties, on the other hand, is the standard of performance that they might confront in the form of a court injunction if they do not negotiate one in a settlement agreement.

Parties to the Action. It generally appears prudent for state regulatory agencies to become actively involved in the negotiation of settlement agreements as well as in the joint pursuit (with the federal government) of judicial decrees. Active participation by state agencies ensures that the identified site will not be delisted without consent of the state agency. "Delisting" could occur if a responsible party and EPA agree to a level of cleanup in settlement of an enforcement action. The agreed upon level of cleanup might be sufficient to remove the threat of imminent harm, resulting in the site not being considered for further funds in spite of considerable local dissatisfaction with the level of cleanup.

Active participation of the state agencies is likely to be well received by alleged responsible parties because the court decree or settlement agreement would then include the two premiere forces in the cleanup action. In some states, introduction of the state agencies to the voluntary agreement may broaden an alleged responsible party's exposure (see for example New Jersey's joint and several liability statute). Responsible parties still may be more receptive to a comprehensive federal/state settlement because their total exposure could be better defined and more effectively limited.

Joining the state agency as an active participant in a settlement agreement will not necessarily foreclose third party litigation by private parties. However, active participation by the federal and state governments in obtaining a judicial remedy or in a properly drafted agreement may help limit the legal exposure of the responsible party to such actions. First and foremost, the active participation of state and federal agencies in the development of a voluntary agreement lends considerable credibility to the reasonableness of the activities undertaken, and especially may demonstrate the reasonableness of the materials left on site following cleanup. Also, state agencies may represent the only party other than the federal government with sufficient resources to pursue remedies under common law. State agencies may find it advantageous to encourage the inclusion of the more-aggrieved private parties in the initial court action or settlement negotiations, closing the books on a site once and forever.

Burden of Proof. The burden of proof for a cause of action under CERCLA and under some state superfund statutes focuses entirely on whether sufficient evidence is available to demonstrate that one is a responsible party, making it strictly liable for cleanup costs. This burden becomes more difficult when there are successive owners and other intervening parties. It may be easier to sustain the burden of proof if the disposal site is at the point of generation, compared to a site removed far from the generation point.

The burden of proof for a private party to maintain a cause of action for personal injury or property damage is very demanding, especially when common law theories of liability are used. Not only must the waste materials be traced to the responsible party, but the waste materials must also be clearly demonstrated to be the legal cause of the injury. This burden is troublesome when contaminants exist in concentrations of a few parts per billion, levels that are generally acknowledged to pervade the human environment. The difficult and costly burden facing private parties in personal injury or property damage litigation may become a significant source of conflict at the state level. Private parties frustrated with the lack of remedy, or dissatisfied with the level of cleanup planned using federal funds, are likely to focus their discontent at the state agencies. Active participation by state agencies and encouragement by state agencies for voluntary agreements to include private parties may diffuse this discontent.

Responsible parties should examine how the private claimant's burden of proof may be affected by its decision to sign a settlement agreement or defend a court action. In the latter case, the federal and state governments will accumulate a data base that may provide a strong evidentary foundation for a private party to later press claims for personal injury and property damage. With a settlement agreement, on the other hand, the data assembled may tend to be less useful to such a third party action and perhaps affirmative steps can be taken in the agreement towards that end.

Limiting Legal Exposure With Negotiated Settlements

The exposure of responsible parties is capable of limitation but not complete elimination. Like standards of performance, limitations on liability are critical factors in the negotiation of a voluntary cleanup agreement. Successful negotiations will to a large extent require a balancing of standards of performance and limitations on liability.

Perhaps the greatest incentive for a responsible party to pursue a voluntarily agreement lies with the potential to negotiate limits on its legal liability. As illustrated in Exhibit 2, many means are available to do this. The extent to which governmental agencies and alleged responsible parties can agree to incorporate these limits may dictate the likelihood of the responsible party seeking a voluntary, rather than an involuntary, cleanup program.

At the outset, inclusion of well-defined standards of performance will enable the responsible party to assess the boundaries of its obligations under the settlement agreement. A principal obstacle to negotiating such standards of performance and other limits on liability is that many of the sites for which the EPA is aggressively pursuing cleanup action have only a limited amount of investigatory work completed. The real extent of the problem at many sites is ill-defined and, therefore, specific standards of performance to adequately address these problems may be very difficult to develop. This lack of information may become a significant stumbling block for a voluntary agreement.

To accommodate this initial absence of information, it may be desirable for governmental agencies and the alleged responsible parties to first pursue site investigation and feasibility studies at the alleged responsible party's expense. Both parties can then reassess their decision to proceed voluntarily, possibly deciding to pursue an involuntary course of action. This flexibility is a distinct advantage of reaching an early settlement agreement. Administrative orders and court judgments do not permit the parties to adjust their respective obligations as easily as if a settlement agreement had been reached.

Consent decrees executed to date have shown some willingness by EPA to insert indemnification, release, nonintervention, and legal assistance clauses in return for the alleged responsible party's commitment of funds or actions. Recent policy statements by EPA's enforcement counsel indicate, however, that such limitations of liability may be far more difficult to obtain in the future.

Another means of limiting liability includes the insertion of a force majeure clause to reflect conditions or activites which are beyond the control of the alleged responsible party. If reasonable penalties for non-compliance with the settlement agreement and a clear and effective procedure for resolving differences encountered in its implementation are set forth, an alleged responsible party may find its liability further limited. Such clauses may be acceptable to the governmental agencies in return for less specific standards of performance or the commitment of larger sums of money. Such clauses may place an additional administrative burden on state agencies who assume management responsibilities.

Both parties to a settlement agreement should assess whether the agreement extends to damages to natural resources in addition to cleanup costs and should assess the extent to which victim compensation is either included or excluded. If private parties are left without legal remedy, they may become a significant source of discontent focused almost entirely at state agencies.

CERCLA, as noted earlier, is silent on the question of apportionment or contribution of cleanup costs among responsible parties. The apportionment and contribution provisions of the Uniform Tortfeasors Act may therefore be available to the extent that one can reasonably allocate the cost of cleanup. Decisions to date suggest that courts will permit the reasonable apportionment of cleanup costs, even if a precise means of division is not readily available. The question of joint and several liability versus apportionment is a particularly important area for negotiation of the settlement agreement, especially when several responsible parties are evident but not involved in the negotiations.

Third Party Claims

Private parties, or even state agencies not involved in federal cleanup enforcement actions, may press claims against responsible parties for personal injury, property damage, public health effects, additional cleanup costs, and other remedies. Although the pursuit of private tort claims for damages is difficult because of burden of proof problems, the cause of action clearly exists. Responsible parties should evaluate their exposure to such claims and negotiate as much protection as possible into the settlement agreement or conduct their defense of government enforcement litigation in a manner that limits such exposure. The decision to settle or litigate the question of cleanup liability at a given site should also reflect the pros and cons of defending such third party claims, giving strong consideration to the data base to be developed for each course of action and its availability and usefulness to support third party claims.

The primary concern of state agencies regarding third party claims is that if these parties are left without redress, state agencies are likely to become the brunt of their dissatisfaction. State agencies should consider involving such third party claimants in the negotiation of settlement agreements to minimize such dissatisfaction.

EXHIBIT 1 THE SUPERFUND DECISION

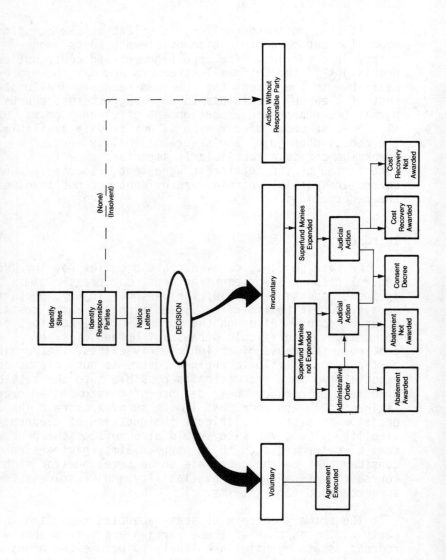

Exhibit 2

THE SUPERFUND DECISION:
A Checklist of Legal Issues

I. EXPOSURE: <u>What is the extent of liability for cleanup under federal and state law?</u>

Strict Liability
o Responsible party under CERCLA?
o Strict liability under state statutes or other federal statutes?
o Strict liability under the common law?
o Potential remedies? (CERCLA strict liability limited to cost-effective cleanup costs and natural resource damages.)
o Sufficiency of evidence?

Other Theories of Liability
o CERCLA administrative order and equitable relief when imminent and substantial endangerment?
o RCRA administrative order and injunctive relief when imminent and substantial endangerment?
o State and other federal statutes?
o Common law negligence, nuisance, and trespass?
o Available remedies? (CERCLA penalties and treble damages, RCRA penalties, injunctive relief backed by contempt power, compensatory and punitive damages, receivership, etc.)

Joint and Several Liability
o One party responsible for total CERCLA cleanup costs and natural resources damages? (U.S. Department of Justice position that CERCLA permits imposition of joint and several liability even though statutory language extracted before enactment.)
o State statute expressly or implicitly provides for joint and several liability?
o State common law requirements for joint and several liability?
o Likelihood of contribution by other joint tortfeasors?

Personal Assets
o Vulnerability of stockholder and officer assets, if corporate resources inadequate? (state law on piercing the corporate veil)

459

Standard of Performance

o Standard of performance for conducting cleanup in order of descending specificity might include:

a. expend a specified sum of money
b. perform specified activities
c. perform specified activities with a reopener clause
d. perform specified activities to meet specific performance criteria
e. meet specific performance criteria
f. "correct the condition"

o Settlement agreement to result in a standard of performance different than a consent decree, administrative order, or court injunction?

Parties to the Action

o Joinder of both federal and state regulatory agencies?
o Intervention by private parties?
o Joint enforcement action by federal and state agencies more formidable than individual actions?
o Inclusion of other defendants to share liability and defense costs?

Burden of Proof

o Elements of cause of action requiring proof?
o Extent of available evidence and difficulty of obtaining additional evidence?
o Remote disposal site or site coterminous with point of generation?
o Partial cleanup actions increase the difficulty of demonstrating threat of future harm?
o Data collection techniques during site investigation capable of supporting litigation? (admissibility under applicable rules of evidence)

II. NEGOTIATION: How can exposure be minimized by settlement?

o Indemnification against the claims of third parties?
o Release from further cleanup liability?
o Promise by regulatory agencies not to pursue further enforcement action?
o Promise of regulatory agencies to assist in defense of claims by third parties?
o Promise not to intervene in subsequent actions?
o Non-admission clause denying liability and factual findings, and prohibiting use of cleanup activities as evidence of responsibility?

o   Limit liability to specified cleanup activities,
    excluding personal injury, property damage, and
    damage to natural resources?
o   Well-defined standards of performance?
o   Force Majeure clause?
o   Reasonable penalties for non-compliance?
o   Procedures for resolution of disputes?
o   Apportionment of cleanup costs among responsible
    parties and Superfund dollars?
o   Settlement agreement fashioned in advance of
    intensive field investigations or litigation,
    avoiding the potentially high costs of discovery?
o   Long-term monitoring and maintenance requirements?
o   Is the site of sufficient priority to reasonably
    anticipate Superfund expenditures considering:
    a. The limited funds available?
    b. The possibility that Superfund may be
        terminated by Congress?
o   Level of cleanup likely to differ when enforcement
    activities lead by state rather than federal
    agencies?
o   Minimize adverse publicity, costs of legal defense,
    and the hidden costs of uncertainty (e.g., chilling
    effects if merger or acquisition planned)?

III.   THIRD PARTY CLAIMS: What exposure exists regarding
                             private claims for personal
                             injuries and property damage?
o   Availability to private parties of federal or state
    superfunds for:
    a. personal injury/property damage
    b. cleanup costs
o   Freedom of state agencies to pursue independent
    causes of action when the EPA signs cleanup
    agreement?
o   Data collected by a responsible party pursuant to
    implementation of diagnostic phase of settlement
    agreement usable as evidence in third party
    litigation?
o   Data collected by federal or state agency in
    preparation for litigation usable as evidence in
    third party litigation?
o   Cleanup actions undertaken under settlement
    agreement or court order sufficient to reduce the
    risks of harm and thus render third party claims
    more difficult to sustain?

NOTES:

(1)   P.L. 94-580, as amended; 42 U.S.C. 6901 _et seq._
(2)   P.L. 96-510; 42 U.S.C. 9601 _et seq._
(3)   "cleanup" is used in this paper as a general term that
      includes "removal", "remedial", and other CERCLA
      authorized actions at a given site.
(4)   CERCLA, Section 105.
(5)   P.L. 93-523, as amended; 42 U.S.C. 300f _et seq._
(6)   P.L. 92-500, as amended; 33 U.S.C. 466 _et seq._
(7)   15 U.S.C. 2601 _et seq._
(8)   Stats. 1981, c. 756; California Health and Safety
      Code, Section 25300 _et seq._
(9)   Kentucky Revised Statutes, Section 224.005 _et seq._
(10) New Jersey Statutes Annotated, Section 58:10-23.11 _et
      seq._

CHAPTER 29

RESPONSIBILITY AND LIABILITY
ATTENDING THE GENERATION,
TRANSPORTATION, STORAGE AND
DISPOSAL OF HAZARDOUS WASTE

Richard T. Sargeant, Esq.
  Eastman & Smith: Counsel, Associated
  Chemical and Environmental Services

INTRODUCTION

Accused polluters are attractive defendants for government lawyers and politicians. Pollution cases create publicity, and, on the whole, have the public's support.

Environmental statutes often create unreasonable obligations and provide for extremely high penalties. Often these penalties are much higher than penalties for similar activity in other industries which pose equal or greater risks to the public's welfare. Although the obligations created by environmental laws may be unreasonable, presently the liability for violations is not particularly arbitrary or unpredictable for companies that make good faith efforts to comply with the applicable statutes and regulations.

The theories of liability pertaining to the disposal of hazardous wastes are still changing, however, and the changes point toward increasing the accountability of all persons involved. It is difficult to predict the standards of liability that will be applied five, ten or fifteen years from now. A decade ago, few persons would have predicted that companies and individuals would now be liable for very significant sums of money because of waste disposed lawfully ten years ago.

This problem attends the disposal of solid waste to a much greater degree than it does other forms of pollution. Hazardous wastes which are deposited in or on the ground are different from most other forms of pollution. Emissions into the air are usually gone, or at least untraceable, soon after they are emitted. Discharges into lakes and streams largely dissipate soon after introduction. Hazardous wastes buried today will however, in all likelihood, be in the same spot in ten years. (If they are not, this alone may constitute a problem for everyone involved.) These wastes thus continue to be a potential source of liability.

STATUTORY SOURCES OF LIABILITY

Resource Conservation and Recovery Act of 1976 (hereinafter "RCRA")

Civil Liability

§3008(a), (c) and (g): The Administrator of the United States Environmental Protection Agency (hereinafter the "US EPA" or "Agency") may issue compliance orders to violators and may assess civil penalties. Penalties of up to $25,000 per day of non-compliance may issue if a violator fails to take corrective action within the time specified in the order.

§7002: The citizen suit provision provides standing to "any person" to bring a civil action against any other person or government instrumentality who is alleged to be in violation of any "permit, standard, regulation, condition, requirement or order which has become effective pursuant to [RCRA]".

§7003: In the case of an imminent and substantial endangerment to health or the environment, the Administrator may sue to enjoin actions contributing to the hazard or "to take such other action as may be necessary". Willful violation or failure to comply with any order issued by the Administrator pursuant to this section may result in fines of up to $5,000 per day.

Criminal Liability

§3008(d): The Act, as amended, provides criminal sanctions for any person who:

(1)   knowingly transports hazardous wastes to
       unpermitted facilities; knowingly treats, stores,
       or disposes of any hazardous wastes identified or
       listed under the regulations without having
       obtained a permit; or knowingly violates a
       material condition of a permit.

(2)   knowingly makes a false material statement or
       representation on any application, label,
       manifest, record, report, permit or other
       document filed for purposes of complying with
       RCRA.

(3)   generates, stores, treats, transports or disposes
       of hazardous wastes and knowingly destroys,
       alters, or conceals any record required to be
       maintained under the regulations promulgated
       under RCRA.

A first conviction for violating (1) above may result
in a fine of not more than $50,000 for each day of violation
and imprisonment not to exceed two years. A first
conviction for violating (2) or (3) above may result in a
fine of not more than $25,000 per each day of violation and
imprisonment not to exceed one year.

Section 3008(e): Under the "knowing endangerment"
provision, a person:

(a)   (i) who knowingly transports, treats, stores or
       disposes of hazardous wastes without a permit
       or in violation of a permit,

       (ii) who withholds material information from its
       application for a permit, or

       (iii) who fails to comply with applicable interim
       status regulations,

(b)   and who knows at the time that he thereby places
       another person in imminent danger of death or
       serious bodily injury, and

(c)   if his conduct manifests either an unjustified
       and inexcusable disregard for human life or an
       extreme indiffference for human life,

is guilty of "knowing endangerment".

Conviction of "knowing endangerment" involving unjustified and inexcusable disregard for human life can result in a fine of not more than $250,000 or imprisonment for not more than two years, or both. Conviction of the more egregious conduct, that manifesting an extreme indifference to human life, can result in a fine of not more than $250,000 and imprisonment for not more than five years.

Comprehensive Environmental Response, Compensation, And Liability Act of 1980 (hereinafter "CERCLA")

Civil Liability

Section 106: Penalties are provided for willful violation of or refusal to comply with an order of the President [or his delegate] made in response to a perceived imminent and substantial endangerment to the public health or welfare or the environment because of an actual or threatened release of a hazardous substance from a facility. The penalty is up to $5,000 per day of violation.

Section 107: Owners and operators of facilities from which there is a release or threatened release of hazardous substances or persons who arranged for the disposal, treatment or transportation of hazardous wastes may be liable for the costs of removal or remedial actions and for damages for resultant loss of natural resources. Subsection 107(c)(3) provides for punitive damages of up to three times the amount of cleanup costs incurred by the government to be assessed against persons who, "without sufficient cause", fail to comply with an order issued pursuant to §§104 or 106.

Section 109: Any person who, after notice and an opportunity for a hearing, is determined to have failed to comply with the financial responsibility provisions in §108 shall be liable for a civil penalty of up to $10,000 per day of violation.

Criminal Liability

Section 103(b): Persons with knowledge of and responsibility to report releases of reportable quantities of hazardous substances, and who fail to do so may be fined not more than $10,000 or imprisoned for up to one year, or both.

Section 103(c): Persons who own or operate, or, at the time of disposal, owned or operated or who transported hazardous substances to a facility with hazardous wastes were required to notify US EPA of the existence of the facilities. Knowing failure to do so can result in a fine or not more than $10,000 or imprisonment of not more than one year, or both.

Section 103(d)(2): Persons who knowingly destroy or otherwise render unreadable or falsify any records required to be maintained by regulations promulgated pursuant to subsection 103(d) may be fined up to $20,000 or imprisoned for not more than one year, or both.

Clean Water Act

Civil Liability

Section 309: Civil penalties of up to $10,000 per day of violation of any permit condition or limitation, or of specified sections of the Clean Water Act, are provided.

Section 311: Liability for cleanup of discharges or spills of oil or hazardous substances into or upon navigable waters is provided. In addition to penalties of up to $5,000 for the discharge itself, §311 provides penalties of up to $10,000 for failing to immediately report such discharges. Subsection (f) assigns liability for the costs of removing the discharged substances.

Section 505: Any citizen may bring an action against any person who is alleged to be in violation of an order of the US EPA Administrator or of an effluent standard or limitation.

Criminal Liability

Section 309(c): Any person who willfully or negligently violates specified sections of the Clean Water Act or any permit condition or limitation implementing said sections may be punished by a fine of not less than $2,500 nor more than $25,000 per day of violation or by imprisonment for not more than one year, or by both. Recidivists may be fined up to $50,000 per day of violation or imprisoned for not more than two years, or by both. (§309(c)(1)).

Persons knowingly making false statements, representations or certifications on any application, record or report filed or required to be maintained or who falsify, tamper with or knowingly render inaccurate any monitoring device are to be punished by a fine of up to $10,000 or by imprisonment for not more than six months, or by both. (§309(c)(2)).

CIVIL VS. CRIMINAL ENFORCEMENT

Civil Enforcement Policy

The overwhelming majority of environmental enforcement proceedings have been civil proceedings. With few exceptions, the government has pursued criminal proceedings only in response to truly egregious conduct.

RCRA hazardous waste enforcement activity began only in 1979[1]. Sixty-one federal judicial hazardous waste actions were filed during the period from July of 1979 through June of 1981. No federal judicial enforcement actions relating to RCRA were filed by the US EPA from June of 1981 through January 1, 1982.

This lull in federal enforcement activity is partially attributable to the change in administrations and subsequent reorganizations of the US EPA enforcement groups. The new administration dismantled the centralized Office of Enforcement in July of 1981 and assigned most of the personnel from that office to individual program offices. Later in 1981, the US EPA formed the Office of Legal and Enforcement Counsel and then reassigned most enforcement attorneys to that office.

Douglas MacMillan, former Director of the US EPA's Office of Waste Programs Enforcement has explained the hiatus by noting that the US EPA is: (a) litigating cases previously filed, (b) utilizing §3008 civil penalty provisions that previously were not available to them, and

(c) implementing CERCLA. Where it is appropriate to do so, CERCLA claims are currently being added to pending judicial actions.

CERCLA enforcement policy was outlined by William Sullivan, US EPA enforcement counsel, in a policy Memorandum dated November 25, 1981, to the Regional Administrators and Regional Enforcement personnel. The US EPA is, wherever practicable, issuing notices to persons that it thinks may be responsible parties. This is consistent with the Agency's policy of negotiating first and suing only if negotiations fail, and with §104(a) of CERCLA which authorizes US EPA to spend CERCLA money unless it determines that timely and effective cleanup will be effected by a responsible party. The recipient of one of these notices is given a short time to agree to undertake necessary field studies, cleanup or remedial measures. If the recipient agrees to undertake the response measures, the US EPA does not spend CERCLA funds for that purpose. If the recipient declines to voluntarily undertake the activities specified, the US EPA then decides whether or not it will spend CERCLA funds for the proposed field studies, cleanup or remedial measures and to seek reimbursement[2].

The US EPA has rejected the argument that its CERCLA enforcement activities should be limited to the 115 sites on the priority list. While it admits that most of its resources will be directed toward these sites, the Agency refuses to limit its activity to the sites on the priority list in order to encourage voluntary cleanup of sites which are not yet on the list or which are not at the top of the list.

A policy Memorandum dated July 7, 1981, issued to Regional Administrators and enforcement personnel[3] ranked RCRA violations and suggested Agency responses to particular violations. The Memorandum separated RCRA Subtitle C violations into three classes.

Class I violations are those which purportedly pose "direct and immediate harm or threats of harm to public health and environment". Examples of Class I violations are the failure to use a manifest, violation of interim status requirements with a resultant discharge, or shipment of hazardous wastes to a facility without a permit. The suggested responses are: (a) an order or lawsuit pursuant to the imminent hazard provision in §7003, or (b) a compliance order or judicial action pursuant to §3008.

Class II violations are those which "involve non-compliance with specific requirements mandated by the statute itself" as opposed to violations of the regulations. Examples of Class II violations include failure to notify as required by §3010 or violation of the § 3005(a) permit requirement. According to MacMillan, the Agency's response to a Class II violation should be: (a) an "interim status compliance letter", or (b) a §3008 compliance order or judicial action.

Class III violations are limited to procedural or reporting violations which "do not pose direct or short-term threats to the public health or environment". Failure to provide for proper training or to submit the proper reports are examples of Class III violations. The response suggested for a Class III violation is a "warning letter".

Political pressure for an increase in hazardous waste enforcement activities has been building. In response, Mr. Sullivan recently said: "'You will see more hazardous waste enforcement in calendar year 1982 than EPA has ever conducted' in the past[4]."

Criminal Enforcement Policy

Since approximately 1978 government officials have been threatening to emphasize the criminal aspect of the enforcement of the pollution control laws[5]. Despite the recent lull in federal hazardous waste enforcement litigation, there is evidence that the Agency intends to increase the number of criminal cases filed.

The Office of Criminal Enforcement has recently hired additional experienced criminal investigators; the office is being upgraded from a staff of two to a staff of twenty-five members. The investigators will be stationed in the various Regions. The US EPA has also agreed with the Federal Bureau of Investigation to refer thirty hazardous waste cases each year to the F.B.I. for investigation.

Enforcement authority has been centralized in Washington, D.C. Previously, criminal investigations were undertaken by the Regions, who then could directly refer the matters to either the Department of Justice or a local U.S. Attorney for judicial action. Now criminal cases must be approved by US EPA in Washington before referral.

The criminal cases that have been filed in the hazardous waste area have, for the most part, involved either surreptitious dumping or false reporting. The Agency has recently indicated that this emphasis will continue, i.e., that the criminal cases will involve egregious violations such as "midnight dumping" or intentional destruction or falsification of records[6].

Civil vs. Criminal

High on the Agency's list for criminal enforcement actions are corporations or individuals who the Agency believes are: (a) acting in bad faith, (b) deliberately "flaunting" the authority of the Agency, or (c) regularly violating the law as a result of "an informed policy decision" made by a corporate official[7].

An example of a situation where the US EPA may decide to bring a criminal enforcement action would be one where the Agency obtains documents that: (a) indicate that high corporate officials with authority to effect compliance were long aware of the violative discharges or disposals, (b) that the discharges or disposals were hazardous, harmful or quantitatively excessive, (c) that it was possible to control or eliminate these discharges or to dispose of the materials properly, and (d) that the cost of discharging or disposing pursuant to the statutes and regulations was more than the individuals or the company wanted to spend[8].

The following considerations are relevant to the government's determination of whether to file a particular case as a civil or as a criminal action[9]:

Requisite Mental State

If the government brings a criminal action, it ordinarily must prove a particular mental state. The requisite mental state varies with the offense.

If the action is brought under RCRA, the government ordinarily must prove that the defendant "knowingly" violated the statute or regulations. While the "knowingly" standard does not include negligent or accidental acts, it has been argued that positive or actual knowledge is not necessary if the defendant was aware of the "high probability" of the facts in question or if steps were taken to avoid actual knowledge[10].

If the government is charging the defendant with "knowing endangerment", the government must not only prove knowledge of the violative actions, but also that the defendant knew that the consequences of the actions would very seriously endanger others. RCRA provides that, for purposes of the crime of "knowing endangerment", "a person is responsible only for his actual awareness or actual belief possessed." Knowledge of others may not be attributed to the defendant for purposes of proving this crime. However facts tending to show that the defendant took "affirmative steps to shield himself from relevant evidence" can be used as circumstantial evidence of actual knowledge[11].

The stated rationale behind the adoption of the "knowing endangerment" provision was to provide a severe penalty for certain life-threatening conduct. It was not intended to result in the prosecution of persons who make difficult business judgments where those judgments were made without actual knowledge of the danger resulting from the particular act[12].

If criminal action is brought under the Clean Water Act, the defendant must either have "willfully or negligently" violated the Act. Remaining to be settled is the issue of whether the government need only prove simple negligence; or whether some more egregious form of behavior such as gross, culpable, or flagrant negligence must be proven. Normally criminal negligence should involve a more serious form of criminal knowledge than civil negligence.

Burden of Proof

The government's burden of proof in a criminal action is "beyond a reasonable doubt". The government must only prove a civil case by preponderance or the greater weight of the evidence. Certainly if the government doubts that it can meet the criminal burden of proof, it may still bring the case as a civil action.

Scope of Discovery

Discovery describes the legal procedures whereby one party in a lawsuit is able to demand and receive information and documents from the other. In a criminal proceeding, discovery is extremely narrow. Liberal or broad discovery in criminal proceedings cannot be reconciled with the constitutional privilege against self-incrimination[13].

In a civil action, the Federal Rules of Civil Procedure permit broad discovery. Any matter, which is not privileged and which is relevant, or which appears reasonably calculated to lead to admissible evidence, is potentially discoverable, even if the information itself may not be admissible[14]. There are exceptions to this statement, but it reflects the general rule.

Seriousness of the Violation

Clearly and understandably, the government is more likely to bring a criminal action in a case where the threat of damage to the environment or human health is serious, than it is in the absence of such circumstances. Accordingly, the Administrator of the US EPA has said that the Office of Criminal Enforcement will emphasize two types of cases: (1) activities which have resulted in substantial environmental harm and/or constitute a hazard to human health, or (2) matters representing willful contempt of court-ordered consent decrees[15].

INDIVIDUAL LIABILITY: WHO IS PERSONALLY LIABLE

Normally the person charged with violating an environmental statute or regulation is a business entity such as a corporation. A corporation is responsible for the acts done by its officers and employees when those persons are acting within the scope of their employment and, in certain cases, even when they are exceeding the scope of their employment[16].

Section 309(c)(3) of the Clean Water Act provides that a "person" as used in the enforcement section of the Act, includes "any responsible corporate officer". Section 1004(15) of RCRA and §101(21) of CERCLA define "person" to include an "individual".

Civil Actions

In civil actions, the US EPA normally proceeds against the corporation when one exists. If there is no corporation or if the operation is small, it will seek enforcement against an individual.

It has also been observed that the agency is likely to proceed against individuals or responsible corporate officials in civil actions when[17]:

(a)   officers or directors have caused their company to enter into a consent decree and thereafter failed to insure that the decree was carried out, or

(b)   the company has failed to move toward compliance in several instances and the agency can see a pattern of conduct and that the individual was involved in the course of such conduct.

The U.S. Supreme Court long ago said[18]:

"A command to the corporation is in effect a command to those who are officially responsible for the conduct of its affairs. If they, apprised of the writ directed to the corporation, prevent compliance or fail to take appropriate action within their power for the performance of the corporate duty, they, no less than the corporation itself, are guilty of disobedience and may be punished for contempt."

Criminal Actions

In the criminal context, a responsible corporate official may be held criminally liable for acts performed in an official capacity[19]. The argument that the act was an official act and therefor binds the corporation, but does not create liability for the individual, will seldom prevail.

As a general rule, where a crime involves criminal knowledge or intent, it is essential to the liability of a particular corporate official that: (a) the official actually and personally do the acts which constitute the offense, (b) that those acts be done under his direction, or (c) that the acts be done with his knowledge and with his authority[20].

Instances where the US EPA may press criminal charges against individuals include the following situations[21]: (a) where intentional corporate non-compliance is the result of "any informed policy decision" made by a responsible corporate official, or (b) where there is evidence of deliberate falsification of records, reports or other required data, or where there is evidence of tampering with monitoring devices. The importance that the US EPA puts on the latter category of violations reflects its need to protect its program. The Agency relies to a large extent on voluntary reporting; actions such as those described in the second category above would destroy the program if they became widespread.

The Dotterweich or Park Doctrine

Some legal writers have discussed and some courts have applied a standard of criminal liability whereby senior corporate officials may be held personally and criminally liable for offenses committed by subordinates even though the officials had not been present at the time and place of the criminal act and had no actual knowledge of the violative activity. Individuals have been found guilty when it was determined that they were responsible officials with the power to prevent or correct the situation leading to the violation[22].

Since 1975 when U.S. vs. Park was decided, most of the cases following the principles enunciated therein have been brought pursuant to the Federal Food, Drug and Cosmetic Act, which does not specifically require criminal knowledge as an element of its criminal violation provisions. The environmental cases which have cited Park have thus far been cases in which the corporate officials were actually aware of the violations and failed to correct them[23], or have involved issues other than those relating to personal criminal liability[24]. The criminal penalty sections of the environmental statutes appear to specifically require either knowledge, willfulness or negligence.

The Park doctrine does not constitute strict criminal liability, i.e. liability without fault[25]. The American system of justice generally requires some personal fault before an individual can be deemed to be criminally liable.

The Park doctrine differs from typical criminal standards in the degree of fault or culpability that must be proven. In Park, the negligence was imputed to the official

475

who had the responsibility for corporate compliance and the power to effect that compliance. He was therefore deemed responsible for the non-compliance despite his lack of actual knowledge of the specific violation with which he was charged.

Culpability at the level contemplated by the doctrine may be more theoretical than real[26]. Culpability at this low level is more consistent with traditional civil liability than it is with criminal liability. Criminal convictions under these facts may violate the public's sense of fair play and reduce the stigma attached to the conviction to something less than that normally associated with criminal convictions.

Some legal writers have suggested that the Park doctrine is properly applied to prosecution for violations of statutes designed to protect the "public health and safety"[27]. To justify the harsh result in these cases, the courts and authors argue that the burden which is imposed upon the corporate official is necessary to protect the innocent and powerless general public. The emphasis thus is not on punishment of the official, but rather on protection of the public[28].

Indeed, one writer has argued that:

"Knowledge, and perhaps negligence, may be imputed to ...an officer ... [who] was in a position to be aware of the company's policies and, having assumed a position of authority in a business enterprise affecting the public's health and well-being, had a high duty of care. A violation of that duty is negligence[29]."

In another article, however, the U.S. Justice Department's arguments in U.S. vs. Park, supra, were criticized as follows:

"Such an authorization for harsh and arbitrary criminal prosecutions under the Act, based on corporate status rather than individual 'wrongful action' and shifting the burden of proving innocence to the accused is offensive to established principles of fairness and justice and can serve no legitimate regulatory objective[30]."

If the doctrine is invoked only in situations involving officials' unjustified failure to act, it may not represent a significant departure from traditional standards. Certainly a person failing to act when he has a duty to act may properly be held to be criminally liable just as one who has acted improperly. "The criminal law may punish 'neglect where the law requires care, or inaction where it imposes a duty'[31]." Implicit in the U.S. Supreme Court's decision in Park is the concept that corporate officials who have the responsibility and power to insure compliance and fail to effect same, have a "responsible relationship" and a "responsible share" in the violations[32]. Those corporate officials had the authority and supervisory responsibility to prevent violations, but failed to do so.

The environmental statutes and regulations concerning hazardous wastes require businesses and responsible corporate officials to establish compliance programs and generally to comply with the provisions of the statutes and regulations. If officials with the power to effect compliance fail to exercise reasonable efforts to comply, the government may be expected to argue that those officials knowingly violated the statutes. With respect to the Federal Food, Drug and Cosmetic Act, the U.S. Supreme Court said:

> "...the Act imposes not only a positive duty to seek out and remedy violations when they occur, but also, and primarily, a duty to implement measures that would insure that violations would not occur. ******* The Act does not ... make criminal liability turn on 'awareness of some wrongdoing', or 'conscious fraud'. The duty imposed by Congress on responsible corporate agents is, we emphasize, one that requires the highest standard of foresight and vigilance, but the Act, in its criminal aspect, does not require that which is objectively impossible[33]."

The present Administration is not expected to strongly urge the application of the Park doctrine to environmental situations. However in the future, the government may be expected to cite the Park doctrine when its proofs of negligence, knowledge or willfulness are somewhat weak or when the Agency's own performance warrants criticism. In these cases the government may argue that the public welfare requires that the Agency prevail or that its errors be overlooked[34]. The Park doctrine will be used by the Agency to tip the scales slightly more in favor of the government and against alleged violators of the statutes regulating the discharge of hazardous substances into the environment.

CONCLUSION

The government has a variety of tools available to it to carry out its responsibilities for the enforcement of the statutes and regulations governing the handling of hazardous wastes.

RCRA provides the US EPA with authority either to sue or to issue compliance orders (with or without penalties) to persons who violate applicable statutes and regulations. The Agency has authority to respond to perceived hazardous waste emergencies pursuant to the imminent hazard provisions of RCRA. Criminal sanctions may issue under RCRA for certain "knowing" violations, including knowingly placing others in danger of serious harm from hazardous wastes.

CERCLA assigns liability for the cleanup costs and damages from facilities from which there are actual or threatened releases of hazardous wastes. Criminal liability under CERCLA is provided for failing to notify the government of actual or threatened releases of hazardous wastes or for failing to maintain certain records.

The Clean Water Act provides penalties for violations of permit conditions and limitations and assigns liability for spills into navigable waters. If the violations are willful or negligent, the Agency may seek criminal penalties.

Very few criminal proceedings have been brought to enforce environmental laws. Recently there have been very few federal judicial enforcement actions brought (either civil or criminal) relating to hazardous wastes. Agency officials have offered a variety of reasons for this lull, but more importantly have given assurances that it will soon end. Indeed, the US EPA has apparently augmented its investigatory capabilities, especially in the criminal area. Most enforcement proceedings, however, will still be brought as civil proceedings. Most of these proceedings in the near future are expected to utilize both RCRA and CERCLA to force the cleanup of the hazardous waste sites on the Agency's priority list.

In the past, when deciding whether to bring an enforcement proceeding as a civil or as a criminal action, the Agency has considered first the nature and seriousness of the violations, and second whether it could sustain the heavier burdens of proof and prevail despite the limited discovery available in criminal proceedings.

Normally enforcement proceedings are directed at a corporation or other business entity. However, where there is evidence of an informed decision by individual officials to conduct business in violation of environmental statutes or regulations, the Agency is more likely to proceed against those individual officials in addition to the business entity.

Some persons are urging that some of the government's traditional burdens and proofs in criminal cases be lessened when the violations involve statutes intended by the legislature to protect the public's health or safety. Specifically they urge that fault be imputed to high corporate officials regardless of whether the official had any knowledge of or personal involvement in the violation.

Existing case law supports the concept that an individual may be criminally liable for failing to act when there is a duty to act. However if criminal liability is assigned in the absence of significant culpability (involvement or knowledge), the resultant liability will be perceived by the public as something less than "criminal", regardless of what it is called. The stated purpose of the above approach is deterrence; however the stigma and deterrent effect of this sort of criminal conviction would lessen should the public perceive this sort of "criminal violation" to involve something less than a true criminal situation. Consequently its adoption is unlikely to produce the desired result, but would, in all likelihood, produce undeserved hardship.

1. MacMillan, D.D., "Hazardous Waste Litigation, 1982", Practicing Law Institute, 196:83 (1982).
2. Sullivan, W. A., Jr., Enforcement Counsel, US EPA, Memorandum to Regional Administrators, et al. Re: Coordination of Enforcement and Fund-Financed Activities Under CERCLA, (November 25, 1981).
3. MacMillan, D.D., Memorandum to Regional Administrators, et al., Re: Guidance on Developing Compliance Orders Under Section 3008 of the Resource Conservation and Recovery Act, (July 7, 1981).
4. "Special Report", Environment Reporter, The Bureau of National Affairs, Inc., 12:1272, 1278 (1982).
5. Moorman, J.E., Address delivered at ALI-ABA Course of Study, "Environmental Law," (February 10, 1978).
6. Environment Reporter (BNA), 12:1214 (1982)).
7. Legrow, S.W., "Memorandum to Regional Administrators, Setting Priorities for Enforcement Actions Concerning July 1, 1977 Violations", reprinted in Environment Reporter (BNA), 8:248 (1977); "Air and Water Act Enforcement Problems - A Case Study," The Business Lawyer, 34:705 (1975); Note: "Federal Enforcement of Individual and Corporate Criminal Liability for Water Pollution," Memphis State University Law Review, 10:596 (1980).
8. The Business Lawyer, supra.
9. Lettow, C.F., "The Environmental Protection Agency's Enforcement Response System," ALI-ABA, Course Materials Journal, 5(4):123 (1981).
10. Weiland, R.A., "Enforcement Under The Resource Conservation and Recovery Act of 1976", Environmental Affairs, 8:673-674 (1980).
11. RCRA Section 3008(f)(2).
12. Joint Explanatory Statement of The Committee of Conference for The Solid Waste Disposal Amendments of 1980.
13. Rules 15 and 16, Federal Rules of Criminal Procedure.
14. Rule 26(b), Federal Rules of Civil Procedure.
15. Environment Reporter, (BNA), 12:1214 (1982).
16. Apex Oil Co. vs. U.S., 530 F.2d 1291 (8th Cir. 1976); U.S. vs. Hilton Hotels Corp., 467 F.2d 1000, 1004 (9th Cir. 1972); U.S. vs. Little Rock Sewer Comm., 460 F. Supp. 6 (E.D. Ark. 1978).
17. The Business Lawyer, supra, 34:669; Lettow, supra, 5(4):121.
18. Wilson vs. U.S., 221 U.S. 361, 376, (1911).
19. U.S. vs. Amrep Corp., 560 F.2d 539, 545 (2nd Cir. 1977); U.S. vs. Gulf Oil Corp., 408 F. Supp. 450, 470-471 (W.D. Pa. 1975); cf. U.S. vs. Sexton Cove Estates, Inc., 526 F.2d 1293 (5th Cir. 1976).

20. Weiland, _supra_, 8:676.
21. Legrow, Memorandum, _supra_, 8:248; _The Business Lawyer_, _supra_, 34:699.
22. _U.S. vs. Dotterweich_, 320 U.S. 277 (1943); _U.S. vs. Park_, 421 U.S. 658 (1975).
23. e.g. _U.S. vs. Frezzo Brothers, Inc._, 602 F.2d 1123, 1130 (3rd Cir. 1979).
24. _U.S. vs. FMC Corp._, 572 F.2d 902, 906 (2nd Cir. 1978); _Hercules, Inc. vs. EPA_, 598 F.2d 91 (D.C. Cir. 1978).
25. cf. Note: "Prosecution of Corporate Officials Under the Federal Food, Drug and Cosmetic Act", _Ohio State Law Journal_, 37:443-444 (1976); Note: "Public Welfare Violations - Imposing Criminal Sanctions With A Strict Liability Standard?", _Univ. of Florida Law Review_, 28:602-603 (1976). See also _U.S. vs. FMC Corp._, _supra_.
26. Note: "Criminal Liability For Public Welfare Offenses:Gambler's Choice", _Memphis State Univ. Law Review_, 10:627 (1980).
27. "Public Welfare Violations", _supra_, 28:596; cf. O'Keefe and Shapiro, "Personal Criminal Liability Under the Federal Food, Drug and Cosmetic Act", _Food Drug Cosmetic Law Journal_, 30(1):5 (1975); "Criminal Liability for Public Welfare Offenses:Gambler's Choice", _supra_, 10:612.
28. O'Keefe and Shapiro, _supra_, 30:31.
29. "Public Welfare Violations - Imposing Criminal Sanctions with a Strict Liability Standard?", _supra_, 28:603.
30. Sethi, S.P. and Katz, R.W., "The Expanding Scope of Personal Criminal Liability of Corporate Executives - Some Implications of United States v. Park", _Food Drug Cosmetic Law Journal_, 32(12): 559-560 (1977).
31. _U.S. vs. Park_, _supra_, syllabus; _U.S. vs. FMC_, _supra_; _U.S. vs. Illinois Central Railroad Co._, 303 U.S. 239, 243 (1938); cf. _Lambert vs. California_, 355 U.S. 225 (1957).
32. _U.S. vs. Park_, _supra_ at 672.
33. _U.S. vs. Park_, _supra_, at 672-673.
34. _U.S. vs. Frezzo Brothers, Inc._, _supra_; _Hercules, Inc. vs. EPA_, _supra_.

CHAPTER 30

HAZARDOUS WASTE AND STRICT LIABILITY, POTENTIAL AND PROTECTION

T. J. Hickey, Esq., P.E.
  Malcolm Pirnie, Inc.

INTRODUCTION

This chapter attempts to highlight some of the threats to the business of the engineer who is practicing or intends to practice his profession in the specialized area of storage, treatment and disposal of hazardous wastes.[1] These threats are real because injury to persons, places and things resulting from hazardous or toxic wastes is real. The injured person wants to be made whole and doesn't particularly care who pays to make him, her or it whole. Very often, injured parties are not satisfied with being made whole: they want vengeance; they want to make a public example of the person or corporation who hurt them; they want the punishment to be such that others will be warned about the dire consequences of such behavior. In short, they want punitive damages and, if possible, imprisonment.

And who can blame a person for seeking revenge if they have lost a spouse, a parent, a child, their family home or their neighborhood because some giant, impersonal manufacturer appeared not to care about the consequences of its action? Whether or not that manufacturer did care or was diligent and responsible in carrying on his business is of no consequence to the grieving parent who has just learned that her six-year old son has suffered irreversible and progressive brain damage due to lead poisoning or who has just buried her husband who was asphyxiated while working in a sewer. As far as that individual is concerned, anyone who had anything to do with this lifelong and unforgiveable hurt should pay. This attitude is being enforced by litigation:

> "(W)e must still insist that the risk of loss fall not upon the innocent victim, but upon those who produce the risk, profit from the risk-producing activity, or are best able to avoid the risk by the use of good safety practices..."

Richard F. Gerry, President of the Association of
Trial Lawyers of American (ATLA), quoted in the
New Jersey Law Journal[2], December 31, 1981

THE PROBLEM

Examples of hazardous waste-caused injuries abound;
witness the following:

1. "The deaths of two men at a waste disposal site
    January 13 (1982) apparently were caused by a
    reaction of two chemicals that caused the release
    of hydrogen sulfide gas...after they reportedly
    opened a valve on the tanker truck.

    "Several others were hospitalized, including the
    driver of the truck...Thirty-seven people, in-
    cluding (the disposal company) employees and
    Shelley Township firefighters, were taken to local
    hospitals for observation."[3]

2. Bunker Hill Company of Kellogg, Idaho, the nation's
    largest refiner of lead, zinc and silver, agreed
    to pay between $6.5 million and $8.7 million to
    nine children in settlement of a claim for poison-
    ing resulting from industrial pollution. Medical
    and psychological experts testified at the trial
    that the children were poisoned by lead emissions
    and lead deposits in the soil from the Bunker Hill
    smelter a few blocks from their home.[4]

3. "Residents of south San Jose are blaming recent
    birth defects in neighborhoood children on chemi-
    cal contamination of drinking water traced to a
    local Fairchild Camera & Instrument Corp. plant.

    "Over an 18-month period in 1980 and 1981, 40,000
    gallons of trichloroethane (TCA) seeped from a
    storage tank into a well 1,500 feet from the
    plant. According to health officials, the con-
    tamination presented no immediate health hazard
    but they are continuing to investigate links
    between TCA and birth defects."[5]

4. Commercial fisherman, boat, tackle, and bait shop
    owners, and marina owners have asserted claims for
    economic harm resulting from a chemical manu-
    facturer's alleged discharge of kepone into the
    James River (Virginia) and Chesapeake Bay.[6]

484

5.   "There is no doubt that (hazardous atmospheres in manholes and sewers) is the most potentially fatal accident cause (to sewer workers)."[7]

Just the briefest pause allows the realization that an engineer was probably involved in some manner in each of the cited problems. The engineer's participation might have been the following:

o   He instructed the disposal site workers in safety procedures.

o   He specified the protective equipment worn by the workers.

o   He specified the wastes that might be placed at the site.

o   He specified the gas detection and warning devices installed at the disposal facility.

o   He designed the slag disposal site.

o   He designed or specified the chemical holding tank.

o   He administered construction of the electronics plant.

o   He designed the valves used in the chemical manufacturing process.

o   He prepared a spill detection and prevention system for the plant.

o   He prepared plans and specifications for the replacement of a length of sewer pipe.

o   He requested that samples be taken of plant influent at certain manholes.

The potentially tragic aspect of all these scenarios is that the engineer need not have been negligent in performing his services in order to be found liable, in some part, for the damages resulting from the cases described. This is so because there is an aspect of the law referred to as strict liability or absolute liability or liability without fault. Originally, such liability was found only in the common law[8] but is now emerging in statutory law.[9] The next section discusses the nature of strict liability.

STRICT LIABILITY

Strict liability may be defined as the principle that in some cases a person may be held liable although he not only has not been charged with a moral wrongdoing but has not even departed in any way from a reasonable standard of intent or care. This principle has frequently found expression where the person's activity is unusual or abnormal in the community and the danger which it threatens to others is unduly great - particularly where the danger will be great even though the enterprise is conducted with every possible precaution. The basis of the liability is the defendant's intentional behavior in exposing those in his vicinity to such a risk.

Generally, the conduct involved does not depart so far from customary social standards as to fall within the traditional boundaries of negligence - usually because the advantages which it offers to the defendant and to the community outweigh even the abnormal risk; but the conduct is allowed to carry it on without making good any actual harm which it does to his neighbors.

The courts have tended to lay stress upon the fact that the defendant is acting for his own purposes and is seeking a benefit or a profit of his own from such activities, and that he is in a better position to administer the unusual risk by passing it on to the public as cost of doing business (insurance) than is the innocent victim.

The classic example is that of the person who conducts blasting operations. It will usually be held that the blaster who injures his neighbor is "at fault" for conducting the operations at all and is privileged to do so only in so far as he insures that no harm shall result. The basis of this liability is the creation of an undue risk of harm to other members of the community.

This doctrine of strict liability for abnormally dangerous conditions and activities is a comparatively recent one in the law. The leading case from which it has developed is Rylands v. Fletcher, decided in England in 1868. The defendants, mill owners in Lancashire, constructed a reservoir upon their land. The water broke through into an unused shaft of an abandoned coal mine and flooded along connecting passages into the adjoining mine of Rylands. The classic and often quoted statement from this case is:

"We think that the true role of law is that the person, who for his own purposes brings on his land and collects and keeps there anything likely to do mischief if it escapes, must keep it at his peril, and if he does not do so is prima facie answerable for all the damage which is the natural consequence of its escape."[10]

On appeal, this statement was limited in the House of Lords to "non-natural" uses of the defendant's land.[11]

Among the "non-natural" uses of land for which strict liability was imposed in England were the storage in quantity of explosives or inflammable liquids, or blasting, or the accumulation of sewage, or the emission of creosote fumes, or pile driving which sets up excessive vibration.

These were deemed to possess the same element of the unusual, excessive and bizarre.

The rule of "Rylands v. Fletcher" then seems to be that the defendant will be liable when he damages another by a thing or activity unduly dangerous and inappropriate to the place where it is maintained, in the light of the character of that place and its surroundings.[12]

In dealing with hazardous wastes[13], the engineer is dealing with substances which may, in many cases, be classified by the law as ultrahazardous or unduly dangerous things. Thus, by his participation in the attempt to rid our environment of toxic and hazardous wastes, he may inadvertently be entering into risks of which his firm is both unaware and ill-prepared to respond to.

DECISION

Critical to the engineer's decision to participate in the process are certain facts which, like gravity, cannot be avoided.

Fact One: The proliferation of hazardous wastes must stop.

Fact Two: Existing sources of exposure to hazardous waste pollution must be cleaned up.

Fact Three: The engineer, private or industrial, must be the leader and doer of the necessary acts to accomplish these ends. Doctors, lawyers, teachers, ecologists and scientists do not have the necessary tools. There is no one else for the job but the engineer.

IMPLEMENTING THE DECISION

The engineer can minimize the threat by taking certain positive steps.

Standard of Care

Normally, when the engineer undertakes an assignment for a client, he implies that he possesses skill and ability sufficient to enable him to perform the required services at least ordinarily and reasonably well; and that he will exercise and apply that skill and ability reasonable and without neglect. However, the engineer does not imply or warrant a satisfactory result. An error of judgment is not necessarily evidence of a lack of skill or care.

When the engineer is dealing with a hazardous substance, the standard of care becomes greater:

No less a degree of care than that commensurate with the apparent danger, or in proportion to the danger reasonably to be anticipated, is reasonable. As is often said, the more imminent the danger, the higher the degree of care. Clearly, when human life is at stake, the rule of due care and diligence requires everything that gives reasonable promise of its preservation to be done, regardless of difficulties or expense.[14]

While no absolute standard of duty in dealing with (hazardous wastes) can be prescribed, it is safe to say, in general terms, that every reasonable precaution suggested by experience and the known dangers of the subject ought to be taken. Where death may be caused by an agency lawfully in use, ordinary care requires that every means that is known, or that with reasonable inquiry would be known, must be used to prevent it.

If, when dealing with a dangerous substance, there is a practical and apparently certain means to determine a safe result, and such means is not used, with injury thus resulting, the person responsible for the failure to use such means is liable.[15]

Clearly, neither the engineer nor his client can deal with hazardous wastes in a "business as usual" manner. A suggested rule of thumb is the following: the engineer's initial determination of the effort required by a client's proposed hazardous waste project should be made without reference to the available funds. If there is a mismatch between the estimated cost of doing the work and available funds, the engineer must be extremely careful, even defensive, in agreeing to reduce or change the degree or nature of the services he will provide. If the engineer does agree to modify his initial estimate of necessary services, the result will be:

1) the services to be provided are adequate because the initial estimate was over conservative or based on a misunderstanding of the project; or

2) the services to be provided are marginally adequate and will probably be sufficient if certain conditions or events occur; or

3) the services are inadequate for the proposed project.

It is suggested that the engineer respond to the three cases as follows:

1) Agree to do the work;

2) Agree to do the work provided his contract with the client stipulates the conditions or events on which the engineer's estimate is based;

3) Walk away from the work.

Contractual Arrangements

Before entering into the performance of services related to a hazardous waste activity, the engineer should be in possession of an executed written agreement which clearly sets forth the mutual duties and responsibilities of the parties.

1) It is suggested that the statement of services to be provided be much more detailed than the language set out in the standard NSPE documents.

2) If possible, a limitation of liability clause should be included. The clause should limit the engineer's liability to his fee or the profit on the fee being paid under the contract. Further, the engineer's liability should be limited to that resulting from his errors, ommisions or negligent acts. He should not agree to be liable for his acts because he is then agreeing to liability which may arise without his fault. Usually, this is an unfair and

489

inequitable transfer of risk by the client to the engineer
and should be firmly resisted.

3) The engineer should avoid indemnification and hold
harmless agreements. In a hazardous waste project, the
engineer should not agree to indemnify the client for anything more than the engineer's negligence. Because of the
potential for strict liability the client should be requested to indemnify the engineer (except for the engineer's
negligence) for injuries and liability arising out of the
client's acts or omissions including prior acts. Without
such a provision, the engineer may find himself unfairly
saddled with the consequences of a situation he had no part
in creating.

Professional Liability Insurance

The engineer would be well advised to ask his insurance
company two questions:

1) Is my firm covered for projects involving hazardous
wastes?

2) Is my firm covered if it is found liable for damages
without having been found negligent in the causation of the
damages? That is, does the insurance cover strict liability?

The latter question is occasioned by the statement of
coverage found in most professional liability insurance
policies: "the insurance afforded by this policy applies to
errors, omissions or negligent acts..." (or similar phrasing).

If the answer to either question is no, attempt to cure
the deficiency by an endorsement to the policy. If this
can't be arranged or is too expensive, evaluate the risk-
benefit equation for your firm. You may find that involvement in hazardous waste projects is not in your firm's best
interests.

Legislative Awareness

The engineering profession will bear a substantial
share of the responsibility for inhibiting the nation's
hazardous waste cleanup program if it doesn't get the message
across to the state and federal legislatures that "We want
to clean up this mess. We're the only ones who know how to
do it. But you won't let us!"

Each engineer, as well as the engineering societies, must take the time to explain to his municipal, state and federal legislative representatives that it is not in the public interest to impose liability without fault on those who are attempting to restore our environment. Any legislator worth his salt will recognize that it is unfair and inequitable to shift the burden of responsibility for hazardous waste problems from those who created and irresponsibly disposed of those wastes to those who are attempting to repair past damage and prevent recurrences of the same.

SUMMARY

By 1950, most engineering organizations had formally endorsed the Fundamental Principles of Engineering Ethics drafted by the Engineers Council for Professional Development. This document stated unequivocally that the engineer "will use his knowledge and skill for the advancement of human welfare."[16]

It will be one of the great ironies, if not tragedies, if the engineering profession is placed in a position where it is forced to say "We won't play the game because we're always IT."

1. This paper does not present the entire scope of potential strict liability for engineers. Such a presentation is far beyond the purposes of the conference and would not serve the practical interests of the conferees.
2. 108 N.J.L.J. Index page 591.
3. Business Insurance, January 25, 1982.
4. Yoss et al v. Bunker Hill Co. et al, No. 77-2030, as reported in the American Bar Association Journal, Vol. 67, December 1981.
5. Business Insurance, February 22, 1982.
6. Pruit v. Allied Chemical Corp. 16 ERC 2014 (D.C. Virginia 1981).
7. Dally, K.A. "Hazards Lurk in Innocent-Looking Manholes," WPCF Highlights 19(2):9(1982).
8. "The common law consists of those principles maxims, usages, and rules founded on reason, natural justice, and an enlightened public policy, deduced from universal and immemorial usage, and receiving progressively the sanction of the courts. Common law is generally used in contradistinction to statue law."

   U.S. v. Miller, 236 F.798, 800 (D.C. Wash. 1916)
9. "We, too, are persuaded that the Legislature's intent in no longer requiring proof of knowledge when it amended the (Water Code) in 1967 was to create a strict liability standard in which no proof of scienter is necessary." American Plant Food Corporation v. State of Texas, 14ERC 1244, 1248 (Tex. Ct. of Crim. App. 1979), emphasis supplied.

   See also Meyer, S., Compensating Hazardous Waste Victims: RCRA Insurance Regulations and a Not So "Superfund" Act, Environmental Law, 11(689), 1981 in which the author advocates the creation of a federal cause of action (toxic contamination tort) for victims of hazardous waste releases based on strict liability. The author further advocates shifting the burden of proof so that, once the plaintiff proves that he or she was exposed to the waste and was injured, the defendant is presumed to have caused the injury and has the burden of providing innocence. The constitutionality of such a presumption is questionable.
10. Fletcher v. Rylands, L.R. 1Ex. 265, 279-80 (1866).
11. Rylands v. Fletcher, L.R. 3H.L. 330, 338 (1868).
12. W. Prosser, The Law of Torts, Section 78 (4th ed. 1971).

13. Hazardous waste defined as a subset of solid waste, which because of its quanity, concentration, or physical, chemical, or infectious characteristics may (1) cause, or significantly contribute to, an increase in mortality or an increase in serious irreversible, or incapacitating reversible illness; or (2) pose a substantial hazard to human health or the environment if improperly managed. Resource Conservation and Recovery Act of 1976, o 1064 (5), 42 U.S.C. o 6903 (5) (1976). EPA classifies materials as hazardous using the following criteria: toxicity, persistence, degradability, potential for accumulation in tissue, flammability, corrosiveness, and other hazardous characteristics. Id. o 3001, 42 U.S.C. o 6921(a). Hazardous waste comes in many forms: solids, powders, liquids, sludges and gases.
14. 57 American Jurisprudence 2d, Section 70.
15. 57 American Jurisprudence 2d, Section 115.
16. Florman, S.C., The Existential Pleasures of Engineering, page 19 (1976).

CHAPTER 31

SUPERFUND - WHERE IS
IT WHEN WE NEED IT?

Steven Stryker
    Battelle Memorial
        Institute

INTRODUCING THE SITUATION

We have all undoubtedly heard a great deal about Superfund in the last few years. We've probably heard or read an
increasing amount since December 1980 when the Superfund
legislation was finally passed as the Comprehensive Environmental Response, Compensation, and Liability Act (CERCLA -
PL 96-510). And especially in the last several months we've
seen and heard a great deal about Superfund, as Congressional
committees have harangued and environmental action groups and
states have sued to prod the Environmental Protection Agency
(EPA) into moving forward on implementation.

With all the attention over the years, there must be
something of interest going on. Indeed there is, and it's
something that all of us involved with environmental engineering will want to be very aware of. We would want to be
aware as citizens and taxpayers in any event, but now we have
another fundamentally important reason: as federal funding
for design and construction of wastewater treatment plants
continues to decline, many observers see the skills no longer
needed there shifting to solving the Superfund problems.
The Superfund authorization is currently $1.6 billion for
five years, in addition to what is spent by industry and the
states. Presumably that amount will solve some, if not all,
of those problems, and presumably it will keep a lot of us
usefully occupied.

THE ORIGINS REVISTED

CERCLA (PL 96-510) was the culminating legislation for
all of the activity that was and is encompassed by the term
"Superfund." In simplest terms, it closes the loop of the
air, water, and solid waste legislation of the 1960's and
-70's by providing federal dollars and national authority

to respond to emergencies and take remedial action at abandoned hazardous waste dump sites. Table I shows the evolution of hazardous and solid waste laws that eventually led to CERCLA, shown along with the other major environmental legislation.

Table I. Evolution of Environmental Legislation

● Solid Waste Disposal Act	1965
● Resource Recovery Act	1970
Clean Air Act	1970
NEPA	1970
● Federal Water Pollution Control Act (a)	1972
Energy Supply and Environmental Coordination Act	1974
TSCA	1976
● RCRA	1976
Safe Drinking Water Act	1976
Clean Water Act	1976
Fuel Use Act	1977
● CERCLA (Superfund)	1980

The first federal law was the Solid Waste Disposal Act of 1965. In that act, Congress recognized the problem (in response to constituent complaints) but gave the federal government only information and research authority leaving regulations and enforcement to the states. Amendments added resource recovery as a goal in 1970, but beyond that, reauthorizations were hardly noteworthy until 1976. That year, of course, was the year of RCRA - the Resource Conservation and Recovery Act.

RCRA was to be the counterpart of CERCLA, which would follow in four years. RCRA provides essentially "cradle-to-grave" regulation of all hazardous wastes and for the first time provides the federal EPA with the teeth of enforcement. RCRA's objective is to control all hazardous waste problems from here on in. CERCLA, as the complement, was enacted to start fixing all the problems that have occurred up to this point. That, of course, is a very large order, and it will involve a great deal of work -- not the least of which will be done within the environmental engineering community. To understand that work and its costly complexity, one has to understand to some extent:

1. the conditions which culminated in the CERCLA legislation; and

2. what has been happening on many fronts to bring Superfund to full implementation.

Because the total impacts are for the most part larger at hazardous waste sites, because the funding is generally going to be larger for site cleanup than for emergency responses, and because (mainly) the environmental engineering community is most likely to be involved in site problems, this discussion will concentrate on the hazardous waste site remediation part of Superfund activity.

WHERE AND WHY SUPERFUND STARTED

Superfund had its major push, if not its birth, at Love Canal. Scientists, engineers, and politicians who had recognized the problems and labored long and hard at solutions have recognized they would have still had years to go had not the Love Canal debacle, assisted by the Valley of the Drums and a few other extremely timely national disclosures, catapulted abandoned hazardous waste sites into headlines and prime time network news. The belief (right or wrong) that potentially deadly disposal sites could exist near any city or town in the country produced pressure and heat on Congress for action. Congress, in its wisdom or for other reasons, chose to deliberate at some length, but finally took action.

In the course of deliberations over abandoned hazardous waste sites, some important information on the scope and extent of the problem (now more widely known) became available for the first time. Per year, the U.S. disposes of about 50 million tons of hazardous wastes. Currently there are an estimated 300,000 industrial operations in the United States which generate hazardous wastes. These operations cut across industry to include agriculture, chemicals, electroplating, leather processing, paper, petroleum, plastics, and many others.

As the name implies, hazardous wastes are not merely unesthetic nuisances. They are either ignitable, corrosive, toxic, reactive, or some combination of these (as defined by EPA). Hazardous wastes may contain acids, bases, toxic

metals, flammables, infectious materials, and explosives.
Some of the worst actors have been pesticides, PCB's, **paint**
sludges, oils, heavy metals, and other chlorinated substances. The reason they are bad actors is because they are
implicated in increased rates of cancers, birth defects, and
other serious human health problems -- without even bringing
up ecological damage considerations.

The single most common way that these hazardous wastes
pose a threat is through insidious migration through or over
the land to public drinking water supplies, particularly
groundwater aquifers. This has happened through past practices of indiscriminate dumping of hazardous wastes in
municipal landfills, on open land, into open pits or lagoons,
and in abandoned private landfills. One of the very first
post-Love Canal remedial actions to be funded through
Superfund is underway at Prices Pit, where chemical wastes
threaten the entire water supply of Atlantic City, New
Jersey. Nearly $1.5 million has been authorized to date, and
that probably will cover only the field investigation, categorization, assessment, feasibility study, and some of the
design for the recommended remedial action option.

For decades there was little or no state or national
regulation, and therefore little or no accountability, for
how hazardous wastes were disposed of in the U.S. And unlike
the case in the industrialized nations of Europe, there was
a perception in the U.S. that the vast expanses of land
equated to a limitless capacity for land disposal. This
legacy has resulted in something like 30-40,000 land disposal
sites in this country with hazardous waste problems. (On the
one hand, reputable organizations such as the Environmental
Defense Fund would put the number higher; on the other hand,
similarly reputable organizations such as the Chemical
Manufacturers Association would have it lower).

The problem, in terms of what Superfund is about, concerns those sites out of the roughly 30-40,000 which are now
abandoned (the official term still seems to be "uncontrolled,"
i.e., the site was private and the owners have long since
vanished into the night, leaving a major problem; the site
owner, unaware of liability problems, or for other reasons,
has gone bankrupt and can't solve the problem; or, for whatever reason no one can be found to accept responsibility for
a problem which is in many cases a clear and present danger
to the public).

By somewhat questionable counts and by questionable, if
well-intentioned, accounting (e.g., the "MITRE Model"), the
real problem set for Superfund is approximately 1-2,000

sites where no responsible party can be identified. Of
these, 115 have been designated as the worst and the next
"dirty" 400 are ostensibly being decided. Whatever the
number and final designation of the really bad actors, one
clear purpose of Superfund was to raise money to fix problems
where a danger existed and no one else could or would act.

In the context of this sort of information, which was
sometimes inflammatory, sometimes valid, rarely scientific,
and all-in-all about on a par with other information-finding
deliberations, the 96th Congress, in the waning days of the
Carter Administration, passed the Superfund legislation.

WHAT SUPERFUND IS SUPPOSED TO DO

The compromise Superfund lesiglation that became the law
of the land on December 11, 1980 requires the following
action:

- EPA will work with other agencies to deal effectively
  with hazardous waste emergencies

- EPA will remedy "uncontrolled" hazardous waste sites

- There will be environmental damage assessment

- Damaged natural resources will be restored

- About $1.6 billion will be raised by taxes levied
  primarily on the chemical industry, to fund federal
  actions to fix the mistakes of the past.

In its barest essentials, with a multitude of important
details and conditions missing, that is what Superfund,
through CERCLA, purports to do.

WHERE ARE WE NOW?

It has now been nearly a year and a half since the passage of CERCLA, and more than $400 million in special taxes
(separate from general revenues) has accumulated in the bank
that is the U.S. Treasury. It is therefor fair to ask, where
is Superfund now? The answer, as may be expected, is complicated. The complications have in truth been political
rather than technical and substantive.

To its credit, the Superfund staff at EPA had been drafting the first National Contingency Plan (NCP) before the
enactment of CERCLA. That plan (again, I am focusing on the
sections dealing with the remedial action at abandoned sites)

called for EPA management of several (most likely three) zone contractors who would be responsible for the complete fixing of all priority abandoned sites in their zone of the country. That would mean, hypothetically, that a firm like Bechtel would manage, for EPA, all investigation, categorization, assessment, feasibility study, design, and construction/cleanup for a zone that might include all the New England states, New York, New Jersey, and all the southern coastal states from Maryland to Florida.

As those plans were being formulated, politics and reality intervened. The U.S. Army Corps of Engineers was about to lose people and money in budget cuts (especially in wastewater treatment grants), and since the Corps designed and constructed things, it seemed natural to EPA Administrator Anne Gorsuch that they should run all of Superfund. And this is what she proposed.

Not surprisingly, the major construction management firms, environmental engineers and their associations, architectural-engineering firms and their associations, and many others protested loudly. The result was that EPA and the President's Office of Management and Budget (OMB) withdrew the Corps super-manager idea and sat on it for nearly nine months.

What came forth in February of 1982 was a compromise: an interagency agreement was reached with Gorsuch and William Gianelli, Assistant Secretary of the Army for Civil Works that EPA would retain overall responsibility and do the up-front work from site identification through recommendation of cure; and the Corps of Engineers would manage all design and construction/cleanup based on EPA's recommendations. Where we stood in the final days of March 1982 was basically on hold until the National Contingency Plan was issued and the EPA - Corps turnover procedures were resolved.

In the meantime, some Superfund activities are moving forward. An estimated $45 million has already been allocated on emergency responses and some priority remedial actions. Three preliminary remedial action zone contracts were awarded over a year ago to Camp-Dresser-McKee, Black & Veatch, and Roy Weston. There are Technical Assistance Teams staffed by Ecology and Environment, Inc. throughout the country ready to assist EPA in emergency response problems, and there are Field Investigation Teams (again, staffed by Ecology and Environment, Inc.) assisting EPA in the preliminary identification and investigation of sites.

WHERE DO WE GO FROM HERE?

To put the whole problem of uncontrolled hazardous waste
in the altruistic context, we all have a job to do for the
good of us all. To put it in a practical, economic context,
we have a job to do for the rest of our careers if we want
it. The real question is, what is it we're supposed to do?
With regard to Superfund, and specifically with regard to
safe and effective cleanup of abandoned sites, what is the
role of the environmental engineering community?

THE PROFESSIONAL "ENVIRONMENTALIST" ROLE IN SUPERFUND

As America now moves to within a month of credible signs
of fullscale implementation of Superfund (which is to say,
allocation of nearly $320 million per year), it is incumbent
upon the environmental engineering community to understand
and be prepared for the roles they may be playing. The
following is a scenario, with very little held back from what
Battelle Memorial Institute is proprietarily thinking about.
It is my best judgment of what is likely to be required by
EPA in Superfund activity.

As Figure 1 shows, there are a number of discrete steps
involved in a Superfund site remediation, regardless of who
draws the diagram. The steps that will most likely involve
environmental engineering are highlighted. With management
of final design and construction/cleanup entrusted to the
Corps of Engineers, the "up front" phase assumes major
importance. What is done up front will have high and wide
visibility. It must be credible, at each site, or nothing
will work. However good (or not good) the up-front work is
that results in design recommendations to the U.S. Army Corps
of Engineers, it is where the basic responsibility lies. So
that job had better be done well, with reputable experts,
and fully documented. The liability question for individuals
and firms looms large. So, one expects, does the question
of professional integrity.

THE CHALLENGE

There will be a great deal of money thrown at the
problem of hazardous waste disposal in the foreseeable few
years. The environmental engineering community will be
gathering a respectable amount of that money. It is incum-
bent on us to consider Superfund as a trust, as it officially
and legally is. In effect, it offers to pay us for a
professional job. We must do a professional job.

501

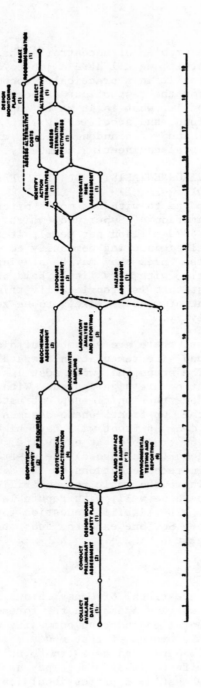

Figure 1.   **SITE REMEDIATION: WORK FLOW FOR CONCEPTUAL DESIGN**

CHAPTER 32

METHODS TO GAIN COMMUNITY
SUPPORT FOR A HAZARDOUS WASTE
FACILITY OR A SUPERFUND CLEANUP

Richard L. Robbins, Esq.
  Chicago

INTRODUCTION

In the last ten years of environmental protection we have all learned that tactics of hostility, litigation and organizing have their place but that they are not the only forms of problem-solving to reduce environmental impact while considering business needs.

Many environmentalists have come to the conclusion that we need less confrontation and more cooperation. The thesis is as applicable to hazardous waste siting as it is to planning for public access to shorefront, solving erosion problems, cleaning up air and water and handling toxic substances.

Citizens have long used workshops, dialogue groups, planning committees and other methods to solve environmental problems. Hundreds of these workshops, dozens of task forces and, perhaps, fifty advisory committees have operated in the Great Lakes states and similar numbers have appeared elsewhere. Industry, government, developers and engineers worked together, exchanged views, translated technical language into layman's language and layman's concerns into technical parameters.

The reason for using these methods is quite obvious. They tend to work better than conventional means of dispute settlement. Cooperation, too, is especially important in the midwest. Environmental problems have been acute and the declining midwestern industrial base has forced us to be selective in our solutions.

SOLUTION NEEDED

Using successful means to solve problems is uniquely
helpful in the hazardous waste area. The midwest has an
annual off site disposal site shortfall of 489 metric tons.
[1] Shortfalls in neighboring regions are often exported to
our region.[2] Chemical spills occur 3,500 times a year in
the country and a large proportion are in the industrialized
midwest.[3] More than 20% of the Superfund cleanup sites
identified last October by EPA are found in the midwest
region.[4] These sites will not only need cleanup but they
also will need safe disposal locations for the waste there
now. Six priority sites alone exist in Ohio. And others of
some modest priority are merely awaiting identication and
action.

Without solution to the hazardous waste problem we will
have industrial barriers to producing such a wide variety
of products as toasters, newspapers, auto paints, computers,
bottled beverages, pipe, steel and plated metal. Even
utilities and universities will have difficulty operating
without hazardous waste sites.

We will also have high costs. Poor treatment, storage
or disposal of hazardous wastes can haunt a generator or
facility operator with quarter million dollar fines under
one federal law, fifty million dollar cleanup costs under
another, strict liability (without any consideration of
fault) to damaged citizens and the potential for legal
action at any time -- not to mention jail sentences for
violators. [5] The community, too, is injured by failure to
solve hazardous waste problems. Economic recovery and
vitality of a region can be retarded by lack of sites.
Heavy fines sap funds from a business. Loss of worker time
and public health care costs are unnecessary burdens to
everyone though they strike a few.

Cooperative solving of siting problems is one means to
reduce these problems and find common solutions to siting
concerns.

One important point which supports this approach is the
present concern of business for the environment -- manifested in general in strong consideration of community
values in the planning and operation of factories and
businesses. Few companies remain who are hostile to air
quality, water quality and land use controls. The environmental community is also aware of the need for a sound
business base. The Open Lands Project in Chicago, for
example, is well on its way toward realizing a national park

and industrial "park" combined along an historic waterway. The Sierra Club in New York is the prime mover behind an intermodal freight terminal covered by parks and housing on New York City's upper west side.[6] The Conservation Foundation leads a group of chemical manufacturers, waste disposers, environmentalists, citizens and political leaders to find joint solutions to waste siting problems.

Cooperation is almost an epidemic. Massachusetts' new waste facility law requires this type of negotiation. Centers for environmental mediation exist in Washington state, Illinois and elsewhere and experts abound.

Cooperation is so important and so valuable because, to some extent, the general public have become environmentalists making business and industry realize that they must deal with communities on a sound basis. In a recent survey 62% of all Americans said they were active or sympathetic to the environmental movement.[7]

In the hazardous waste area cooperation is required because almost the entire community subject to hazardous waste turns out to be environmentalists -- the Kiwanis, the Rotary, the League of Women Voters, sometimes even the banks and the Chamber of Commerce. The environmental position is taken by almost everyone.

The case studies of hazardous waste siting quickly reveal this situation. There are good examples (and some bad examples) of siting, community involvement and the community's environmental perspective at Sturbridge, Massachusetts; Cottage Grove, Minnesota; Wright City, Missouri; Wes-Con in Idaho; Allied Chemical at Rossville, Maryland; IT Corporation at Brentwood, California; SCA/ Earthline at Wilsonville, Illinois and Resource Recovery Corporation at Pasco, Washington.[8]

Local concern is evident in state laws which govern siting. A number of states override the prevalent local opposition. They include Alabama, Arkansas, Arizona, California, Maryland, Minnesota and Massachusetts.[9] Even so, locals have prevailed.[10]

It is clear from these situations and the state laws that cooperation can be helpful in the location of a site, the specific planning of the site, construction, operation, maintenance and closure. Cooperation is also just as important when remedial measures are proposed to solve a problem, resupply water to an affected household or to clean up a site or a spill.

The best way to get at the essence of "cooperation" is to look at some case situations and to try and extract from these real-life facility sitings some information as to the effect of cooperative activity.

Much of the case study material is taken from EPA reports. The techniques described are derived from activities undertaken principally by the author in the Great Lakes region and by others throughout the country who have used cooperative approaches to bridge the gaps between industry, citizens and government.

3M CHEMOLITE, MINNESOTA

In 1978 3M's Chemolite Division in Minnesota discovered that they needed a site to dispose of hazardous waste they had previously been discarding in a landfill. 3M proposed use of a site adjacent to their plant in Cottage Grove. The site was unique in that 3M had successfully opposed plans for a regional hazardous waste landfill in the same year. 3M had testified against the site because of potential for leakage into groundwater used for production and the company's need for future expansion space. A more difficult situation could not be imagined.

According to EPA, 3M launched a "low key public relations campaign," consulted privately with local officials and gave a strong presentation to the local zoning commission.

3M relied upon its reputation as "competent and responsible" and the fact that 3M "would be there long after the site was filled." 3M is, of course, a major employer in the community and the region and the site was used for locally generated wastes.[8]

The Chemolite success suggests that local reputation and employment coupled with a "low key public relations campaign" and appropriate consultations with officials that are well designed are valuable in securing public approval.

KANSAS INDUSTRIAL ENVIRONMENTAL SERVICES (KIAS), KANSAS

KIAS is a regional facility rather than one that serves one business or nearby industry. Local citizens were "fighting mad" when they heard about the site and a planned public meeting. EPA says the meeting did little to "allay public concerns". However, a second meeting stressed the tough state permit that would be issued. That permit proved the selling point for the site. It contained such criteria

506

as approval of all transportation routes by the state, high liability insurance and annual permit review and approval.

KIAS maintained that permit and public good will with a "good neighbor" policy of maintaining local roads in winter, plowing snow, pulling residents' cars out of snow drifts, using local suppliers and providing an onsite employee to handle complaints. KIAS made friends with the community and responded to their needs.

Strong points in KIAS favor were the "ability" to include state and local concerns in their plans, the "good neighbor" policy and the tough state permit standards which reassured the public.[8]

WES - CON, IDAHO

Wes-Con, at Bruneau and Grandview, Idaho, was well known in the area proposed for a site. They welcomed public inquiry and concern, invited citizenry to see the operation, took into account public concerns over nuclear wastes and nerve gases and barred them from the site and accepted waste from local ranchers and public facilities in the area without charge.

An important part of WES-CON's efforts included local control of the operation. Wes-Con "organized a local corporation with native and long-time residents in ownership positions" to control the site and met extensively with civic leaders.[8]

OTHER SITES

SCA/Earthline's proposal at Bordentown, New Jersey failed. Reasons included an "inability to talk with citizens", debates by their local attorney and adverse publicity at another site. SCA wasn't credible or communicating according to EPA.[8]

Allied Chemical at Rossville, Maryland, was seen as insensitive to the community's self image and priorities. They failed also because they withheld detailed presentations at a zoning hearing and their piecemeal attempts to answer public questions made the public suspicious.[8]

Even sound public relations and extensive presentations to community associations were unable to overcome technical (geologic fault) objections at an IT site in Brentwood, California.

WHAT COOPERATIVE ELEMENTS ARE IMPORTANT?

The literature of public involvement in environmental disputes adds significantly to our ability to abstract from these situations some findings on "cooperation", dealing with a community and other variables in the siting process.

One expert points to certain attitudes and considerations which are important to success:

1. Design of the involvement activities reflects the value placed on public concerns by the company and is sensed by the public.

2. Visibility of the activities - documenting the relationship between concerns and action taken to meliorate them - is critical.

3. The extent to which the plans are subject to change, the open attitude, is important.

4. Company consideration of values _and_ professional expertise is necessary.

5. Use of professional expertise to create options and not foreclose them is helpful.

6. Communication in language understandable by the public is needed.[11]

From Chemolite we learn that consultations with officials are important and the reputation of the company critical. In KIAS including public concerns in the plan were important. Wes-Con points to actual partnership with the community. Other hazardous waste siting approaches suggest that "ability to talk with citizens" and extensive presentations to community groups are important as well.

Experience in the Great Lakes region suggests that projects are accepted by a community when the community is informed at an early stage, given clear pathways to provide their ideas, see their ideas considered by technical staff and are provided with a wide variety of information by the proponent of any project. Both information and information transfer are required.

It seems clear that new methods of receiving and providing information are important to the siting process. But also important is the very information itself. What is it

508

the public wants to know? And what is it the public wants to change?

A short list of the kinds of information citizens and communities want will be helpful in the design of a communication program. The suggestions listed here were largely gleaned by participation in a number of hazardous waste disputes in the Great Lakes areas; but mainly from participation in a Conservation Foundation hazardous waste dialogue group referred to earlier. The results of that group's work will be published soon in a far more comprehensive form by the Conservation Foundation.

WHAT INFORMATION DOES A COMMUNITY WANT?

1. First, industry will need to know what <u>background the community has on the subject</u>. Important information here relates to recent controversies over hazardous waste or other environmental or health issues.

2. Industry also needs to know the <u>technical expertise</u> that the community has or can call upon to evaluate a site to determine how best to present data.

3. The public wants to know about the <u>need</u> for the site, the type of technology, the types of wastes, current disposal practices.

4. The public needs to know reasons for the <u>particular location</u>, regulatory barriers to other sites, compatibility with state or federal plans, the potential for resource reduction, recycling, reuse and recovery.

5. Also important is the <u>operator's experience</u>, activities in similar areas, assurance of financial stability, level of competency of the staff, general provisions to protect the public health, perimeter control, emergency procedures, potential paths of escape and potential impacts on drinking water or public health.

6. The public wants to know about specific <u>transportation</u> routes, carrier qualifications, routes, schedules, time of day, packaging methods, methods of on-site storage, handling of reactive and ignitable wastes, monitoring wells and plans for expansion.

7. The public wants to know about <u>processes used to treat waste</u>, control of processes, disposal of residues; special disposal effects on groundwater, technologies to be used, monitoring of disposal effects. (Everyone

knows that nothing is forever impermeable and assertions to the contrary have only spread suspicion.)

8. The public wants to know about the _operational plan_ for the facility, how it will be operated, how staff will be trained and monitored, how waste input controlled.

9. For landfills the public wants to know _ground and surface water locations_, geologic structure, design of burial cells and liners, leachate collection systems and groundwater monitoring.

10. For surface impoundments _odors and other emissions_ are important to the public as is wildlife (read domestic dogs) control.

11. For incinerators the public is interested in _controls on routine emissions_, accidental release prevention and disposition of residue.

12. Communities are interested in _general effects on health_, _economic benefits and costs to a community_ such as jobs created or maintained, costs of fire and police service, effects on taxes and property values.

13. The public also wants information on the current and future potential for _effective regulatory programs_ - the permits, approval, testing, enforcement system.

14. _Long term closure_ is important to the public who want to know about trust funds, insurance and future uses of the site.

Although much of this information can probably be found in a permit application, it is clear that citizens of communities near facilities find that this source of information is not very useful. Citizens need a system that communicates in their language and is responsive to their questions.

In addition, though, and even more importantly, it is not the information itself that is important, it is the community's ability to change that information to make it conform to their needs and concerns that is far more critical. The case studies described above and much citizen action in the last decade tend to bear this out.

The next section describes methods to answer these questions and to alter and improve the plan to meet community needs and secure local support and needed approval.

HOW INDUSTRY CAN COMMUNICATE INFORMATION TO CITIZENS AND
LISTEN TO THEIR CONCERNS?

Citizens have serious concerns about hazardous waste
sites and many of these concerns relate to communications
difficulties. These include:

1. Distrust of technical solutions and a poor record
    of disposal (Love Canal, Valley of the Drums,
    Hudson River PCBs).

2. Failure to understand technical language or
    concepts.

3. Resentment of powerful outside business leaders,
    lawyers and consulting firms.

4. Little trust in state or federal agencies or
    these agencies persistent or forceful monitoring
    of a project or activity. Agency budget cuts
    are a serious problem.

5. Past frustration in dealing with agencies and
    developers.

6. Tendencies of fringe elements to mislead the
    public and leadership by irrational opposition.

7. Confusion by advocates who unquestionably support
    all actions and proposals.

8. Failures of facility operators, agencies and
    others to consider community needs.[1]

What can be done to surmount these problems?

An operator or developer should not send in know-it-all
experts, mount a slick public relations campaign, withhold
important information or defend the project plans without
opportunity for alteration. Nor should any operator or
developer plan, design or operate the site from a remote
location or distant corporate headquarters and fail to consider local public attitudes toward the company, the regulating agencies or the hazardous waste issue as a whole.[12]
These actions will generally make the problems worse since
they fail to recognize attitudes and local needs.

What can be done involves bridging the information gap
with some new mechanisms.

These should

   --- aim at problem-solving
   --- contain two-way communication methods
   --- use negotiation, cooperation and compromise
   --- aim at consensus building
   --- focus on clearly delineated areas of disagreement
   --- find underlying concerns and resolve them

The mechanisms should go beyond that required by state or federal law. Federal law, mainly the Resource Conserva-tion and Recovery Act of 1976, is not especially helpful.[12] Superfund requirements for cleanup embodied in a proposed National Contingency Plan (§300.61, 47 Fed. Reg. 10990 March 12, 1982) are not especially helpful even though administrative guidance suggests that tougher requirements than "keep the local community informed" may soon issue. State law requiring participation will need to be examined to determine state minimums. However, in general, these laws do not adopt cooperative approaches but stress formal hearings and comment methods which often encourage adversarial situations.

Appropriate bridging of the information gap has been successful in the past when the following activities were utilized.

1. A target group for communication is carefully selected.

2. That group is provided with a variety of communication methods to talk with government regulators and facility operators.

3. A small group is turned into a task force which deals with controversial issues and takes an important part in planning, design, operation and monitoring of a facility site.

4. Communication is two-way and occurs as early as possible in the process. Industry avoids using the media as the sole source but "panders" to media needs when necessary.

TARGET PUBLIC

The target public should include those concerned about the waste facility and those who are community leaders. It should include:

--- nearby landowners
--- those using groundwater or surface water or in
    the nearby airshed
--- local officials and zoning and planning staff
--- leaders of key civic groups, conservation groups,
    business groups
--- key elected officials such as state legislators
--- generators of hazardous waste (who are "closet"
    advocates)
--- public health and safety interests
--- local scientists and engineers
--- unions, farmers, ranchers and other important
    sectors.

Of course, the exact list will depend on the area, the
mixture of waste and the strength of interests and interest
groups.

VARIETY OF COMMUNICATION METHODS

Though the public hearing is often selected as the means
to describe a facility plan and to solicit advice, it is
often one of the least valuable methods of providing information or obtaining information.

--- It comes at a vary late stage in most proceedings.
--- It is adversarial in nature -- citizens are often
    antagonistic, proponents are defensive, lawyer-
    types often dominate the process.
--- Some of the public, often the most supportive,
    cannot speak well or feel uncomfortable speaking.
--- The hearing lends itself to "grandstanding",
    political posturing and use as a platform to
    build organizational strength and plays to media.
--- The hearing is not a dialogue.[1]

Though the hearing can be improved upon [4] other methods
are needed to supplement the hearing, provide for the needs
of those without speaking ability, those with specific technical ability and others with special needs and concerns.

The enterprising facility developer can use an extensive
number of mechanisms to talk with the community.[13]
These include:

1. Newsletters, fact sheets, telephone "hot lines"
    which describe specific information about a
    project, explain technical details in citizen
    language. (A fact sheet, for example, might
    carefully detail how a site will be operated,

2. Small group sessions, briefings, workshops, presentations to special committees. (These can be
   used to learn about the concerns, say, of the
   local fire department, or the Kiwanis or the
   Audubon Society; or they can be convened to consider a particular aspect of a site such as
   transportation routes and times of operation but
   they must be led by trained, experienced professionals.)

3. Opinion polls, information surveys (by telephone,
   mail, or house-to-house survey). One facility
   developer is reputed to have visited every house
   in the area of the site to discuss the project
   with residents.

THE TASK FORCE OR ADVISORY COMMITTEE

Advisory committees have long been used in state and
local environmental programs. Composed of representatives
of interest groups (who generally function as individuals
rather than delegates), the advisory committees have made
suggestions to agencies and served as a foil against which
to test proposed programs.

Within the task force, discussion is used to build consensus out of compromise among the interest groups. For
example, an environmental leader might want a state water
program to upgrade oxygen levels in lakes used as fisheries.
The environmentalist is joined by a fishermans' group but
opposed by local governments and industrial representatives.
A compromise where certain industries on certain lakes are
required to upgrade treatment in prime fishing areas emerges
and the government agency considers that in new rules that
are published.

In the facility siting situation, however, the interest
groups are not so clearly defined; nor do they have a larger
agenda (statewide) on which to bargain -- any use of advisory
committees must be carefully structured to reflect this and
to reflect some of the lack of local experience in working
on such groups.

A task force, carefully structured and well led, can be
very useful in facility siting. If appropriately designed
it can

--- gain the support of community leadership
--- turn opponents into participants
--- fully expose public concerns

--- provide a professional type dialogue situation
where issues between community and developer
can be discussed and solutions derived.

How does a task force operate?

Selection of a task force is important. The task force
should be a small group -- not more than twenty so that
about 12 to 15 attend each meeting. Members should include
persons who represent the interest groups concerned with
the site. The members should be able to lead their constituent groups; but should not be true "representatives"
needing to return to the group for support.

Organized this way the task force recognizes the
importance of special interests, the efficiency of dealing
with only a few people and the considerable attachment,
enthusiasm and eventual "ownership" that a small group
develops for a facility where they have negotiated over
many of the parameters.

To insure that the task force achieves the appropriate
objectives it must be professionally led by someone who
knows how to facilitate discussion, develop and maintain
agendas and achieve consensus.

The task force must be technically capable of dealing
with facility siting issues. This can be achieved by
placing scientists and engineers on the task force, assign-
ing government or industry staff to assist the task force or
educating the task force in a series of workshops or course
sessions. The course approach has worked well in some Great
Lakes issues. In one case, over 150 citizen leaders and
local officials spent ten four-hour sessions learning about
Lake Michigan.[14] When the task force is specifically
invested with a particular responsibility, the incentive to
learn is higher.

Generally, a task force should be as self selected as
possible. A developer or a community might discuss potential members with community leaders and from that develop
the task force itself -- by recommendation.

SPONSORSHIP AND FUNDING

Considering the limited funding for state and federal
programs it is likely that industry itself will need to pay
for the task forces, workshops, polls and other processes
that go beyond a mandated hearing. Such funding is seen as
unusual to the public though it is only another cost of a

515

permit and usually less expensive than lawyers and litigation. It could, in some cases, make the operation and the outcome suspect. Another approach would be to provide a grant to a local government or civic group to carry out the activities insulating industry from direct control. Costs of such programs can vary from $500 per workshop to $20,000 for a series of task force meetings, technical consultants, workshops, opinion surveys and informative fact sheets.

SUMMARY

Experience in environmental management programs and hazardous waste siting situations suggest that one key to gaining community support is a sound program of "bridging" the communication gap between citizen and developer. "Bridging" involves two steps -- focusing on the appropriate aspects of siting, operation and closure and developing new methods of carrying out that discussion.

Key points to consider include the following:

1. Entire communities act as environmentalists with respect to hazardous waste and must be considered as such.

2. The cooperative approach to siting is a necessary and an important technique for successful siting.

3. Experience in facility siting shows that failure to plan along with a community is often fatal.

4. Programs should not be sold. They should be cooperatively planned considering community needs.

5. The public has a clear list of concerns that they want adequately considered.

6. The community will need education and will need technical language translated into their own language. Technical experts will need to consider "feelings" and "emotions" as well as hard technical facts.

7. Communication methods selected must consider general community fear and distrust, irrational proponents and opponents and the

often justified concern that government agencies will not police a site.

8. The public hearing cannot be relied upon as the sole means to inform the public and listen to concerns. Where necessary, though, it can be improved by pretraining citizens before the hearing, using a trained hearing examiner and scheduling testimony so that busy people can attend.

9. Other methods of communication such as newsletters, small workshops to elicit concerns and deal with them, and opinion surveys can improve industry response and public understanding.

10. The task force of key interest group representatives aimed at developing a consensus on siting, planning, design, operation, monitoring and closure is one successful means to build community support and draw out major concerns.

11. Full information should be provided to the community through access to the site and regular reporting.

12. A "good neighbor" policy of assisting the community, insuring that fire and police costs are not increased and any damage is compensated for is an important part of a successful project.

REFERENCES

1. Robbins, Richard L., Working with the Public, Local Officials and Nearby Landowners to Reduce Opposition to Hazardous Waste Facilities in Robbins, editor, Limiting Liability for Hazardous Wastes (Chicago-Kent College of Law, 1981).
2. U.S. Environmental Protection Agency, Hazardous Waste Generation and Commercial Hazardous Waste Management Capacity, SW-894 (1980).
3. Environmental Emergency Response Act, Senate Report No. 848, 96th Cong., 2d Sess., (1980).

4. 114 Dumps Ranked by Risks to Health, <u>New York Times</u>, October 24, 1981, p. 4.
5. Robbins, Richard L., <u>Liability for Hazardous Materials Spills</u> (Unpublished Paper, 1982).
6. Sierra Club, Urban Waterfront Project, <u>Sixtieth Street Railroad Development</u> (Sierra Club, New York, 1982).
7. U.S. Council on Environmental Quality et al., <u>Public Opinion on Environmental Issues</u> (1980); Friends of the Earth, <u>New York Times</u>, February 2, 1982, p. A13 (67% of Americans were against any relaxation of environmental laws - CBS/New York Times poll).
8. U.S. Environmental Protection Agency, <u>Siting of Hazardous Waste Management Facilities and Public Opposition</u>, SW-809 (1979).
9. Tarlock, A. Dan, Anywhere But Here: An Introduction to State Control of Hazardous Waste Facility Location in Robbins, editor, <u>Limiting Liability for Hazardous Wastes</u> (Chicago-Kent College of Law, 1981).
10. New England Town Rises Up to Block a Toxic Waste Plant, <u>New York Times</u>, October 17, 1981, p. 2.
11. U.S. Department of the Interior, Water & Power Resources Service, <u>Public Involvement Manual</u>: Involving the Public in Water and Power Resources Decisions (1980).
12. Stanford Environmental Law Society, <u>Hazardous Waste Disposal Sites</u>: A Handbook for Public Input and Review (1981).
13. U.S. Department of Transportation, Federal Highway Administration, <u>Effective Citizen Participation in Transportation Planning</u>, Volumes I and II (1976) (FHWA/SES - 76/09)
14. Robbins, Richard L., Public Participation in Great Lakes Decisions in Kaplan, editor, <u>Decisions for Lake Michigan</u> (Purdue University, 1979).

CHAPTER 33

HAZARDOUS WASTE
DELISTING PETITIONS

K. A. Frato
ERM-Midwest, Inc.

INTRODUCTION

The United States Environmental Protection Agency (EPA) promulgated the first phase of regulations on May 19, 1980, implementing the hazardous waste management system established by Subtitle C of the Resource Conservation and Recovery Act of 1976 (RCRA). RCRA regulations have defined wastes as "hazardous" if: (1) they meet the Subpart C characteristics of ignitability, corrosivity, reactivity, or Extraction Procedure (E.P.) Toxicity, or (2) they are listed under Subpart D and contained in Section 261.31 or 261.32 of the regulations.

The wastes listed under Subpart D were named because they usually contain one or more potentially hazardous constituents, based on data collected by EPA. Individual plants may have waste streams that differ from the industry norm because of differences in raw materials, processes, or equipment. Therefore, the regulations (contained in Section 260.22) provide that a waste generator may petition EPA to exclude (i.e., delist) a specific waste from the Subpart D lists.

The regulations address other types of petitioning procedures. Specifically Section 260.20 establishes procedures for petitioning EPA to modify or revoke any provisions in Parts 260 through 265 of the regulations including characteristics of listed hazardous wastes and establishes EPA's action on such petitions. Examples of some of the petitions EPA has received in this area are:

- Request from the American Iron and Steel Institute for industry-wide exclusion of lime treated spent pickle liquor sludge from the hazardous waste rules.

- Request from the Department of the Army for a modification of the ignitability characteristic to encompass sustained burning rather than flash point.

Another form of petition addressed in the regulations are those petitions for equivalent testing methods. A person may petition the Administrator of EPA to add new testing or analytical methods to 40 CFR Parts 261, 264 or 265 if the person demonstrates that the proposed method is equal to or superior to the corresponding method prescribed in the regulations. An example of this type of petition would be a request to use neutral water instead of acidic treatment in the EPA Toxicity Test Procedure when testing wastes to be disposed of in neutral or alkaline environments.

DELISTING PETITIONS - BACKGROUND

Of the three major types of petitions, this paper will focus on petitions for delisting. This procedure is only available for listed waste from individual generators or wastes discharged from hazardous waste management facilities and the exclusion is obtained on a site-specific basis. The December 7, 1981 and March 3, 1982 Federal Registers list 288 petitioners and the wastes being considered for delisting. A simple bar chart shown as Figure 1 and prepared from information in the Federal Register, illustrates those wastes which seem to be the most popular candidates for the delisting process. Over 100 petitions have been submitted which contain F006 (wastewater treatment sludges from electroplating operations). At least 60 petitions for K062 (spent pickle liquor from steel finishing operations) are contained on the Federal Register lists. Many of the petitions submitted to EPA were for K063 (sludges from lime treatment of pickle liquor) but on November 12, 1980, EPA removed K063 from the hazardous waste list. However, since these lime treatment sludges are generated from K062, a listed hazardous waste, they are still considered hazardous and are candidates for delisting.

Wastes from the petroleum refining industry account for a significant number of delisting petitions. Over twenty petitions have been received by EPA for K051 (API separator sludge) with petitions for K048 (Dissolved air flotation (DAF) float) and K049 (Slop oil emulsion solids) typically included in the request for exclusion.

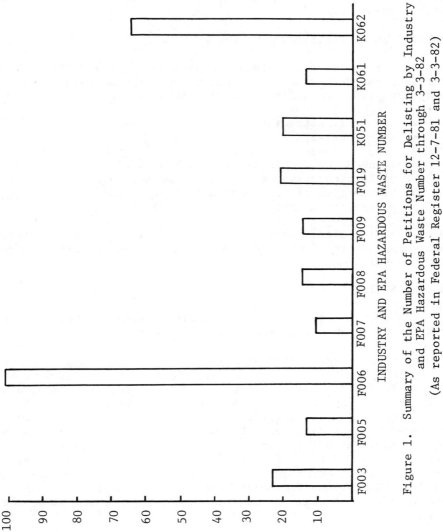

Figure 1. Summary of the Number of Petitions for Delisting by Industry and EPA Hazardous Waste Number through 3-3-82 (As reported in Federal Register 12-7-81 and 3-3-82)

Similarly, Figure 2 shows the number of temporary exclusions granted by EPA for selected hazardous wastes through December 1981. F006 and K062 have received the largest proportion of temporary exclusions. Of 266 petitions filed as of mid-December 1981, 72 have received temporary exclusions. EPA officials administering the petitioning process have indicated that a very high percentage of the petitions submitted will be granted.

ADVANTAGES OF DELISTING

The incentive for delisting a hazardous waste is that it can be managed as a non-hazardous waste. The advantages will vary from plant to plant but could include:

- Lower off-site disposal costs
- Less paperwork
- No groundwater monitoring
- No need for double liners and leachate collection system for earthen impoundments
- Elimination of the Superfund closure and post-closure taxes ($2.13/dry ton).

One of ERM's clients estimates reduction of disposal costs from $66/ton to $6/ton (equivalent to annual savings of $180,000/year) through delisting.

DELISTING PROCEDURES

The overall procedure for delisting wastes is shown schematically in the delisting logic diagram (Figure 3).

Review of Reference Material

An industry should begin the delisting procedure by obtaining the EPA Listing Background Document and key support references from which the specific Listing Background Document was prepared. These Background Documents are available for review at the Regional EPA library (where they may be copied), at the Headquarters EPA library in Washington, or from the National Technical Information Service (NTIS) as Document Number 81-190035. Background Documents support the regulations and provide EPA's response to public comments. In particular, the documents include:

- Summary of basis for listing
- Profile of the industry including a description of the processing operations and which waste streams are considered hazardous

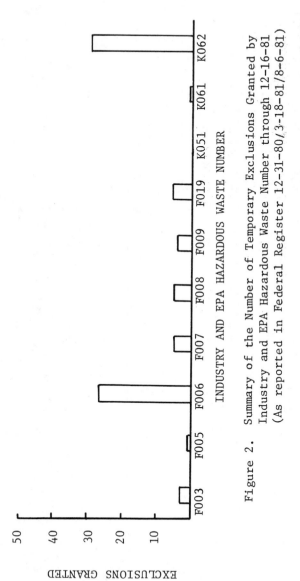

Figure 2. Summary of the Number of Temporary Exclusions Granted by Industry and EPA Hazardous Waste Number through 12-16-81 (As reported in Federal Register 12-31-80/3-18-81/8-6-81)

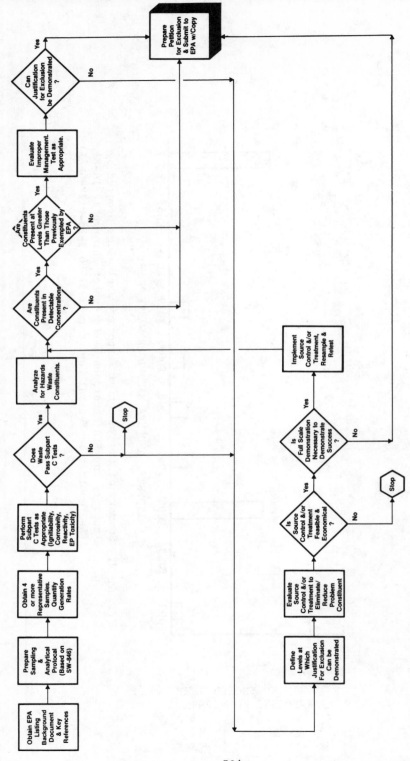

Figure 3. Generalized Steps for Hazardous Waste Delisting (from ERM Resources Bulletin – Spring 1981)

524

- Current disposal practices
- Hazards posed by the waste including health and ecological effects.

Copies of exemplary petitions may be obtained by contacting the RCRA Docket Clerk (Deneen Shrader - Office of Solid Waste at (202)755-9173). Additional EPA requirements for specific wasts may be obtained by requesting a copy of "Informational Requirements for Temporary Exclusions" from the RCRA Docket Clerk. For instance, for F006 (a popular delisting candidate) EPA lists eleven information requirements. The following requirements are included in this list:

- Description of the waste treatment system
- Disposal scenario used for waste generated prior to November 19, 1980, and scenario proposed for waste if delisted
- Total analysis of the sludge for total cyanide
- E.P. Toxicity test results for cyanide using the E.P. extraction procedure substituting distilled water for acetic acid.

The documents listed above should be reviewed carefully in order to understand what EPA considers to be the normal concentrations of hazardous constituents in the listed waste, the reason it is of concern, and the proper management scenario for the waste.

Preparation of Sampling & Analytical Protocol

The second step is to prepare a written sampling and analytical protocol. This protocol, based on EPA's "Test Methods for Evaluating Solid Waste, Physical/Chemical Methods" otherwise known as SW 846, should outline in detail the source of the waste, procedures for obtaining and preserving representative samples, and the analytical procedures to be used.

Collection of Samples

The third step is to obtain representative samples of the waste. EPA regulations require that an adequate number of samples be taken (in no case less than four) over a period of time sufficient to represent the variability or the uniformity of the waste. Chain of custody documentation should be followed along with documentation of the sampling and analytical protocol. A bound notebook (with consecutively numbered pages) in which field data can be recorded during sampling procedures is highly recommended. Sampling personnel should record the following entries in the log book:

- Location of sampling point
- Date and time of collection
- Collector's sample identification number
- Sampling methodology (i.e., was the sampling protocol followed?)
- Field observations or measurements as needed.

It is suggested that the sampler sign the top and bottom of each page and cross out unused areas of pages to prevent back-entry of information. The standard rule of thumb whenever one is involved in a sampling exercise is to record sufficient information so that someone can reconstruct the sampling without reliance on the collector's memory. Of critical importance to this step is the acquisition of a laboratory that has the necessary equipment and quality control procedures required by EPA for hazardous waste testing and also will provide the necessary documentation of chain of custody of the samples. A visit to potential laboratories with a tour of the facilities and a first hand review of capabilities and quality control procedures is a logical first step in choosing a reputable laboratory.

Sample Analysis

Once the samples have reached the chosen laboratory, the Subpart C tests should be performed. The waste must pass the Subpart C tests in order to be considered further. Analysis for all the hazardous waste constituents is normally conducted at the same time. Using analytical procedures listed in SW 846, it needs to be demonstrated that constituents are not present in detectable concentrations. If they are not, then a petition may be prepared by the generator presenting the required information.

Petition Preparation

If the listed waste passes the Subpart C tests but hazardous constituents are present, the levels in the Extraction Procedure (EP) and the constituent analysis should be compared with other wastes which have been previously exempted. If they are lower than the established precedent, the petition can be prepared on this basis. A call to EPA (RCRA Hotline 800-424-9346) at this point would be helpful. They are able to give general guidelines (informally only) for petitions that might be successful. For example:

Inorganics: If the EP toxicity is greater than 30 x NIPDWS then EPA probably won't delist the waste. For EP toxicity between 20 to 30 x NIPDWS EPA will take a hard look. For concentrations less than 10 x NIPDWS there should be no problem obtaining EPA approval for the petition.

Cyanide: Total cyanide should be less than 20 ppm. If greater than 10 ppm then EPA will look at the disposal technique. EPA has raised a red flag over some wastes containing complexed cyanides. They have data indicating that complexed cyanides may photodecompose to free cyanide in the presence of sunlight. As an interim measure they have proposed a management method which involves covering the waste on a daily basis (i.e., land-filling). EPA has also devised a laboratory test to aid in determining if photo-degradation of complexed cyanides could be a problem. This test involves exposure of a sample of the waste to a mercury arc lamp and subsequent quantifications of free cyanide.

Nickel: Nickel should be less than 20 ppm. If greater than 15 ppm then EPA will take a closer look. Originally the interim nickel leachate level was 10 ppm, but based on additional toxicity data the level was raised to 20 ppm.

    If tests fail Subpart C tests or if the hazardous waste constituents are equivalent to normal industry ranges then source control or treatment may be opted for. This procedure becomes a standard engineering investigation designed to determine the optimum cost solution. EPA personnel have indicated a willingness to discuss laboratory procedures (i.e., various types of extractions, leachate tests, etc.) which they believe are representative tests of improper management.

    Once the analytical results have been obtained they should be reviewed carefully to determine if the samples collected provided adequate representation of the waste. The variability of the analytical results between samples may impact the sample size. The upper confidence level (UCL) should be calculated and compared with EPA threshold values. If the upper bound of this interval (i.e., the UCL) is below the applicable threshold then the waste can be considered non-hazardous. If the UCL is above the threshold, due to high standard deviation and small sample size, then it cannot be concluded that the waste is not a hazardous waste and additional samples may need to be collected and analyzed.

    When a petition is justified, the following requirements must be met in the submission:

- Names and qualifications of the person sampling and testing the wastes.
- Dates and times of sampling and testing.

- Description of the waste and an estimate of
  monthly and annual quantities of waste covered
  by the demonstration.
- Pertinent data on and discussion of the factors
  delineated in the Listing Background Documents
  discussed earlier.
- Description of methodology and equipment used
  to obtain representative samples.
- Description of the sample, handling and preparation techniques, including techniques used
  for extraction, containerization and preservation of the samples.
- Description of the tests performed including
  the results, names, and model numbers of instruments used in performing the tests.

EPA Actions on Petitions

The time period to gain a final exclusion (i.e., petition
approval) for a waste can run up to six months because of the
backlog at EPA and the public comment cycle. However, EPA has
a mechanism for granting informal approval. As soon as EPA
conducts its review of a petition, and if it appears to merit
the exclusions requested, EPA will issue a memorandum of
"enforcement discretion" to the cognizant EPA region advising
them that in Headquarters' view, the waste may be considered
non-hazardous. At this point the company filing the petition
is also notified that the Regional Office has received the
memorandum. Once concurrence with the Regional EPA Office
has been established then the waste may be disposed of as
non-hazardous. Experience to date suggests that this can be
achieved within 30 days after petition submission if a completely acceptable petition is submitted the first time. In
many instances, however, attempts have been made to short-
circuit certain details of the petition process, prompting
EPA to request additional information, necessitating the submission of amendments, and considerably lengthening the process.

Exclusions granted by EPA apply to situations where the
waste is under Federal control and subject to the hazardous
waste management system established under RCRA. States which
have received interim authorization to administer and enforce
a hazardous waste management program may establish delisting
requirements which differ from the Federal program. For this
reason, it is advisable to maintain close coordination with
both State and Federal officials responsible for the delisting
process to assure that specific provisions of the responsible
regulatory agency have been complied with. In those cases
where a State has interim authorization, EPA has reviewed

528

the petitions and granted exclusions which apply when the waste is transported to a non-authorized State or is transported in interstate commerce.

Summary

The delisting process detailed in the above discussion represents an alternative to the complex and costly procedures required by EPA regulations implementing the Resource Conservation and Recovery Act. Delisting represents a cost-effective process which provides a means for companies to significantly reduce waste disposal costs. Successful delisting petitions are based on implementation of a program consisting of a series of logical steps with close coordination and review with EPA and State officials.

CHAPTER 34

REGIONAL HAZARDOUS WASTE
DISPOSAL FACILITIES IN EUROPE
AND IN THE UNITED STATES

Wilbert L. O'Connell
    Battelle Columbus Laboratories

The Resource Conservation and Recovery Act is changing
the way many industries have been disposing of their hazardous wastes. The formerly common practice of dumping the
waste in a landfill is losing favor as the mandated standards
for a secure landfill have decreased the availability and increased the cost and risk of liability of that type of disposal. As U.S. industry has looked around for other disposal
options, it has found that incineration is an increasingly
attractive alternative for disposing of burnable hazardous
wastes. Many of the larger chemical and oil companies that
produce large quantities of burnable wastes have had their
own incinerators for years, but many smaller firms that produce small to moderate amounts of burnable wastes have found
that sending the waste to a commercial incinerator is preferable to building and operating a small incinerator of their
own.

    There are and have been quite a few small commercial in-
cineration facilities available in the U.S., but they
frequently have been limited in the types and quantity of
wastes they could accept. The trend is now towards large
disposal facilities providing a variety of treatment methods
which are capable of treating and disposing of a wide variety
of wastes. Because of the size of these facilities, they are
rather widely spaced and draw wastes from up to several hundred miles away.

    Those studying the subject have noted that there seem to
be more of these regional facilities in Europe than there are
here. Is this really so, and if it is, why, and what can we
learn from it? First, let us look at the more prominent facilities of this type in Europe. (Table I) This list may not
be complete, but it probably covers the more prominent facilities now in service. Those facilities which have not yet

TABLE I.   EUROPEAN FACILITIES

Owner	Location	Year Opened
Gesellschaft Zur Beseitigung Von Sondermull	Schwabach, WG	1968
Gesellschaft Zur Beseitigung Von Sondermull	Schweinfurt, WG	1972
Kommunekemi	Nyborg, Denmark	1975
Sidibex	Sandouville, FR	1977
Plafora	St. Vulbas, FR	1978
Gesellschaft Zur Beseitigung Von Sondermull	Ebenhausen, WG	1979
Entsorgungsbetriebe Simmering GES.m.b. & Co. K.G.	Vienna, AU	1980
Hessischen Industriemull GmbH	Biebesheim, WG	1981

opened have been omitted. Eight facilities have been identified, all of which have started up since 1968, and there are
at least one more in Sweden and one in Germany which are
scheduled to start up soon. The regional facilities in the
U.S. (Table II) consist of at least five such facilities,
and there are at least five others in the planning or construction phases. An integrated facility may contain all or
only some of the treatment processes shown in Table III.
All of the facilities covered in this discussion contain at
least blending and storage facilities, a rotary kiln, and
afterburner, and wastewater treatment.

TABLE II.   AMERICAN FACILITIES

Owner	Location	Year Opened
Rollins Environmental	Deer Park, TX	1970
Rollins Environmental	Baton Rouge, LA	1972
Rollins Environmental	Bridgeport, NJ	1969
Ensco	Eldorado, AR	
Cincinnati MSD	Cincinnati, OH	1979

TABLE III.   HAZARDOUS WASTE DISPOSAL FACILITY COMPONENTS

Reception and Unloading

Oil and Solvent Reclamation

Dewatering

Blending and Storage

Incineration

    Feed Equipment

    Rotary Kiln

    Secondary Combustion Chamber

    Heat Recovery and Power Generation

    Air Cleaning

Wastewater Treatment

Residue and Inorganic Waste Land Disposal

The services that are offered by the European facilities are indicated in Table IV. In Europe most facilities offer a wide range of services including incineration of organics, aqueous waste treatment, solvent or oil recovery, land-filling of residues and heat recovery. In the U.S. (Table V) the facilities don't seem to offer such a wide range of services at a single site. That does not mean that the services are not available; they are just not available at a single site.

TABLE IV.   SERVICES AVAILABLE AT EUROPEAN FACILITIES

Facility	Incin	Landfill	Oil Rec	Inorganic	Heat Rec
Schwabach	x	x	x	x	x
Schweinfurt	x				
Kommunekemi	x	x	x	x	x
Sidibex	x		x	x	x
Plafora	x			x	
Ebenhausen	x	x	x	x	x
Vienna	x				x
Biebesheim	x		x		x

TABLE V.   SERVICES AVAILABLE AT USA FACILITIES

	Incin	Landfill	Oil Rec	Inorg	Heat Rec
Rollins, TX	x	x		x	
Rollins, LA	x	x		x	
Rollins, NJ	x				
Ensco, AR	x				
Cincinnati, OH	x				

There are reasons for these differences. Europe is more developed and more densely populated than the U.S. so that land for landfilling is hard to find and, when available, expensive. Because of its scarcity and expense, landfilling has been reserved for the more difficult to treat wastes rather than being used as freely as in the USA. Energy has always been more expensive in Europe than in the USA so there has been more incentive to search out methods of recovering energy wherever it was available such as in burnable wastes. These two factors are obvious to a causual observer. It is not as obvious that in Europe most of these facilities have been built by or with the cooperation of some branch of government (Table VI). Only two of the European facilities were built by private industry. Consortiums of industry participated in three of them, leaving only one facility built by a single company. In the USA only one facility has been built by a government body (Table VII). In the USA economic factors and governmental attitudes have directed waste disposal operations towards the use of landfills as the primary disposal method. Governmental attitudes are becoming less favorable towards landfills and more encouraging towards incineration so increased use of incineration can be expected.

TABLE VI. ORGANIZATION OF EUROPEAN FACILITIES

Facility	% of Participation		
	Public	Consortium	Private
Schwabach	100		
Schweinfurt	70	30	
Nyborg	100		
Sidibex		100	
Plafora		100	
Ebenhausen	70	30	
Vienna	100		
Biebesheim	26		74

TABLE VIII. ORGANIZATION OF USA FACILITIES

Facility	% of Participation		
	Public	Consortium	Private
Rollins, TX			100
Rollins, TX			100
Rollins, NJ			100
Ensco, AR			100
Cincinnati, OH	100		

One of the older European facilities is Kommunekemi at
Nyborg, Denmark. It was built by the government in 1975 to
take all of the hazardous wastes from Denmark. It consists
of two separate incinerators, one liquid burner for chlorinated solvents capable of handing about 7000 tons/year, and
one rotary kiln for solids and non-halogenated liquids which
is capable of handling about 29,000 tons/year. Steam is
generated from the rotary kiln's boiler for use in district
heating. The plant also has an oil recovery system which
recovers fuel oil for resale, an inorganic waste treatment
system and a landfill for disposing of the process residues.
The plant, as is common in Europe, is served by a series of
about 21 remote pickup stations which receive wastes for bulk
shipment to the plant. Kommunekemi has its own fleet of
trucks and railcars which deliver the waste from the collection centers to the plant. They are now in the process of

535

installing a secondary rotary kiln incinerator which will be capable of handling chlorinated wastes.

The Sidibex plant was constructed in 1977 by a consortium of industries. It contains systems for oil recovery, inorganic waste treatment and incineration. The incinerator is a rotary kiln which can handle about 50,000 tons/year of non-halogenated organics and recovers steam for sale to the local industries and power for internal use.

At Schwabach, ZVSMM has an older rotary kiln which can handle 16,000 tons/year of non-halogenated wastes, an inorganic waste treatment facility, waste oil separation and a landfill.

A more modern and larger facility was constructed by GSB at Ebenhausen in 1979. This facility consists of two rotary kilns capable of handling over 63,000 tons/year of wastes. These units have scrubbers and can handle halogenated wastes. An inorganic waste treatment plant is also on site and their landfill is located nearby. The incinerators produce steam and power for internal uses, but apparently none is sold. In 1972 they purchased an older grate-type incinerator at Schweinfurt and in 1973 purchased a solvent reclaimer at Geretrsried. Forty percent of the company is held by the State of Bavaria, 30 percent by local cities, and 30 percent by various industries.

A slightly different plant was constructed at about the same time in Vienna, Austria. The wastes available at Vienna included about 1.2 million tons per year of sewage sludge, so the plant includes facilities for disposing of it. The plant consists of two rotary kilns and two fluid bed sludge incinerators. There is a sludge dewatering and drying facility, an organic waste treatment plant, and a landfill for residues. Power is generated for internal use, but none is sold.

The most modern facility was recently started up at Beibesheim with the help of the German State of Bavaria. The facility consists of two rotary kilns with heat recovery and flue gas scrubbers, offsite inorganic treatment facilities at Frankfurt and Kassel, waste oil recovery and an offsite landfill. This plant is unique in that the scrubbers are a new type developed by Ciba Geigy which combines wet and dry scrubbing in a way that produces no aqueous effluent from the scrubbers.

The United States has not developed the large regional waste disposal facilities that Europe has. For the reasons previously discussed, landfills have been emphasized with smaller, more specialized commercial facilities. Rollins Environmental was one of the earliest firms to construct integrated regional facilities in the USA. They opened their first facility in 1969 in Bridgeport, New Jersey, followed by similar facilities in Deer Park, Texas, in 1970, and Baton Rouge, Louisiana, in 1972. All three facilities have similar

incinerators (Figure 1) consisting of a rotary kiln and a
liquid incinerator sharing the same secondary combustion
chamber and scrubber system. The units are capable of burning halogenated materials, and the Texas unit has a permit
for burning PCBs. To get this type of permit, they had to
prove that they would destroy 99.9999 percent of the PCBs
fed to it and remove 99 percent of the HCl from the stack
gas. Inorganic waste treatment and landfills are available
at Louisiana and Texas, but are no longer available at New
Jersey.

Energy Systems Company (ENSCO) also operates a commercial incinerator at Eldorado, Arizona. The incinerator is a
rotary kiln with a very large secondary combustion chamber
(Figure 2). It also has received a permit to burn PCB, and
it is reported that its capacity is committed for several
years ahead for PCB disposal.

The only commercial incineration facility owned and
operated by a government body that has been found is the
Cincinnati Metropolitan Sewer District facility in Cincinnati,
Ohio (Figure 3). This unit consists of a short rotary kiln
and a liquid burner sharing a common secondary combustion
chamber and scrubber system, making the unit suitable for
burning halogenated wastes. The unit presently burns mostly
grit from the adjacent sewage treatment plant in the kiln,
although it can also handle other solids by grinding them
prior to incineration.

SCA Chemical Services, Inc., provides the variety of services available at European plants except that they are not
all available at any one site. They have solvent reclaiming
operations and an incinerator for non-chlorinated liquids at
their Braintree, MA, site inorganic waste treatment and solvent recovery at Newark, NJ, aqueous waste treatment, and
landfilling at the Model Cities site near Niagara Falls, NY,
where they plan to add incineration in the future. Sometime
this spring they will start up a modern incineration facility
at the Old Hyon site in Chicago, IL, with a capacity of about
60,000 tons/year of wastes.

Other companies are also planning regional waste disposal facilities. Waste Technologies Industries has applied
for an RCRA permit for a facility to be located in East
Liverpool, OH. IT Corporation has also applied for a permit
for a facility to be located near Baton Rouge, LA, and the
Gulf Coast Waste Disposal Authority has plans for a site near
Houston, TX.

So we can see that with the facilities planned for the
USA in the near few years, the number of regional facilities
in the USA might approach those in Europe.

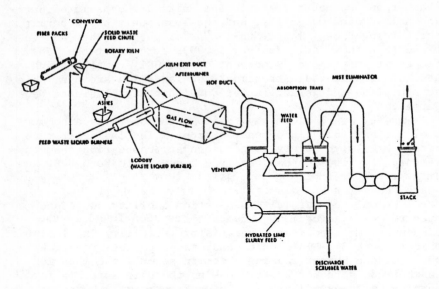

FIGURE 1.   ROLLINS ENVIRONMENTAL DEER CREEK, TX

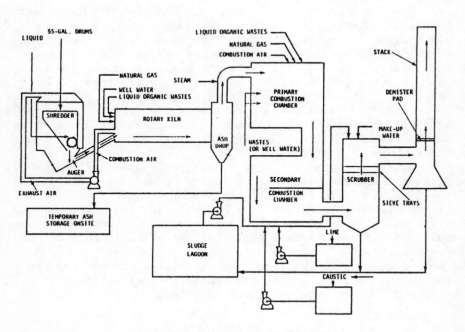

FIGURE 2.   ENSCO ELDORADO, ARK.

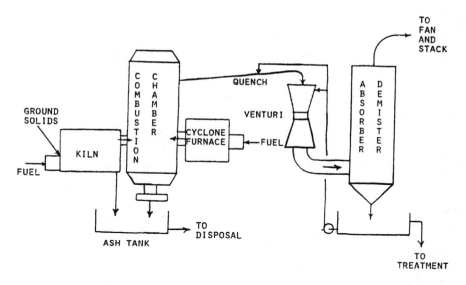

FIGURE 3.   CINCINNATI MSD

Benzene, 162, 260, 304,
   395-397, 400, 431
Benzoyl chloride, 207, 224
Biebesheim, West Germany
   site, 532-533, 535-536
Biller-Biemann algorithm,
   237
Biochemical oxygen demand,
   416-418
Biological detoxification,
   362-364, 376, 384, 421,
   431-434
Biscayne Aquifer site, 244
Bordentown, New Jersey, 507
Borings, 87, 90, 136, 147,
   150, 155-156, 159, 174,
   181, 183, 185, 191-196,
   260, 262, 360, 366, 373,
   392-394, 404-406, 413
Braintree, Massachusetts,
   537
Brentwood, California, 505,
   507
Bridgeport, New Jersey,
   244, 532, 534-536
Bridgeport Rental and Oil
   Services site, 244
Brominated materials, 229,
   (*see also* Halogenated
   materials, specific
   compounds)
Bromine, 233
Bromophenoxybenzene, 237
Bruneau, Idaho, 507
Butane, 46, 53-54

Cadmium, 33, 50, 71-72,
   124-125, 225, 260, 262,
   409-411, 415
Calcium carbide, 224
Calcium chloride, 300
Calcium hypochlorite, 220
Calcium salts, 301-302, 304
   (*see also* specific com-
   pounds)
Calcium sulfide, 70
California, 447-448, 505,
   507
California Hazardous Sub-

stances Act, 447
Camphor, 433
Carbon tetrachloride, 201,
   206
Carbonyl chloride, 49
Carbonyl sulfide, 54
Carcinogens, 225, 228, 251,
   365, 376, 498
Catalyst, spent, 59, 61-63,
   65, 67-69, 76-78, 256
Caustic, 42, 62, 65, 67
   (*see also* Bases, spe-
   cific compounds)
Cement, 127, 254, 293, 340
Cerium, 213, 225
Certification, 118, 120,
   126-128
Chain of custody, 121-123,
   126, 525-526
Chem-Dyne site, 14,
   199-203, 205-209, 211,
   213, 219, 225, 243
Chemical Control Corpora-
   tion site, 199, 277-278,
   280-287
Chemical Manufacturers
   Association, 498
Chemical oxygen demand,
   412, 416-419
Chemicals and Minerals Rec-
   lamation site, 243
Chemical Transportation
   Emergency Center, 178
Chesapeake Bay, 484
Chicago, Illinois, 537
Chlorides, 122, 124-125,
   415
   (*see also* specific compounds)
Chlorinated materials, 221,
   224-225, 227, 229, 236,
   251, 435, 498, 535-536
   (*see also* Halogenated
   materials, PCB, specific
   compounds)
Chlorine, 233, 235, 254
Chloroform, 206, 260, 262,
   364
Chlorophenol, 364

76, 247, 289, 536
Dichlorobenzene, 386-387
Dichlorophenol, 395
Dichlorophenoxy acetic
  acid,
  (see 2, 4-D)
Diesel Fuel, 231
Dimethyl ether, 54
Dinitrotoluene, 281
Discovery, 472-473, 479
Drain layers, 320, 322,
  325-329
Dredging, 81-82, 89, 93

Earth resistivity,
  (see Resistivity)
East Liverpool, Ohio, 537
Ebenhausen, West Germany
  site, 532-533, 535-536
E. coli, 434
Eldorado, Arkansas,
  532, 534-535, 537-538
Electric arc furnace, 33-34
Electromagnetic subsurface
  profiling,
  (see Ground penetrating
  radar)
Electron capture detection,
  232
Electroplating industry,
  422, 497, 520
Elizabeth, New Jersey, 199,
  277
Encapsulation, 267, 270-275
  290, 315-316, 337,
  339-340
Endrin, 50
Energy Supply and Environmental Coordination Act,
  496
Enesco site, 255, 532,
  534-535, 537-538
Engineers Council for Professional Development,
  491
Environmental Defense Fund,
  498
Environmental Impact Statement, 92

EP Toxicity,
  (see Toxicity)
Ethanol, 110
Ethyl acetate, 133, 211
Ethylene dichloride, 304
Ethylene glycol, 377
Europe, 531-537
Explosives, 103, 220,
  224-225, 228, 248-249,
  252, 263, 277, 280-281,
  283-285, 487, 498

Facility decontamination,
  180-181, 241, 243-244,
  246-247, 249-264,
  265-276, 315, 357-358,
  362, 441, 443-445,
  449-454, 456, 458-461,
  469, 497, 499, 500-502,
  505
Facility inspection,
  173-180, 183-197,
  199-226, 228, 241,
  243-244, 246-247, 252,
  254, 259-263, 456,
  500
Facility, regional, 506,
  531-539
Facility security, 247-258
Facility sites,
  (see specific sites)
Facility siting, 9-10,
  18-23, 82, 84, 87, 89,
  92-93, 95, 405, 423,
  503-517
Fatty acids, 431
Fecal coliform, 124-125
Federal Bureau of Investigation, 470
Federal Food, Drug and Cosmetic Act, 475, 477
Federal Water Pollution
  Control Act, 496
Ferric chloride, 31
Ferrous sulfate, 304
Field Brook site, 243
Filter cake solids, 61, 68,
  73, 256
Financial responsibility,

PCB, 81-82, 84, 87, 89, 91,
93, 184, 223-224, 229,
230-234, 244, 247,
254-255, 266, 277, 283,
376, 431, 446, 498, 511,
537
Pennsylvania, 243, 322,
326-327
Pentachlorophenol, 136
Perchloroethylene, 224
Permeability, 85, 88-91,
154, 259, 289, 293,
296-305, 307-313, 322,
326, 329-330, 332, 339,
341, 343, 346-348,
351-352, 355, 358, 360,
373, 379, 404-405, 411,
413
Permitting, permits, 5, 9,
14, 95, 104, 246, 253,
444, 446, 464-465, 469,
507, 510, 516
Peroxides, 281
Personal injury, 441, 443,
447-448, 455, 461
Personal liability,
(see Liability)
Personal protective
devices,
(see Safety)
Pesticides, 124-125, 157,
220, 223, 257, 304, 376,
390, 395, 498
(see also specific pesti-
cides)
Petroleum industry, 49, 57,
422, 497, 520
Pharmaceutical waste, 225,
422
Phenol, 122, 124-125, 301,
304, 395, 415, 417
Phosphate, 416, 418-419
Phosphorus trioxide, 224
Photoionization detection,
219, 261
Pickle liquor, 29-30,
42-43, 519-520
Picric acid, 110, 281
Pittsburgh, Pennsylvania,

322, 326-327
Platinum, 77
Pneumatic recovery, 364
Pollution Abatement Ser-
vices site, 241, 244-249,
252-254, 257-260, 262-264
Polychlorinated biphenyl,
(see PCB)
Polychlorinated diphenyldi-
oxin, 225
Polymeric waste, 205, 219,
221, 224, 226, 256, 497
Polynuclear aromatics, 238
Positive displacement, 359
Post closure, 3, 6, 118,
168, 243, 332, 341, 425,
444, 522
Potassium, 416, 418
Potassium fluoride, 65
Potassium hydroxide, 65
Price Landfill site, 244,
498
Propane, 46, 53, 67, 76-78
Property damage, 441, 443,
447-448, 455, 461
Propyl acetate, 377
Propylene, 54
Pseudomonas, 433
Pseudomonas putida mt-2,
432, 434-435
Public participation, 84,
92, 184, 503-517
(see also Citizen groups,
Citizen's suits)
Pulp and paper industry,
422, 497
Punitive damages, 452, 459,
466, 483
Pyrophoric waste, 225

Radar,
(see Ground penetrating
radar)
Radioactive waste, 19, 103,
124-125, 154, 213, 225,
247-249, 257-258, 315,
317, 333-336, 341, 422,
507
Radium, 124-125